AF862943

The Human Locomotor System

Thompson Sarkodie-Gyan • Huiying Yu

The Human Locomotor System

Physiological and Technological Foundations

Springer

Foreword

Huiying Yu and Thompson Sarkodie-Gyan have co-authored a book that presents the biomechanical and physiological principles underlying the human locomotor system. The book covers cutting-edge technologies for, for example, biosensors and wearable devices that enable such technologies as brain-machine interfaces and explores the role of artificial intelligence and computational modeling in enhancing the performance of the human locomotor system.

This book provides a comprehensive foundation for understanding both the normal and pathological aspects of the human locomotor system. It is a valuable resource that can serve as a basis for further investigation in fields such as medicine, rehabilitation, neurology, degenerative diseases, and physiology. By applying the knowledge gained from this book, researchers and practitioners can integrate the information in this book to help guide work toward developing innovative solutions to improve human health and wellbeing.

At this critical juncture, there is an urgent need to address the challenges faced by older adults and individuals with disabling diseases, as well as to harness technological advancements to enhance their quality of life and independence. This book focuses on the physical, perceptual, and motor aspects of the human locomotor system, with a special emphasis on neuroscience and cognitive science research for progressive degenerative diseases and age-related cognitive decline. It highlights the importance of measurement and instrumentation in healthcare technology, as they enable accurate diagnosis, monitoring of patient conditions, and efficient administration of therapeutic treatments.

This book will serve as a helpful resource for a wide range of professionals, including professors, graduate students, rehabilitation engineers, neuroscientists, sensorimotor scientists, space biomedical researchers, human motion analysts, and biomedical engineers specializing in prosthetics, assistive technologies, and other medical devices. Its interdisciplinary approach makes it an important reference for anyone working in the fields of human locomotion and healthcare technology.

FISA Foundation – Paralyzed Veterans of America Distinguished Professor
University of Pittsburgh,
Pittsburgh, PA, USA

Rory A. Cooper

Preface

The human locomotor system is a very complex multisystem. It involves components including, but not limited to, physical, perceptual, and motor aspects. This complexity further translates into the interactions among muscles, sense organs, motor pattern generators, and the central nervous system (CNS) toward creating intelligence in mobility. The CNS controls all body activities, from heart rate and sexual functions to emotion, learning, and memory. It shapes our thoughts, hopes, dreams, and imagination and is what makes us human. The neuromusculoskeletal system (neural control of the musculoskeletal system) enables intelligent behaviors, including walking, swimming, running, crawling, reaching, capturing, and grasping, among many other locomotive and manipulation activities.

However, musculoskeletal conditions are the most common cause of severe long-term pain, functional limitations, physical disability, and social and economic implications, affecting most people worldwide.

Furthermore, several illnesses, infections, and injuries that cause damage to the brain and its functions may lead to serious disorders that affect memory, cognition, movement, or consciousness or cause conditions such as chronic pain. The chronic pain and physical disability brought about by musculoskeletal conditions affect social functioning and mental health, further diminishing the quality of the patient.

There is a significant thrust within the scientific community seeking to understand how the muscles, sense organs, motor pattern generators, and the brain interact toward creating mobility and other functions. Understanding of the neuromusculoskeletal system may assist to ameliorate human health problems that may involve the design of prosthetics, the restoration of movement following brain or spinal cord injury, as well as the design, actuation, and control of mobile robots.

Current medical interventions remain marginal and unable to offer prevention and effective treatment for most chronic musculoskeletal conditions because their etiology and pathogenesis are unknown. This is a result of our marginal understanding of the functions of the locomotor system in both healthy and unhealthy states.

Sensorimotor physiologists have attempted to dissect the musculoskeletal system into simple elementary mechanisms to derive an integrative understanding of locomotor behavior. Furthermore, there have been attempts to elucidate the mechanisms underlying intelligent adaptive behavior in the locomotor system. These attempts have involved the direct capture of the activities of the neuronal system in human locomotion. They do not, however, clarify how the nervous system adaptively functions as a dynamic system and how it effectively coordinates adaptive interactions with the musculoskeletal system during locomotion.

Other studies have tried to artificially emulate locomotion by using mathematical models and robots based on control of a robot simulation model. This does not lead to understanding biological locomotor mechanisms, as the control laws are artificially constructed solely based on an engineering perspective independent of actual biological mechanisms.

This book is currently at a pivotal point for recognizing the need to respond to the needs of the aging and disabling diseases, coupled with technology initiatives for maintaining independence and reasonable quality of life; and neuroscience and cognitive science research into progressive degenerative diseases and age-related cognitive decline. The book addresses the human locomotor system's physical, perceptual, and motor aspects. It introduces the science of measurement and instrumentation as a significant aspect of the technology initiatives because it grants us good diagnosis, monitors patient condition, and administers therapeutic treatments both reliably and efficiently.

This book introduces the biomechanical and physiological foundations of the locomotor system. Furthermore, the book introduces technologies for intervening in the brain, technologies including biosensors and wearable biosensors, and the application of artificial and computational intelligence in the human locomotor system.

Following the introduction of the physiological and technological foundations, the authors have exposed the challenges and associated opportunities of the locomotor system to the readership.

This book may serve as a foundation for describing the normal and abnormal human locomotor system. It will allow scientists to build on this knowledge to create advances in different fields, including medicine, rehabilitation, neurological, degenerative, and physiological, toward ameliorating human health problems.

The book will be of utmost interest to professors and lecturers, graduate students, rehabilitation engineers, neuroscientists, sensorimotor scientists, space biomedical researchers, surveillance experts, gait analysts, and biomedical engineers for amputation, prosthetics, spinal cord injuries, and other conditions.

El Paso, TX, USA Thompson Sarkodie-Gyan
Las Cruces, NM, USA Huiying Yu

Contents

Chapter 1
The Human Locomotor System: Physiological and Technological Foundations

Introduction to the Human Locomotor System

The human locomotor system constitutes many physical, perceptual, and motor aspects.

Physical Aspects of the Human Locomotor System

The human locomotor system (also known as the musculoskeletal system) consists of the skeleton, skeletal muscles, ligaments, tendons, joints, cartilage, and other connective tissues. The locomotor system involves a human body system that provides the body with movement, stability, shape, and support.

The system is subdivided into muscular (which consists of all types of muscles in the body, including tendons that attach the muscles to the bones) and skeletal (which are the ones that act on the body joints to produce movements).

The human locomotor system involves the interaction of the muscles, sense organs, motor pattern generators, and the brain toward the production of coordinated movement both in diverse and complex environments. Coordinated movement is an adaptive human behavior driven by the intelligent sensory-motor functions toward balance, posture, and stability of the body. Hence, the human dynamic behavior in space is a very complex process because it involves many physical, perceptual, and motor aspects. Figure 1.1 depicts the physiological/biomechanical factors that affect the dynamics of locomotion.

The physiological factors that affect the dynamics of the locomotor system include neural control, muscle function, and postural control; however, more subtle alterations in underlying physiology, including cardiovascular changes and mental health, may also influence the variability of locomotion [1].

© The Author(s), under exclusive license to Springer Nature Switzerland AG 2023

T. Sarkodie-Gyan, H. Yu, *The Human Locomotor System*,
https://doi.org/10.1007/978-3-031-32781-0_1

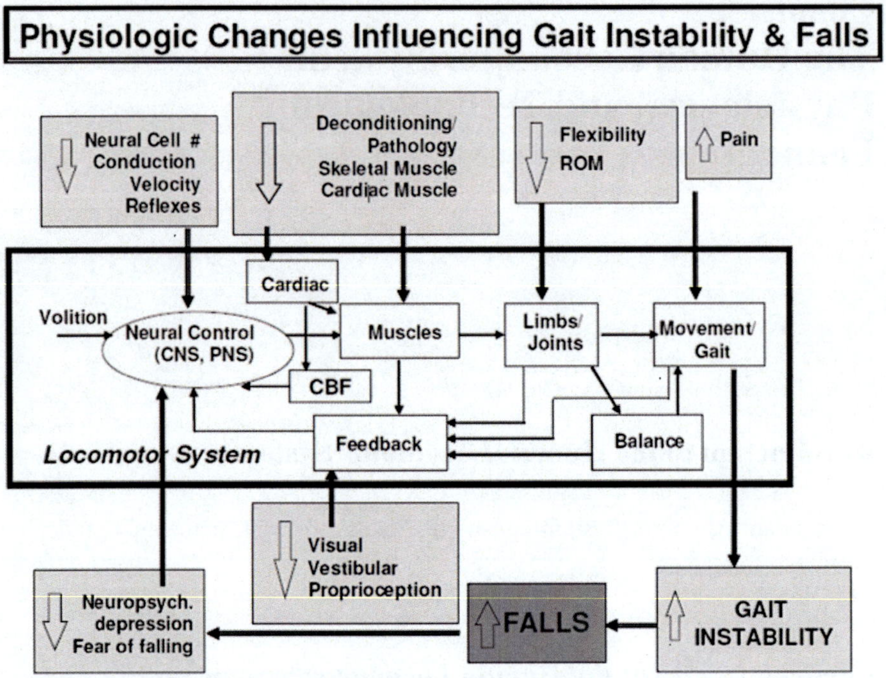

Fig. 1.1 Simplified block diagram of the locomotor system

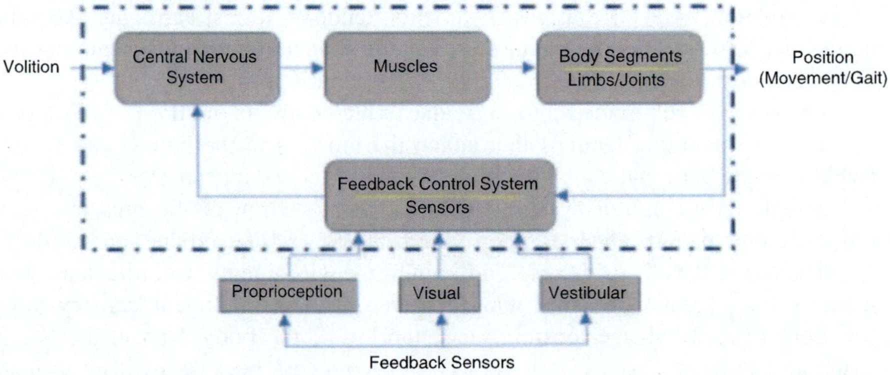

Fig. 1.2 Simplified conceptual scheme of the human locomotor system

Furthermore, Fig. 1.2 illustrates a simplified conceptual scheme of the human locomotor system.

From this conceptual scheme of the human locomotor system, as illustrated in Fig. 1.2, a linkage system (body segments), actuators (muscles), feedback sensors (consisting of proprioception, visual, and vestibular systems), and the central nervous system (CNS) are depicted.

The scheme illustrates a feedback control system in which a position or set point is generated within the central nervous system (CNS). This set point may also be understood as the posture/balance of the human body. This set point is then compared with the actual position of the limb/posture, and the CNS will send a neural signal to the muscles. The muscles exert forces on the bones of the skeletal system, which will start moving if it is not constrained by the environment. Notably, the interaction between the body and environment imposes some constraints on the redundant degree of freedom in the total dynamical system. The movements are detected by the sensors, which are all over the body, in the muscles, joints, and skin as well as the visual and vestibular sensory systems. The sensory receptors in the muscles, joints, and skin provide the brain with essential information about the position and motion of the body and limbs. Using these sensory signals, the actual position of the skeletal system can be reconstructed.

The skeletal system of the musculoskeletal system is illustrated in Fig. 1.3.

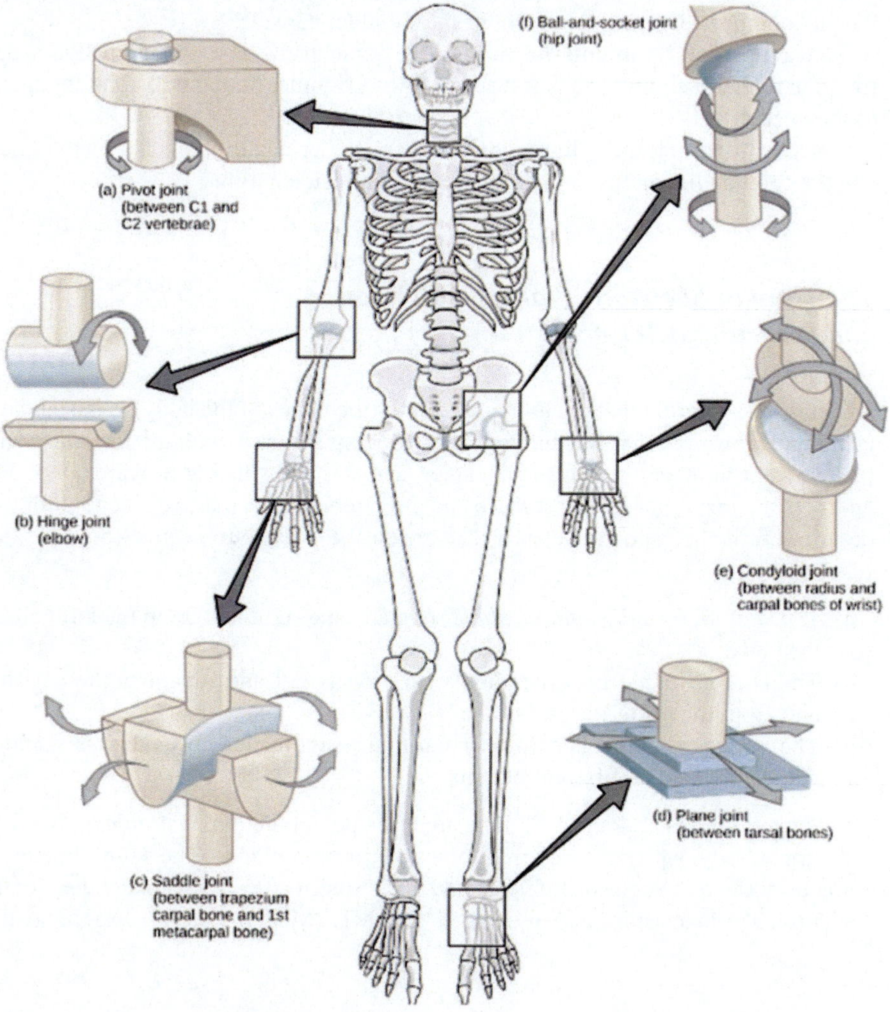

Fig. 1.3 Illustration of the skeletal system

The musculoskeletal system is made up of bones, cartilage, ligaments, tendons, and muscles, which form a framework for the body. Tendons, ligaments, and fibrous tissue bind the structures together to create stability, with ligaments connecting bone to bone and tendons connecting muscle to bone.

The skeleton is built of bone tissue. Joints, or articulations, are the intersections between bones. Ligaments connect bones at the articulations, thus reinforcing the joints. The skeleton consists of approximately 20% of the total body weight. The skeletal system is generally broken down into axial and appendicular skeletons. Bone tissue performs many functions, including support, attachment sites, leverage, protection, storage, and blood cell formation.

A **ligament** is a short band of tough fibrous connective tissue that binds bone to bone and consists of collagen, elastin, and reticulin fibers. The ligament usually provides support in one direction and often blends with the capsule of the joint. Ligaments can be capsular, extracapsular, or intra-articular.

Capsular ligaments are simply thickenings in the wall of the capsule, much like the glenohumeral ligaments in front of the shoulder capsule.

Extracapsular ligaments lie outside the joint itself. The collateral ligaments found in numerous joints are extracapsular (for example, fibular collateral ligament of the knee).

Finally, **intra-articular ligaments**, such as the cruciate ligaments of the knee and the capitate ligaments in the hip, are located inside a joint.

The Human Skeleton: Planes of Reference and Directional Terms

Under the assumption that the human skeleton (the bones of the body) is in standard anatomical position, that is, standing erect, looking forward, with the feet close and parallel to each other, the arms at the sides, and the palms facing forward (Figs. 1.4 and 1.5), the main planes of reference and directional terms that are essential for the description of the human skeleton that divide the body into sections include the following:

(i) The *sagittal* (or *midsagittal) plane* separates the right half from the left half of the body.
(ii) The *coronal plane* is perpendicular to the sagittal and separates the anterior half from the posterior half of the body.
(iii) The *transverse plane* is perpendicular to the sagittal and coronal planes, and it may be located at different heights.

As depicted in Figs. 1.4 and 1.5, there are three perpendicular anatomical planes in the human body, which are referred to as the cardinal planes. The sagittal plane is formed by the SI (Superior/Inferior) and AP (Anterior/Posterior) axes, the frontal (or coronal) plane is formed by the SI and ML (Medial Lateral) axes, and the

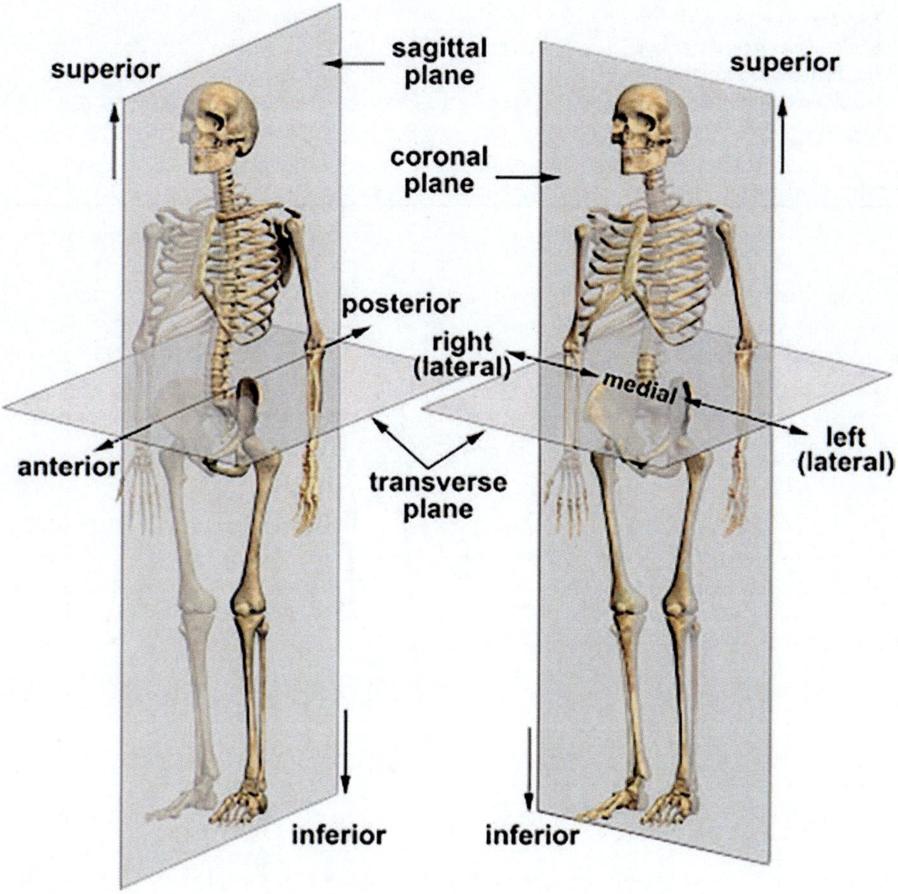

Fig. 1.4 Planes of reference and directional terms of the human skeleton

transverse plane is formed by the AP and ML axes (Figs. 1.4 and 1.5). Therefore, a three-dimensional analysis is necessary for a complete representation of human motion.

From Figs. 1.4 and 1.5, the main directions for parts of the body are superior, inferior, anterior, posterior, medial, and lateral, whereas the terms proximal and distal are more appropriate for the limbs [3].

Superior is toward the head, **inferior** toward the feet, **anterior** toward the front of the body, **posterior** toward the back of the body, **medial** toward the sagittal plane, and **lateral** away from the sagittal plane. For the limbs, **proximal** lies toward the trunk of the body, and **distal** lies away from the trunk. Terms that are often used for the hands and feet include **palmar**, which is the palm side of the hand; **plantar**, which is the sole side of the foot; and **dorsal**, that is, the top side of the foot or the back side of the hand, Fig. 1.6. Note that when the terms **right** and **left** are used,

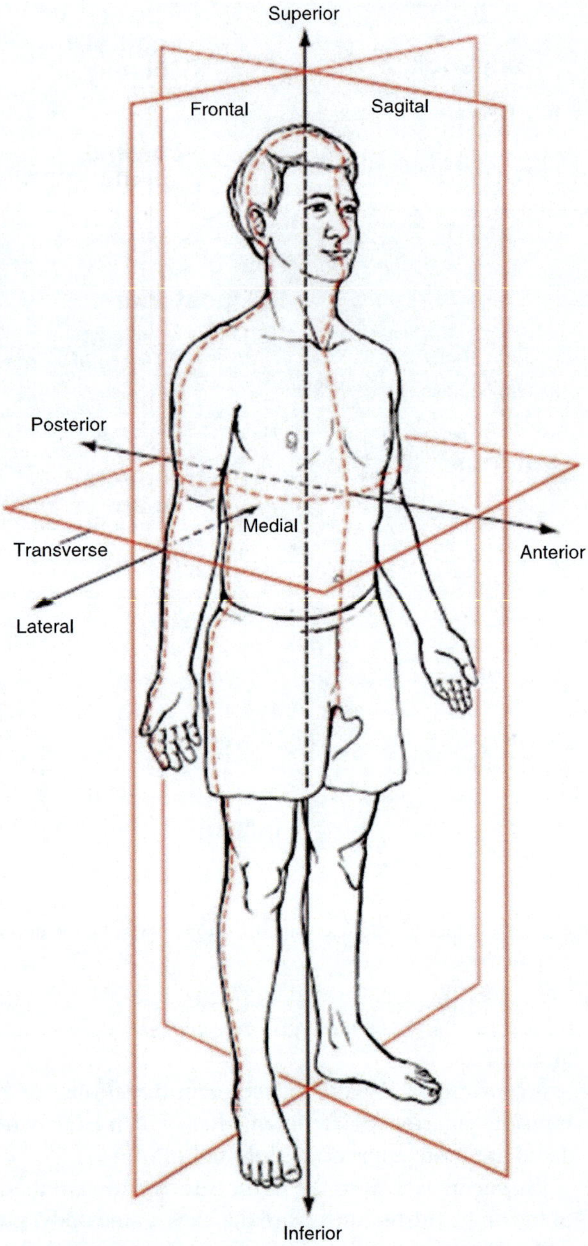

Fig. 1.5 The cardinal planes as a reference frame for the three-dimensional representation of the human body

they refer to the sides of the individual being studied and not to the sides of the observer [3].

A three-dimensional analysis is necessary for a complete representation of human motion. The motion of any bone can be referenced with respect to either a

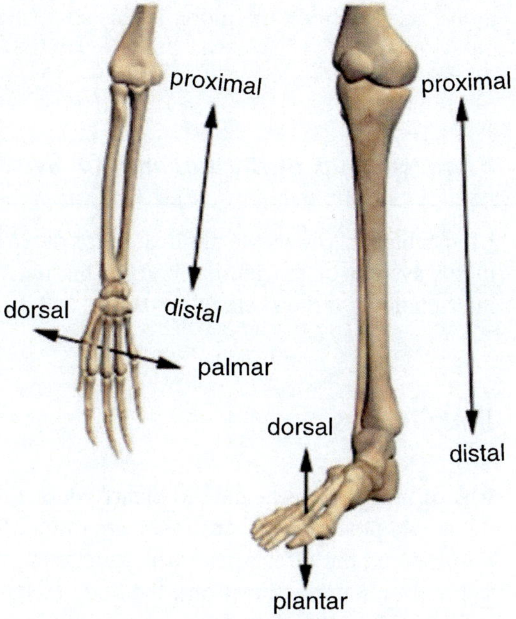

Fig. 1.6 Directional terms for the upper and lower limbs

local or global coordinate system. For example, the motion of the tibia can be described by how it moves with respect to the femur (local coordinate system) or how it moves with respect to the room (global coordinate system). Local coordinate systems are useful for understanding joint function and assessing range of motion, while global coordinate systems are useful when functional activities are considered [3].

Forces in the Musculoskeletal System

The musculoskeletal system is responsible for generating forces that move the human body in space as well as preventing unwanted motions. Understanding the mechanics and patho-mechanics of human motion requires an ability to study the forces and moments applied to, and generated by, the body or a particular body segment.

Some of the more common force generators with respect to the musculoskeletal system include muscles/tendons, ligaments, friction, ground reaction, and weight.

The distinction between the mass and the weight of a body is that the amount of matter composing that object constitutes the mass, whereas the weight is the force acting on that object due to gravity and is the product of its mass and the acceleration due to gravity. So while the mass of an object is the same on Earth as it is on the moon, its weight on the moon is less since the acceleration due to gravity is lower on the moon.

Biomechanics of the Locomotor System

Biomechanics *a*llows the application of the laws of mechanics to describe the locomotor system of the human body. This means that biomechanics establishes the interrelations of the skeleton, muscles, and joints.

Body Planes

It is of utmost significance to clearly depict the specific body planes of reference (anatomic position), because they are often used in the description of the structural position and the directions of the functional movement. In the standard position of reference (anatomic position), the body faces forward, the hands are at the sides of the body, the palms face forward, and the feet point straight ahead. The body planes are derived from dimensions in space and are oriented at right angles to one another [3]. The sagittal plane is vertical and extends from the anterior to the posterior (or from the front to the back). Its name is derived from the direction of the human sagittal suture in the cranium. The median sagittal plane (also called the midsagittal plane) divides the body into right and left halves, Fig. 1.7a. The coronal plane is vertical and extends from side to side. Its name is derived from the orientation of the human coronal suture of the cranium. It may also be referred to as the frontal plane, and it divides the body into anterior and posterior components, Fig. 1.7b. The transverse plane is a horizontal plane and divides a structure into upper and lower components, Fig. 1.7c.

In the midsagittal plane, movements of flexion and extension take place in the sagittal plane.

In the coronal plane, movements of abduction and adduction (lateral flexion) take place in the coronal plane.

In the transverse plane, movements of medial and lateral rotation take place in the transverse plane.

The human skeleton is a system of bones joined together to form segments or links. These links are movable and provide for the attachment of muscles, ligaments, and tendons to produce movement. Whereas the bones form the levers, the ligaments that surround the joints establish hinges, and the muscles provide the forces for moving the levers about the joints. An articulation constitutes the junction of two or more bones and can be classified according to function, position, structure, and the degrees of freedom for movement they allow.

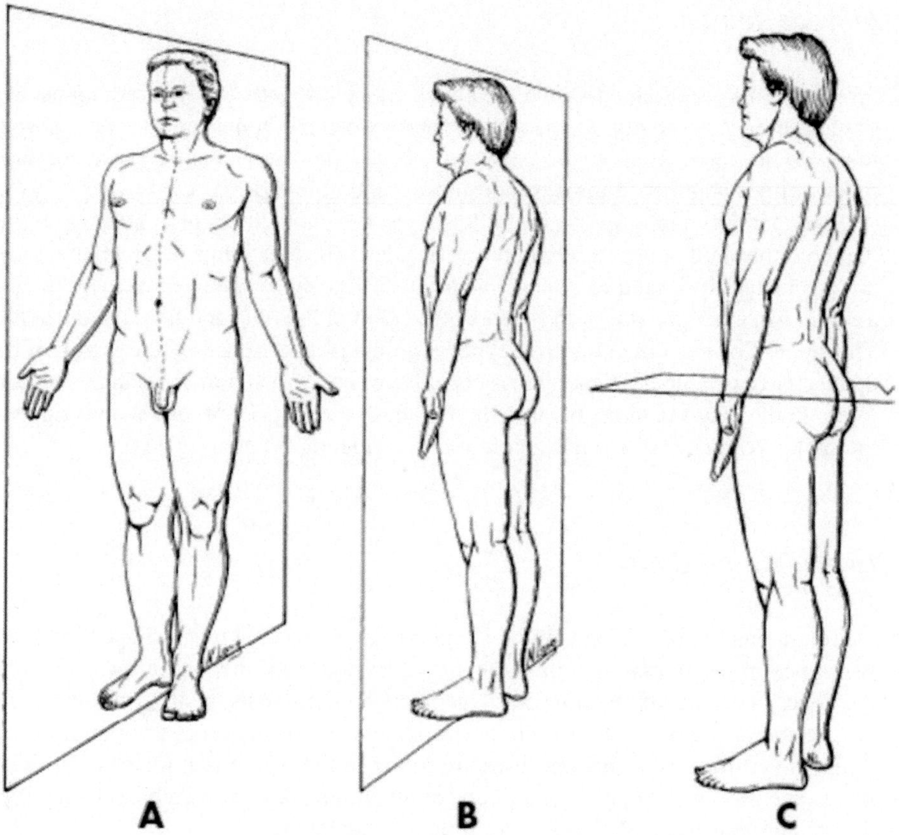

Fig. 1.7 (**a**) Midsagittal plane (**b**) Coronal plane (**c**) Transverse plane

Joints

Joints are locations at which bones of the skeleton connect with one another. A joint is also called an articulation. The majority of joints are structured in such a way that they allow movement. In fact, not all joints allow movement [11]. Of all those joints that do allow movement, the extent and direction of the movements they allow also vary.

Joints may be classified as either structural or functional. Whereas the structural classification of joints depends on the manner in which the bones connect to each other, their functional classification depends on the nature of the movement the joints allow. There is significant overlap between the two types of classifications because function depends largely on the structure.

Furthermore, the structural classification of joints is based on the type of tissue that binds the bones to each other at the joint. Hence, there are three types of joints in the structural classification; they include the fibrous, cartilaginous, and synovial joints.

Fibrous Joints

Fibrous joints are joints in which the bones are held together by fibrous connective tissue. There is no cavity, or space, present between the bones, and so most fibrous joints do not move at all or are only capable of minor movements. There are three types of fibrous joints: sutures, syndesmoses, and gomphoses.

Sutures are found only in the skull and possess short fibers of connective tissue that hold the skull bones tightly in place. Syndesmoses are joints in which the bones are connected by a band of connective tissue, allowing for more movement than in a suture. An example of a syndesmosis is the joint of the tibia and fibula in the ankle. The amount of movement in these types of joints is determined by the length of the connective tissue fibers. Gomphoses occur between teeth and their sockets; the term refers to the way the tooth fits into the socket like a peg. The tooth is connected to the socket by a connective tissue referred to as the periodontal ligament.

Cartilaginous Joints

Cartilaginous joints are joints in which bones are connected by cartilage. There are two types of cartilaginous joints: the synchondroses and the symphyses. Whereas the bones are joined by hyaline cartilage in a **synchondrosis**, hyaline cartilage covers the end of the bone, but the connection between bones occurs through fibrocartilage in **symphyses.** Synchondroses are found in the epiphyseal plates of growing bones in children. Symphyses are found at the joints between vertebrae. Either type of cartilaginous joint allows for very little movement.

Synovial Joints

Synovial joints are the only joints that have a space between the adjoining bones and are characterized by a fluid-filled space. This space is referred to as the synovial (or joint) cavity and is filled with synovial fluid, called a synovial cavity, between the bones of the joints. A typical synovial joint is illustrated in Fig. 1.8. Synovial fluid lubricates the joint, reducing friction between the bones and allowing for greater movement. The ends of the bones are covered with articular cartilage, a hyaline cartilage, and the entire joint is surrounded by an articular capsule composed of connective tissue that allows movement of the joint while resisting dislocation. Articular capsules may also possess ligaments that hold the bones together. Synovial joints are capable of the greatest movement of the three structural joint types; however, the more mobile a joint, the weaker the joint. Knees, elbows, and shoulders are examples of synovial joints.

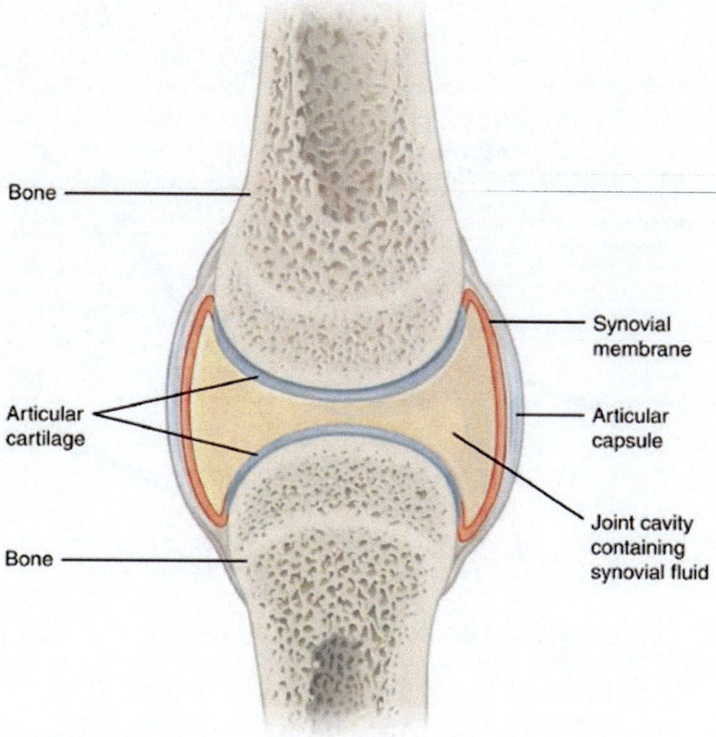

Fig. 1.8 The synovial membrane, articular capsule, joint cavity with synovial fluid, articular cartilage, and bone ends are the main components of a typical synovial joint

Movable Joints

Movable joints can be classified according to the type of movement they allow [11]. There are six classes of movable joints. These six classes of movable joints include: pivot, hinge, saddle, plane, condyloid, and ball-and-socket joints. Figure 1.9 illustrates an example of each class as well as the type of movement it allows.

Categories of Synovial Joints

Synovial joints are classified into six different categories on the basis of the shape and structure of the joint. The shape of the joint affects the type of movement permitted by the joint, Fig. 1.9. These joints can be described as planar, hinge, pivot, condyloid, saddle, or ball-and-socket joints.

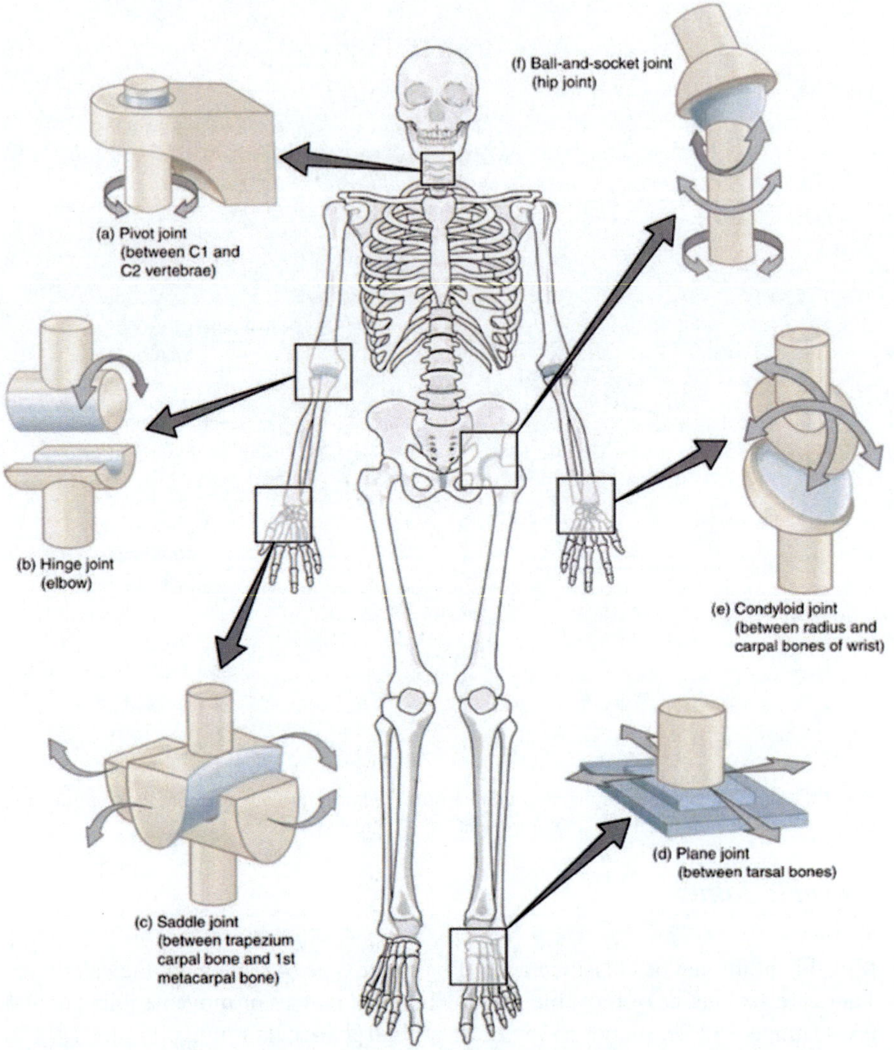

Fig. 1.9 Diagram illustrating the six different classes of movable joints in the human body. Planar, hinge, pivot, condyloid, saddle, and ball-and-socket are all types of synovial joints

Planar Joints

Planar joints have bones with articulating surfaces that are flat or have slightly curved faces. These joints allow two bones to exhibit gliding movements over one another, and so the joints are sometimes referred to as gliding joints. The joints between the tarsals in the ankles and between the carpals in the wrists are mainly gliding joints. In the wrist, this type of joint allows the hand to bend upward at the wrist and also to wave from side to side while the lower arm is held steady. Planar joints are found in the carpal bones in the hand and the tarsal bones of the foot, as well as between vertebrae, Fig. 1.10. The range of motion is limited in these joints and does not involve rotation.

Hinge Joint

A hinge joint allows back-and-forth movement like the hinge of a door. The slightly rounded end of one bone fits into the slightly hollow end of the other bone. In this way, one bone moves while the other remains stationary, like the hinge of a door. An example of a hinge joint is the elbow. This joint allows the arm to bend back and forth. The knee is sometimes classified as a modified hinge joint, Fig. 1.11.

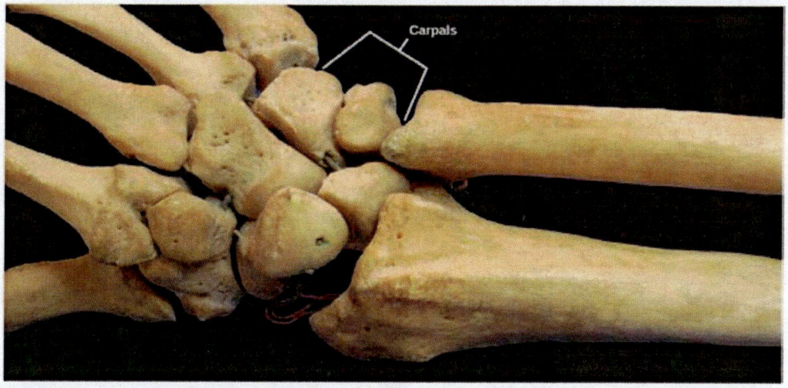

Fig. 1.10 Examples of planar joints include the joints of the carpal bones in the wrist

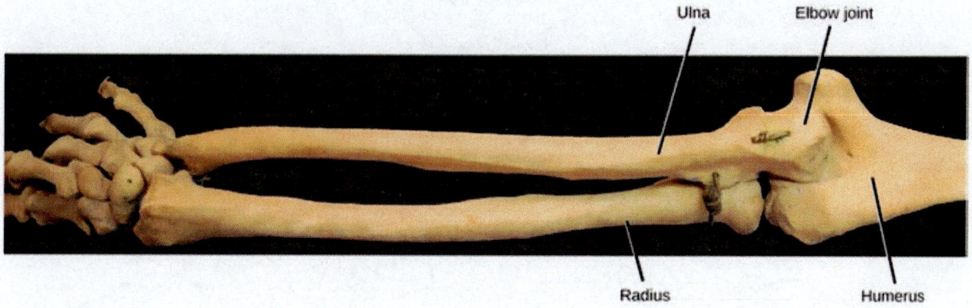

Fig. 1.11 The elbow joint, where the radius articulates with the humerus, is an example of a hinge joint

Pivot Joint

A pivot joint allows one bone to rotate around another. It consists of the rounded end of one bone fitting into a ring formed by the other bone. This structure allows rotational movement as the rounded bone moves around its own axis. An example of a pivot joint is the joint between the first two vertebrae in the spine. This joint allows the head to rotate from left to right and back again, Fig. 1.12.

Condyloid Joint

A condyloid joint is one in which an oval-shaped head on one bone moves in an elliptical cavity in another bone, thereby allowing movement in all directions except rotation around an axis, Fig. 1.13. This type of joint is also sometimes called an ellipsoidal joint. It allows angular movement along two axes, as seen in the joints of the wrist and fingers, which can move both side to side and up and down.

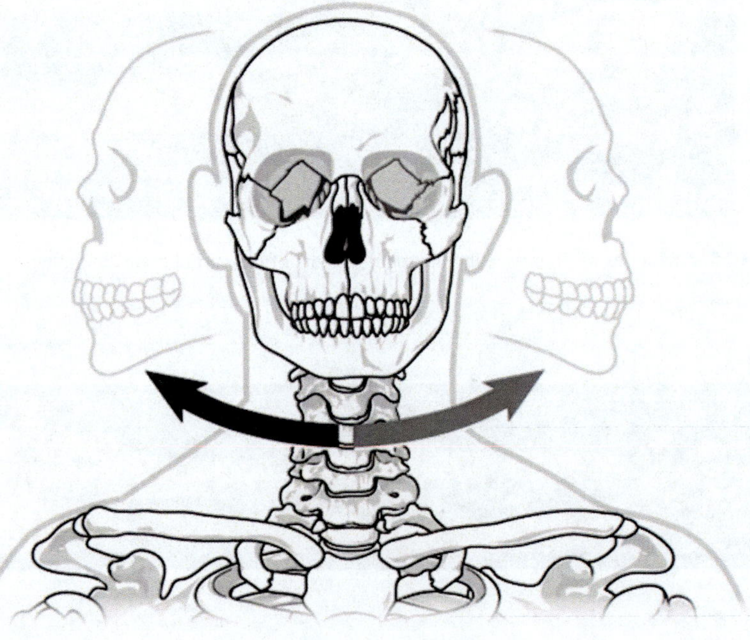

Fig. 1.12 An example of a pivot joint is in the neck that allows the head to move back and forth

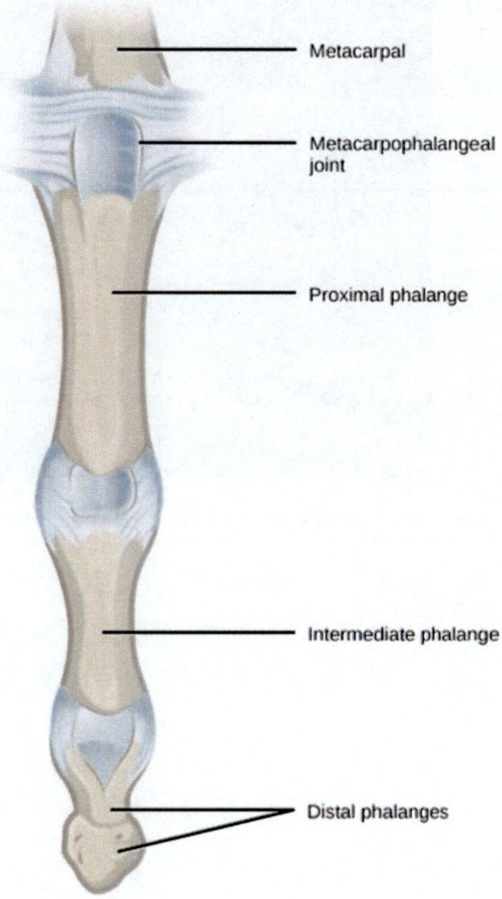

Fig. 1.13 Examples of condyloid joints are the metacarpophalangeal joints in the finger

Saddle Joints

Saddle joints are so named because the ends of each bone resemble a saddle, with concave and convex portions that fit together. Saddle joints allow two different types of movement. They allow angular movements similar to condyloid joints but with a greater range of motion. An example of a saddle joint is the thumb joint, which can move back and forth and up and down but more freely than the wrist or fingers, Fig. 1.14.

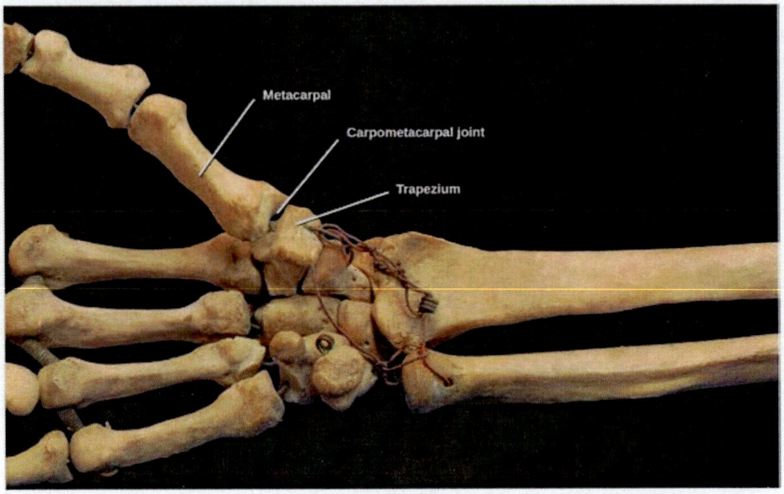

Fig. 1.14 Examples of saddle joints are the carpometacarpal joints in the thumb

Fig. 1.15 An example of a ball-and socket joint is the shoulder joint

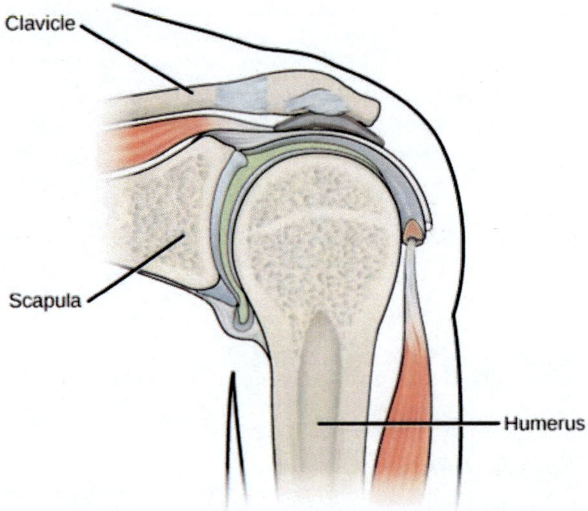

Ball-and-Socket Joints

Ball-and-socket joints possess a rounded, ball-like end of one bone fitting into a cup-like socket of another bone. This allows the greatest range of movement of any movable joint in all directions. It allows forward and backward as well as upward and downward motions. It also allows rotation in a circle. The hip and shoulder are the only two ball-and-socket joints in the human body, Fig. 1.15.

Mapping Between Upper and Lower Limb Joints

Human locomotion is a complex process that shows some inherent synergies and coordination, also called inter-joint coordination, between the upper and lower limbs.

Coordinated, stable, and adaptive human locomotor movement is dependent upon the integrated coordination between the lower and upper limbs, as well as the trunk. There emerges a coupled inter-limb control of diagonally opposite limbs moving synchronously during a walking task. This coupled inter-limb control could represent an evolutionary remnant of quadrupedal locomotion. Movements of the upper limbs are coordinated with those of the hindlimbs through propriospinal pathways that also exist in humans, as in quadrupeds, and connect cervical and lumbar spinal circuits, resulting in alternating rhythmic activation of the corresponding central pattern generators (CPGs). Corticospinal connections also contribute to arm muscle activity during adult upright walking.

Research reports suggest that arm movements coordinated with leg movements form an integral part of human bipedal locomotion and help reducing overall energy expenditure [6, 20]. Furthermore, there are some important physiological differences in the control of the inter-limb coordination between quadrupeds and humans, presumably in relation to the different control of balance and the different use of the forelimbs.

Walking is the most common daily functional activity among people. When people walk, the swinging motion of their arms along with bipedal action is a natural human behavior whereby one leg must support the body in stance while the other limb moves through the swing phase to facilitate forward progression [6, 20]. Positional and rotational movements of the hand and leg joints occur in a coordinated fashion between the upper and lower body segments.

Dynamics of the Upper Extremity

The upper limb consists of the upper arm, forearm, and hand and is connected to the axial skeleton by the shoulder girdle. The shoulder complex has the greatest mobility of all joints in the human body, whereby each articulation contributes to the movement of the arm through coordinated joint actions [7, 8]. There occurs free motion and coordinated actions between all four articulation joints of the shoulder complex, namely the scapulothoracic, sternoclavicular, acromioclavicular (AC), and glenohumeral joints [9].

The **sternoclavicular joint** serves as the only point of skeletal attachment of the upper extremity to the trunk.

The clavicle is connected to the scapula at its distal end via the **acromioclavicular joint.**

The scapula interfaces with the thorax via the **scapulothoracic joint.** In fact, this articulation just connects bone to bone. However, it is a physiological joint that

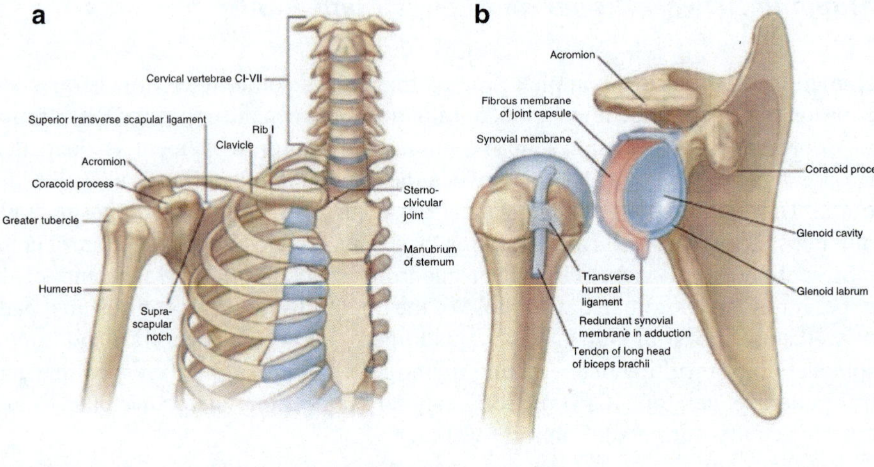

Fig. 1.16 Illustration of the anterior view of the bony framework of the shoulder

contains neurovascular, muscular, and bursal structures that allow for smooth motion of the scapula on the thorax.

Motions at the **glenohumeral joint** (also called the shoulder joint) are represented by the movements of the arm. This is a synovial ball-and-socket joint that offers the greatest range of motion and movement potential of any joint in the body. Figure 1.16 illustrates the anterior view of the bony framework of the shoulder.

The shoulder girdle is formed by the clavicle and scapula. The scapula forms a joint with the arm at the humerus (glenohumeral joint), and the proximal point of the clavicle forms a joint with the thorax at the sternum (sternoclavicular joint), while the distal point of the clavicle articulates with the acromion of the scapula, forming the acromioclavicular joint. The shape of the glenohumeral joint combined with its loose articulation gives the shoulder a high range of motion and makes it the most mobile of all the joints in the human body [7, 8]. The humerus is the only bone that supports the upper arm and articulates with the bones of the forearm—the ulna and the radius—at the elbow joint. The elbow joint consists of three articulations: the humeroradial joint, the humeroulnar joint, and the radioulnar joint.

The principal muscles responsible for joints actions at the shoulder joint include the following:

Abduction
Deltoid
Supraspinatus
Serratus anterior and trapezius assist abduction past the horizontal plane

Adduction
Pectoralis major
Latissimus dorsi
(aided by terres major and subscapularis)

Extension
Lasissimus dorsi
Teres major
Posterior deltoid fibers

Flexion
Pectoralis major
Anterior fibers of deltoid
Coracobrachialis biceps brachii (long head)

Lateral rotation
Infraspinatus
Teres minor (aided by posterior fibers of deltoid)

Medial rotation
Subscapularis
Teres major
Pectoralis major
Lattismus dorsi (aided by anterior fibers of deltoid)

Joints of the Elbow

The elbow is a critical element for a functional upper extremity (the upper extremity consists of a mechanically linked system between the shoulder, elbow, wrist, and hand). It is a stable joint, with structural integrity, good ligamentous support, and good muscular support [11]. It has three joints allowing motion between the three bones of the arm and forearm (humerus, radius, and ulna). Movement between the forearm and the arm takes place at the **ulnohumeral** and **radiohumeral** articulations, and movements between the radius and the ulna take place at the **radioulnar articulations.** The primary functions of the elbow are to position the hand in space, act as a fulcrum for the forearm, and allow for powerful grasping and fine motions of the hand and wrist.

The ulnohumeral joint is the articulation between the ulna and the humerus and is the major contributing joint to flexion and extension of the forearm, Fig. 1.17. The radiohumeral joint is the other joint that participates in flexion and extension of the forearm. The radioulnar joint is the articulation that establishes movement between the radius and the ulna in pronation and supination [7, 8, 11]. There are actually two radioulnar articulations: the superior in the elbow joint region and the inferior near the wrist, respectively.

The principal muscles responsible for joints actions at the elbow joint include:

Extension
Triceps (aided anconeus)

Flexion
Brachialis
Biceps brachii (especially when forearm is supinated)
Aided by brachioradialis and other muscles in flexor forearm compartment

The principal muscles responsible for joints actions at the radioulnar joint include:

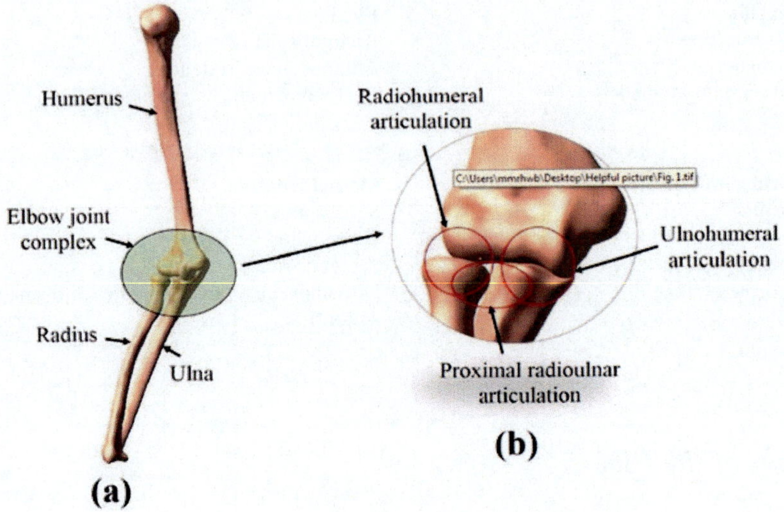

Fig. 1.17 (**a**) View of right upper extremity and (**b**) the articulations of the elbow joint

Supination	**Pronation**
Supinator	Pronator quadratus
Biceps aids when there is resistance	Pronator teres aids when there is resistance

Wrist Joint

The human wrist joint, which is also called the radiocarpal joint (is also a diarthrodial joint), is a ball-and-socket joint that allows for flexion, extension, abduction, and adduction movements, Fig. 1.18. It is a condyloid synovial joint of the distal upper limb that connects and serves as a transition point between the forearm and hand.

The wrist joint consists of eight carpal bones, which are interposed between the forearm (radius and ulna) and the five metacarpal bones [9]. Furthermore, the wrist consists of two rows of carpal bones, namely the proximal carpal row (PCR) that includes the three carpals that participate in wrist joint function: the scaphoid, lunate, and triquetrum, including the pisiform bone that sits on the medial side of the hand to serve as the site of muscular attachment. The distal carpal row (DCR) includes four carpals: the trapezium that interfaces with the thumb at the saddle joint, the trapezoid, the capitate, and the hamate [9].

The articulation between the two rows of carpals is called the midcarpal joint, and the articulation between a pair of carpal bones is referred to as an intercarpal joint.

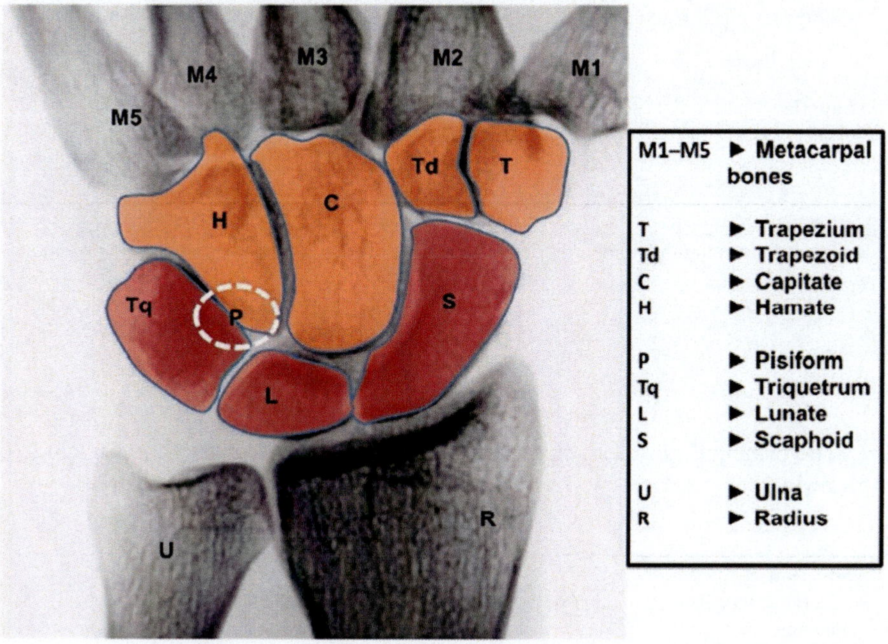

Fig. 1.18 Bones of the wrist from dorsal

The soft tissue structures surrounding the carpal bones include the tendons that cross the carpus (the carpus is divided into the proximal and the distal rows) or attach to it, the ligamentous structures that connect the carpal bones to each other, and the bony elements of the hand and forearm.

The principal muscles responsible for joints actions at the wrist joint include the following:

Abduction
Flexor carpi radialis
Extensior carpi radialis ulnaris
Longus and brevis

Adduction
Flexor carpi ulnaris
Extensor carpi

Extension
Extensor carpi radialis brevis and longus
Extensor carpi ulnaris
(Aided by extensor digitorum)

Flexion
Flexor carpi radialis
Flexor carpi ulnaris
(aided by flexors of the fingers)

The principal muscles responsible for the movement of the fingers include:

Abduction Dorsal	**Adduction** Palmar interossei
Extension Extensor digitorum (Aided by lumbricalis and interossei)	**Flexion** Flexor digitorum superficialis and profundus. Lumbricals (the Lumbricals act to flex the metacarpo-phalangeal joint and extend the interphalangeal joints.

The principal muscles responsible for the movement of the thumb include the following:

Abduction Adductor pollicis longus and brevis	**Adduction** Adductor pollicis
Extension Extensor pollicis longus and brevis	**Flexion** Flexor pollicis longus and brevis.
Opposition Opponens pollicis.	

The Lower Extremity

Human Hip Joint

The hip joint, also referred to as the acetabulofemoral joint, is the joint between the head of the femur and the acetabulum of the pelvis. It has three degrees of freedom and is a ball-and-socket joint consisting of the articulation between the acetabulum on the pelvis and the head of the femur. It is a diarthroidal joint formed at the interface of the proximal head of the femur and the acetabulum, or socket, of the pelvis [10].

The hip joint, Fig. 1.19, connects the upper and lower halves of the body. Its primary function is to support the weight of the body in both static (e.g., standing) and dynamic (e.g., walking or running) postures. Articulation at the hip joint facilitates a variety of movement patterns, including flexion and extension, abduction and adduction, medial and lateral rotation, and circumduction [10]. The external rotation of the hip accompanies the forward pelvis, while the internal rotation accompanies the backward pelvic side.

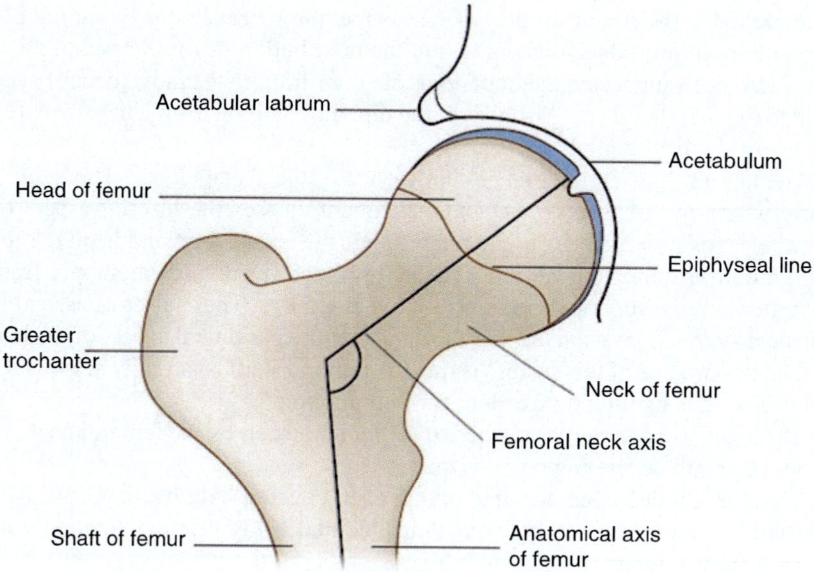

Fig. 1.19 Bony anatomy of the hip

The hip is a ball-and-socket joint and consists of the four principle features of a synovial or diarthrodial joint, namely:

- A joint cavity
- Joint surfaces that are covered with articular cartilage
- A synovial membrane that produces synovial fluid
- A ligamentous capsule [3]

The femur is the longest bone in the body, extending from the hip joint proximally to the knee joint distally. The rounded head of the femur interfaces with the acetabulum, forming the ball-and-socket connection that facilitates a large rotational range of motion in healthy hips.

Distal to the femoral head, the femoral neck extends infero-laterally and bridges the head to the femoral diaphysis (i.e., shaft). Bony outcroppings in the superior portion of the femoral shaft, known as the greater and lesser trochanters, are the attachment sites of tendons/muscles and ligaments, which are important for stabilization, rotation, and abduction/adduction of the hip. Distally, the femur ends in rounded condyles that interface with the tibia at the knee joint.

The acetabular labrum is a fibrocartilaginous structure that spans the length of the acetabular rim. At the inferior gap of the horseshoe shaped acetabulum (i.e., at the acetabular notch), the labrum becomes contiguous with the transverse acetabular ligament. The labrum can increase the surface area of the acetabulum by up to 27% and is thought to assist with load distribution and stabilization of the hip joint [10, 11, 13].

Attached to the rim of the acetabulum is the fibrocartilaginous labrum. It plays a role in normal joint development and in the distribution of forces around the joint. It has also been suggested that it plays a role in restricting the movement of synovial fluid to the peripheral compartment of the hip, thus helping exert a negative pressure effect within the hip joint.

The labrum runs around the circumference of the acetabulum, terminating inferiorly where the transverse acetabular ligament crosses the inferior aspect of the acetabular fossa. It attaches to the bony rim of the acetabulum and is quite separate from the insertion of the capsule. The labrum receives a vascular supply from the obturator and the superior and inferior gluteal arteries. These arteries ascend in the reflected synovial layer on the capsule and enter the peripheral aspect of the labrum.

The importance of this feature is that the femoral shaft is laterally displaced from the pelvis, thus facilitating freedom for joint motion.

The head of the femur is attached to the femoral shaft by the femoral neck, which varies in length depending on body size.

The femoral head and acetabulum are each covered with hyaline cartilage that, assisted by intra-articular synovial fluid, facilitates low-friction rotational movement of the hip joint.

Surrounding Musculature

The movement of the hip joint is governed by muscles. Muscles controlling hip motion can be grouped into the gluteal (superficial and deep) region and the anterior, posterior, and medial compartments of the thigh.

Superficial gluteal muscles, including the gluteus maximus, gluteus minimus, gluteus medius, and tensor fascia latae serve to stabilize the hip and produce extension, medial rotation, lateral rotation, and abduction of the femur. Muscles deep in the gluteal region (i.e., deep hip rotator muscles) are the piriformis, superior gemellus, obturator internus, inferior gemellus, and quadratus femoris. These muscles originate at different points on the pelvis but share attachments at the greater trochanter, thereby producing lateral rotation of the hip [10–12].

In the anterior compartment of the thigh, the iliopsoas (iliacus and psoas major), sartorius, and quadriceps (rectus femoris, vastus lateralis, vastus medialis, and vastus intermedius) serve as the major flexors of the hip. Muscles of the medial thigh compartment are primarily hip adductors and include the pectineus, adductor longus, adductor magnus, adductor brevis, gracilis, and obturator externus. Finally, hamstring muscles, including semitendinosus, semimembranosus, and biceps femoris (long and short heads), make up the posterior thigh and are the major extensors of the hip [12, 13]. Contributions from these muscles not only produce rotational movements but also level and stabilize the pelvis during ambulation. The rectus femoris and hamstring muscles are biarticular, which means that they span the hip as well as the knee, and make important contributions to the rotation and stabilization of each joint.

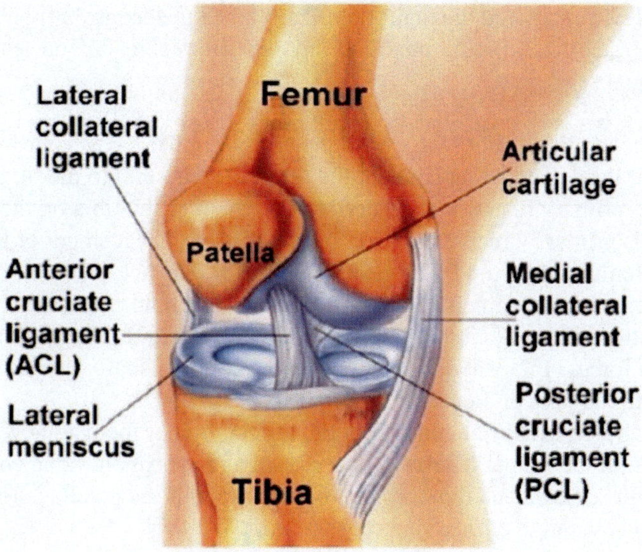

Fig. 1.20 The knee joint anatomy of the human with visible cruciate ligaments

Knee Joint

The knee joint is a hinge-type synovial joint that mainly allows for flexion and extension, including a small degree of medial and lateral rotation, Fig. 1.20. It consists of tibiofemoral and patellofemoral articulations [13]. The tibiofemoral joint is the articulation between the two longest and strongest bones in the body (the femur and the tibia). It is the weight-bearing component of the knee joint. And the patellofemoral joint consists of the articulation of the patella with the trochlear groove on the femur.

The knee supports the weight of the body and transmits forces from the ground while allowing a great deal of movement between the femur and the tibia. It allows locomotion with minimum energy requirements from the muscles, provides stability during the activities of daily life, and also accommodates different terrains and/or environments [12, 13].

Ligaments are fibrous bands of tissue that connect bone to bone and provide support to joints.

The knee is reinforced by two collateral ligaments, one on the medial side and another on the lateral side, as well as two stronger ligaments (the cruciate ligaments) that prevent excessive anterior, posterior, varus, and valgus displacement of the tibia in relation to the femur [14].

The *anterior cruciate ligament (ACL)* provides the main stabilization for the knee and enables smooth and stable flexion and rotation of the knee, as well as anterior movement of the tibia relative to the femur.

The ACL is considered the main stabilizer of the knee, contributing to about 85% of the knee stabilization and enabling smooth and stable flexion and rotation of the knee.

The posterior cruciate ligament (PCL) is located inside the knee, just behind the anterior cruciate ligament (ACL). It connects your upper leg to your lower leg. It is one of several ligaments that connect the femur (thighbone) to the tibia (shinbone). It offers the primary restraint to posterior movement of the tibia on the femur.

The ACL primarily resists anterior and rotational displacement of the tibia relative to the femur, while the PCL prevents posterior displacement.

Both the cruciate ligaments stabilize, limit rotation, and cause sliding of the condyles over the tibia in flexion. They both also offer some stabilization against varus and valgus forces. In a standing posture, with the tibial shaft vertical, the femur is aligned with the tibia and tends to slide posteriorly.

The patellofemoral joint consists of the articulation of the patella with the trochlear groove on the femur. The patella is a triangular sesamoid bone encased by the tendons of the quadriceps femoris. The primary role of the patella is to increase the mechanical advantage of the quadriceps femoris [15–17].

The patella is connected to the tibial tuberosity via the strong patellar tendon [16]. It is connected to the femur and tibia by small patellofemoral and patellotibial ligaments that are actually thickenings in the extensor retinaculum surrounding the joint.

The lateral meniscus (external semilunar fibrocartilage) is **a** fibrocartilaginous band that spans the lateral side of the interior of the knee joint. It is one of two menisci of the knee, the other being the medial meniscus. It is nearly circular and covers a larger portion of the articular surface than the medial.

The medial meniscus is a fibrocartilage semicircular band that spans the knee joint medially and is located between the medial condyle of the femur and the medial condyle of the tibia. It is also referred to as the internal semilunar fibrocartilage [16, 17].

The menisci are important in the knee joint. The menisci enhance stability in the joint by deepening the contact surface on the tibia. They participate in shock absorption by transmitting half of the weight-bearing load in full extension and a significant portion of the load in flexion. In flexion, the lateral meniscus carries the greater portion of the load. By absorbing some of the load, the menisci protect the underlying articular cartilage and subchondral bone. The menisci transmit the load across the surface of the joint, reducing the load per unit area on the tibiofemoral contact sites.

The lateral collateral ligament (LCL) is a thin band of tissue running along the outside of the knee. It connects the thighbone (femur) to the fibula, which is the small bone of the lower leg that runs down the side of the knee and connects to the ankle.

The *medial collateral ligament (MCL)* is a band of tissue that runs along the inner edge of your knee. It helps to connect your shin and thigh bones to keep your knee stable and working properly when you move.

Articular cartilage is the smooth, white tissue that covers the ends of bones where they come together to form joints. Healthy cartilage in our joints makes it easier to move [15–17]. It allows the bones to glide over each other with very little friction. Articular cartilage can be damaged by injury or normal wear and tear.

Foot and Ankle

The foot and ankle constitute a complex anatomical structure that comprises 28 bones, 33 joints, 112 ligaments, and 13 extrinsic and 21 intrinsic muscles acting on the segments. All of the joints must interact harmoniously and in combination to achieve a smooth motion [11, 13].

The ankle joint (or the talocrural joint), as shown in Fig. 1.21, is a synovial hinge joint that consists of the articulation of three bones: the tibia, the fibula (bones of the lower leg), and the talus (bone of the foot). The articulations occur between the talus and the tibia, and the talus and the fibula. The tibia and fibula form a deep socket, a concave surface, for the trochlea of the talus, creating a mortise [13, 15].

Ankle Joint (Talocrural Joint)

The ankle complex consists of three articulations: (1) the talocrural joint, (2) the subtalar joint, and (3) the distal tibiofibular syndesmosis. These three joints function in concert to allow coordinated movement of the hind foot.

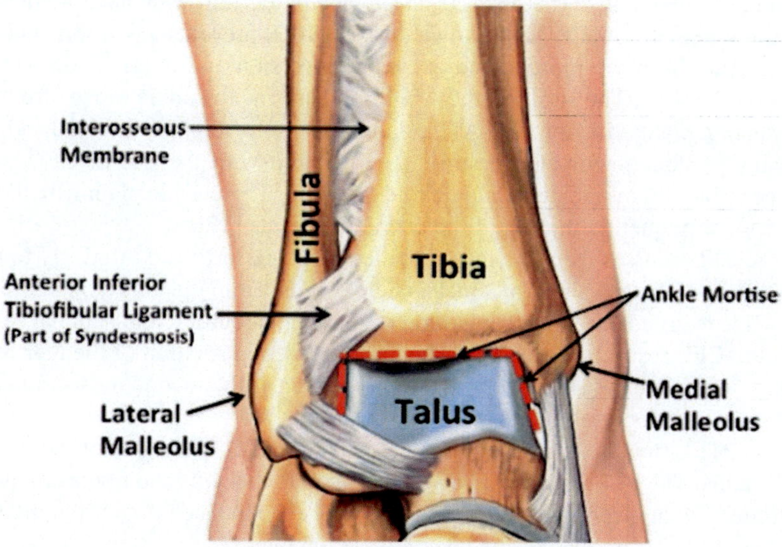

Fig. 1.21 Figure of the ankle joint anatomy

With reference to the feet, each big toe has two joints: (1) the metatarsophalangeal joint at the base of the toe and (2) the interphalangeal joint just above it. The other four toes have three joints each: (1) the metatarsophalangeal joint at the base of the toe, (2) the proximal interphalangeal joint in the middle of the toe, and (3) the distal phalangeal joint closest to the tip of the toe [14, 15].

Anatomy of the Talocrural Joint (Ankle Joint)

The talocrural joint (the ankle joint) is formed by the articulation of the dome of the talus, the medial malleolus, the tibial plafond (the distal and inferior aspects of the tibia are also known as the plafond), and the lateral malleolus [14, 15].

The distal and inferior aspect of the tibia is connected to the fibula via tibiofibular ligaments, forming a strong mortise that articulates with the talar dome distally. It is a hinge joint and allows for dorsiflexion and plantarflexion movements in the sagittal plane.

When the ankle complex is fully loaded, the articular surfaces are the primary stabilizers against excessive talar rotation and translation; however, the contribution of the ligaments to talocrural joint stability is crucial. The talocrural joint receives ligamentous support from a joint capsule and several ligaments, including the anterior talofibular ligament (ATFL), posterior talofibular ligament (PTFL), calcaneofibular ligament (CFL), and deltoid ligament. The ATFL, PTFL, and CFL support the lateral aspect of the ankle, while the deltoid ligament provides medial support [14, 15].

The ATFL lies on the dorsolateral aspect of the foot and courses from the lateral malleolus anteriorly and medially toward the talus at an angle of approximately 45° from the frontal plane. The ATFL is an average of 7.2 mm wide and 24.8 mm long. In vitro kinematic studies have shown that the ATFL prevents anterior displacement of the talus from the mortise and excessive inversion and internal rotation of the talus on the tibia. The strain in the ATFL increases as the ankle moves from dorsiflexion into plantar flexion. The ATFL demonstrates lower maximal load and energy to failure values under tensile stress as compared with the PTFL, CFL, anterior inferior tibiofibular ligament, and deltoid ligament. This may explain why the ATFL is the most frequently injured of the lateral ligaments [15].

The CFL courses from the lateral malleolus posteriorly and inferiorly to the lateral aspect of the calcaneus at a mean angle of 133° from the long axis of the fibula. The CFL restricts excessive supination of both the talocrural and subtalar joints. In fact, the CFL restricts excessive inversion and internal rotation of the rear foot and is most taut when the ankle is dorsiflexed. The CFL is the second most-often injured of the lateral talocrural ligaments [14].

The PTFL runs from the lateral malleolus posteriorly to the posterolateral aspect of the talus. The PTFL has broad insertions on both the talus and fibula and provides restraint to both inversion and internal rotation of the loaded talocrural joint. It is the least commonly sprained of the lateral ankle ligaments.

Anatomy of the Subtalar Joint (Tolocalcaneal Joint)

The subtalar joint (also known as the talocalcaneal joint) is formed by the articulations between the talus and the calcaneus, and, like the talocrural joint, it converts torque between the lower leg (internal and external rotation) and the foot (pronation and supination). The subtalar joint allows the motions of pronation and supination and consists of an intricate structure with two separate joint cavities [14, 15]. The posterior subtalar joint is formed between the inferior posterior facet of the talus and the superior posterior facet of the calcaneus. The anterior subtalar, or talocalcaneonavicular joint, is formed from the head of the talus, the anterior-superior facets, the sustentaculum tali of the calcaneus, and the concave proximal surface of the tarsal navicular. This articulation is similar to a ball-and-socket joint, with the talar head being the ball and the anterior calcaneal and proximal navicular surfaces forming the socket in conjunction with the spring ligament.

The anterior and posterior subtalar joints have separate ligamentous joint capsules and are separated from each other by the sinus tarsi and canalis tarsi. The anterior joint lies farther medially and has a higher center of rotation than the posterior joint, but the two joints share a common axis of rotation. This discrepancy results in an oblique axis of rotation of the subtalar joint.

Anatomy of the Midtarsal Joint

The midtarsal joint is also known as the transverse tarsal joint or Chopart's joint. It is an S-shaped joint when viewed from above. It consists of two joints: the talonavicular joint and calcaneocuboid joint.

The talonavicular (TN) joint is formed between the talar head and the concavity on the navicular. It does not have its own capsule but rather shares one with the two anterior talocalcaneal articulations [14, 15].

The calcaneocuboid (CC) joint is formed between the anterior facet of the calcaneus and the posterior cuboid. Both articulating surfaces present a convex and concave surface with the joint being convex vertically and concave transversely. Very little movement occurs at this joint.

Anatomy of the Tarsometatarsal (TMT) Joint Complex

This joint is also known as Lisfranc's joint. This complex divides the midfoot from the forefoot.

The distal tarsal rows, including the three cuneiform bones and cuboid, articulate with the base of each metatarsal to form the TMT complex [14, 15]. It is an S-shaped joint and is divided into three distinct columns:

- Medial – composed of 1st metatarsal and medial cuneiform
- Middle – composed of 2nd and 3rd metatarsals, intermediate and lateral cuneiforms, respectively
- Lateral – composed of 4th and 5th metatarsals and the cuboid

Anatomy of the Metatarsophalangeal (MTP) Joints and Interphalangeal (IP) Joints

The Metatarsophalangeal (MTP) joints are formed between the metatarsal heads and the corresponding bases of the proximal phalanx [14, 15]. The interphalangeal joints of the toes are formed between the phalanges of the toes. Each toe has proximal and distal IP joints, except for the great toe, which only has one IP joint.

The Foot (Categories)

The foot consists of three categories: the forefoot, which contains the five toes (phalanges) and the five longer bones (metatarsals); the midfoot, which is a pyramid-like collection of bones that form the arches of the feet (cuboid, navicular, and three cuneiforms); and the hindfoot, which forms the heel and ankle (talus and calcaneous). These categories consist of 26 irregularly shaped bones (seven tarsals, five metatarsals, and 14 phalanges) and six joints (ankle, subtalar, midtarsal, tarsometatarsal, metatarsophalangeal (MTP), and interphalangeal (IP) joints). The structure of the foot is illustrated in Fig. 1.22.

Bones in the foot may be divided into three categories based on where they are located [13–15]. These categories (regions of the foot) are illustrated in Fig. 1.23.

Forefoot

The forefoot is the very front part of the foot that includes the toes and the ball of the foot. It is made up of several parts. These parts include:

Phalanges: These are the toes. They are made up of a total of 14 bones: two for the big toe and three for each of the other four toes.

Metatarsals: These are five long bones that extend from the base of each toe to the midfoot. The first metatarsal bone leads to the big toe and plays an important role in propulsion (forward movement). The second, third, and fourth metatarsal bones provide stability to the forefoot.

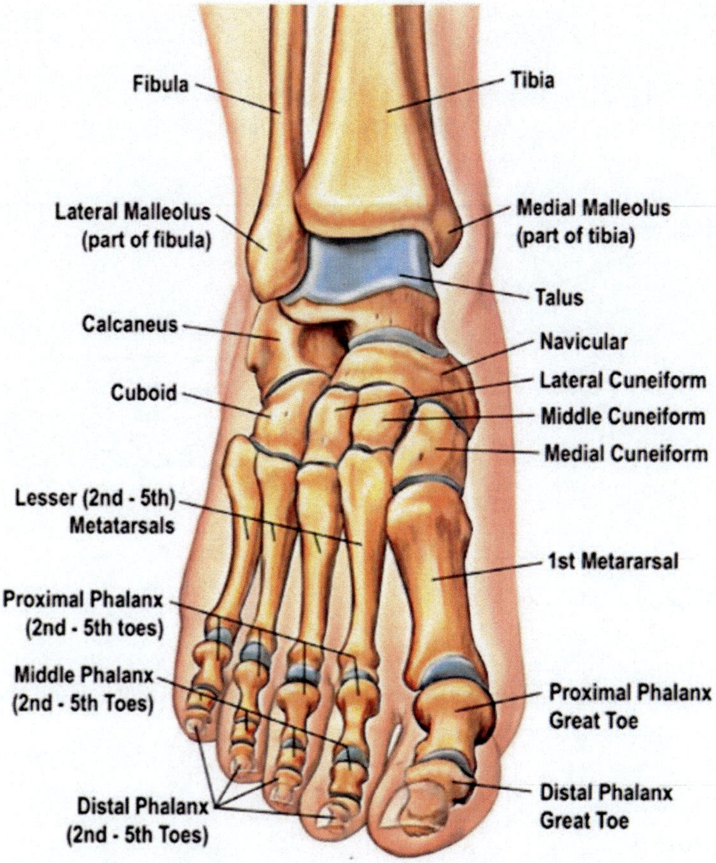

Fig. 1.22 Illustration of the foot

Sesamoid bones: These are two small, oval-shaped bones beneath the first metatarsal on the underside (plantar surface) of the foot. It is embedded in a tendon at the head of the bone (the part nearest to the big toe). Its role is to reinforce and reduce stress on the tendon.

Midfoot

The midfoot is made up of five of the seven tarsal bones: the navicular, cuboid, and medial, middle, and lateral cuneiforms. The junction between the hind and midfoot is termed Chopart's joint, which includes the talonavicular and calcaneocuboid joints.

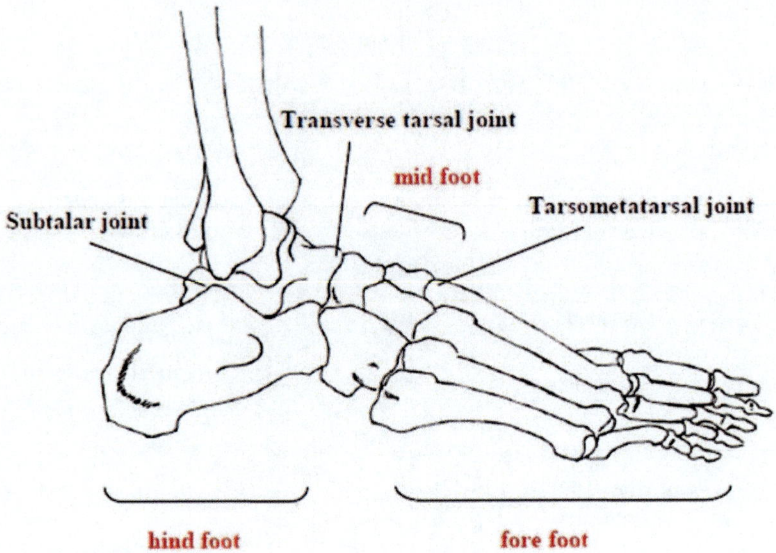

Fig. 1.23 Regions of the foot

Together, the tarsals form the arch of the foot. The arch plays a key role in weight-bearing and foot stability. The articulation of the midfoot and the forefoot forms the Lisfrane joint.

Hindfoot

The hindfoot is the most posterior aspect of the foot. It consists of the talus and calcaneus, two of the seven tarsal bones.

Talus: This is the bone that sits between the calcaneus and the two bones of the lower leg (the tibia and fibula). It helps transfer weight and pressure across the ankle joint.

Calcaneus: This is the large foot at the heel of the foot, also known as the heel bone. Its main function is to transfer most of the body weight from the legs to the ground.

The talus and calcaneus articulation is referred to as the subtaler joint, which has three facets on each of the talus and calcaneus.

The hindfoot motion is often defined as occurring in the cardinal planes as follows: sagittal-plane motion (plantar flexion-dorsiflexion), frontal-plane motion (inversion-eversion), and transverse-plane motion (internal rotation-external rotation). The hindfoot motion, however, does not occur in isolation in the individual planes; rather, coordinated movement of the three joints allows the hindfoot to move as a unit about an axis of rotation oblique to the long axis of the lower leg. The hindfoot motion does not occur strictly in the cardinal planes because the talocrural and subtalar joints each have oblique axes of rotation. Coupled hindfoot motion is best described as pronation and supination. In the open kinetic chain, pronation consists of dorsiflexion, eversion, and external rotation, while supination consists of plantar flexion, inversion, and internal rotation. In the closed kinetic chain, pronation consists of plantar flexion, eversion, and external rotation, while supination consists of dorsiflexion, inversion, and internal rotation [13, 14].

The talocrural, or tibiotalar, joint is formed by the articulation of the dome of the talus, the medial malleolus, the tibial plafond, and the lateral malleolus. The shape of the talocrural joint allows torque to be transmitted from the lower leg (internal and external rotation) to the foot (pronation and supination) during weight-bearing. This joint is sometimes called the "mortise" joint and, in isolation, may be thought of as a hinge joint that allows the motions of plantar flexion and dorsiflexion. The axis of rotation of the talocrural joint passes through the medial and lateral malleoli. It is slightly anterior to the frontal plane as it passes through the tibia but slightly posterior to the frontal plane as it passes through the fibula. Isolated movement of the talocrural joint is primarily in the sagittal plane, but small amounts of transverse- and frontal-plane motion also occur about the oblique axis of rotation [15, 16].

Muscles Controlling Foot Movements

The muscles that control the movements of the foot originate in the lower leg and are attached to the bones in the foot with tendons.

The main muscles that facilitate movement in the foot include:

- Tibialis posterior: The muscle that supports the foot's arch
- Tibialis anterior: The muscle that allows the foot to move upward
- Peroneus longus and brevis: The muscles that control movement on the outside of the ankle
- Extensors: The muscles that raise the toes to make it possible to take a step
- Flexors: The muscles that stabilize the toes and curl them under

Foot-Aided Tendons

Tendons are fibrous connective tissues that attach muscles to bones. There are three major tendons that help facilitate foot movement, including flexion (the forward bending of the foot) and dorsiflexion (the backward bending of the foot):

Achilles tendon: This is the most notable tendon of the foot that runs from the calf muscle to the heel. It is the strongest and largest tendon in the body that makes it possible to run, jump, climb stairs, and stand on your toes [16].

Tibialis posterior: This tendon attaches the calf muscle to the bones on the inside of the foot and supports the arch of the foot.

Tibialis anterior: This tendon runs from the outer bone of the lower leg to the tarsals and first metatarsal, which enables dorsiflexion.

Primary Ligaments of the Foot

Ligaments are fibrous connective tissues that connect bone to bone. These are the primary ligaments of the foot:

Plantar fascia: This is the longest ligament of the foot that runs from the heel to the toes to form the arch. The plantar fascia provides strength for walking and assists with balance [15, 16].

Plantar calcaneonavicular: This is the ligament that connects the calcaneus to the talus. Its role is to support the head of the talus.

Calcaneocuboid: This is the ligament that connects the calcaneus to the tarsal bones. It helps the plantar fascia support the arch of the foot.

Muscle Physiology of the Human Locomotor System

A muscle is a soft tissue, a complex biochemical plant that transforms chemical energy into mechanical energy. It is controlled by neural input from the central nervous system (CNS). Therefore, muscles are the actuators of the musculoskeletal system. They generate forces between two end-points either actively through contraction or passively through their resistance to stretch [4, 11].

The muscle belly is connected to the bones or other muscles at points called origin and insertion by tendons. Muscles and tendons are surrounded by connective tissue, holding them together and separating them from their neighborhood. They also allow the muscle to slide inside this hull during movement and guide it along a predefined path, preventing displacement to the side. The muscle belly itself consists of muscle fibers that are grouped together into muscle fiber bundles (fascicles) as illustrated in Fig. 1.24. The fibers may have a length of approximately 15 centimeter and a diameter between 10 micrometers and 100 micrometers, respectively, in human. From Fig. 1.16, each muscle fiber bundle consists of muscle fibers that consist of a chain of sarcomeres. During muscle activation, a biochemical process increases the overlapping region of the actin and myosin filaments and pulls the z-membranes closer together, which results in muscle contraction [11, 16].

Each fiber is made up of so-called myofibrils that constitute some tightly packed filaments that go from one end of the muscle to the other. These myofibrils form the contractile elements of the muscle and exhibit an approximate diameter of one micrometer.

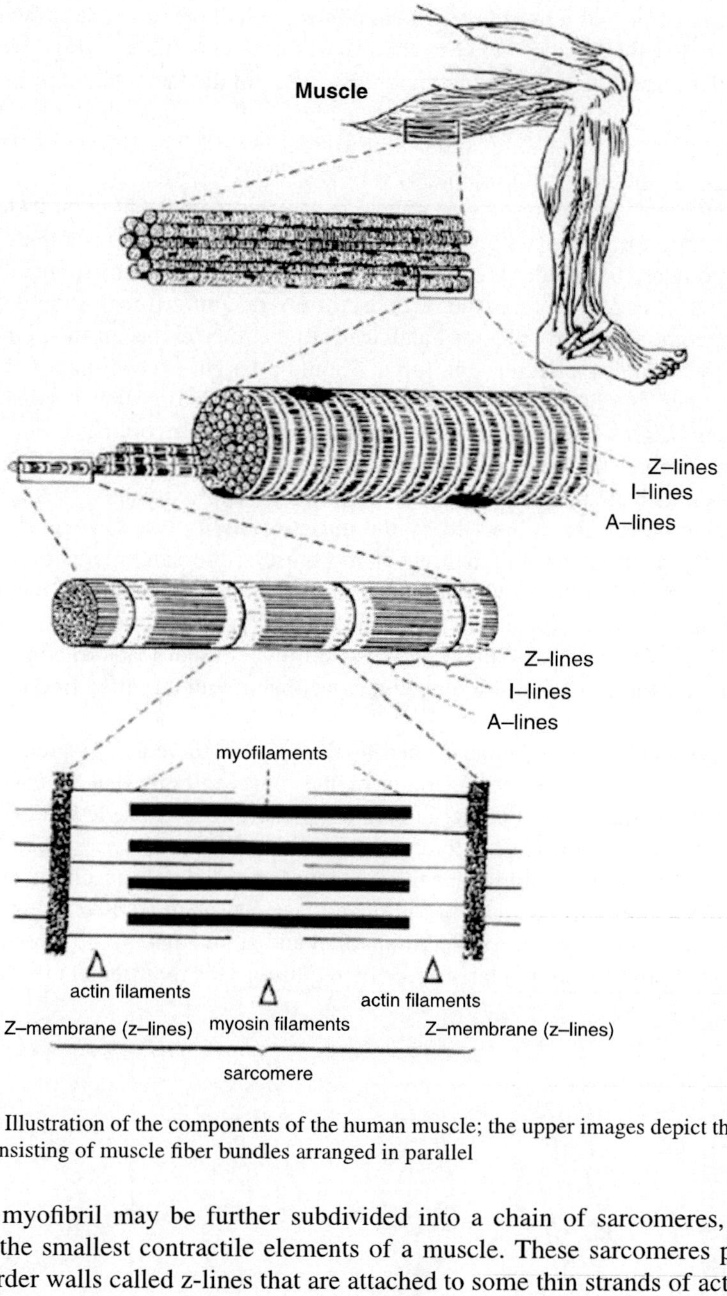

Fig. 1.24 Illustration of the components of the human muscle; the upper images depict the whole muscle consisting of muscle fiber bundles arranged in parallel

Each myofibril may be further subdivided into a chain of sarcomeres, which exhibits the smallest contractile elements of a muscle. These sarcomeres possess some border walls called z-lines that are attached to some thin strands of actin filaments. The distance between these z-lines may be approximated between 2 and 3 micrometers, depending upon the level of contraction of the sarcomere. The myosin filaments are situated between the actin filaments.

The activation of a muscle results in a biochemical process that allows the actin filaments to glide deeper between the myosin filaments (cross-bridge binding), in order to increase the overlapping region. By virtue of the fact that the actin filaments are attached to the z-lines results in the decrease in the distance between the z-lines, the sarcomere shortens, and so does the whole muscle. The shortening of a sarcomere could amount to approximately 60% of its rest length.

The force a sarcomere may be able to generate is dependent upon its length that means, upon the distance between the z-lines. This stands to reason that the longer the sarcomere, the smaller the overlapping region of the actin and myosin filaments, respectively, and, therefore, the decrease in the resulting force. Should the sarcomere become shorter on the one hand, leads to the interference of the filaments, and hence, resulting in a decrease in force. Should a length exceed that of the optimal muscle fiber length, results in the emergence of a passive force that steadily increases with length. This force is a result of the elasticity of the myofibrils.

Furthermore, the optimal muscle fiber length for the active force-length curve decreases with an increase in muscle activation by approximately 15%. The muscle force may be further influenced by the muscle velocity (the change of length per unit time). During the lengthening of the muscle, the sarcomere force increases mainly by virtue of the stretch of the elastic elements. On the other hand, the sarcomere force decreases to zero in a hyperbolic fashion during shortening. This occurs because the filament movement of actin and myosin that leads to contraction follows a repeated process of binding and detachment with a limited frequency, called the cross-bridge cycling.

Based on the task to be performed by the muscle, there are apparent discrepancies in the macroscopic arrangement of the muscle fibers; that means, the more muscle fibers that work together in parallel, the stronger a muscle becomes, and the longer the muscle fiber, the more sarcomeres are linked in series, thus leading to faster contractions. Should a specific volume for a skeletal muscle be available, it may result in the arrangement of optimized fascicles as illustrated in Fig. 1.25.

The fascicles may be arranged in parallel and at an angle to the direction of pull should the production of a large force be required. This requirement demands more

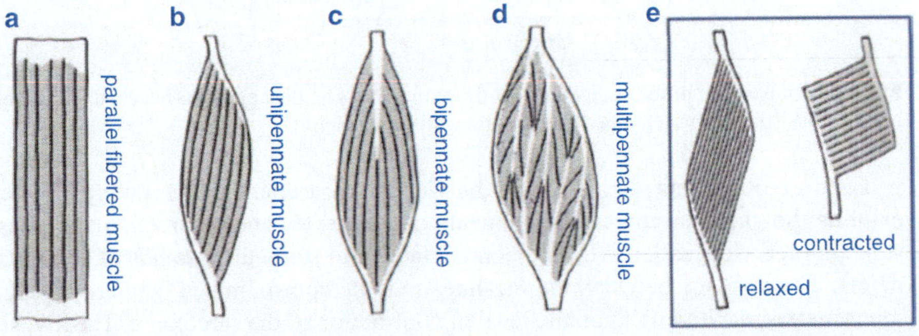

Fig. 1.25 Illustration of different muscle structures (Fig. 1.25e (Adapted from [12]) shows the change of the pennation angle in unipennate muscles through muscle contraction. (Adapted from [UHI96])

fibers in parallel within the same volume by sacrificing a larger range of contraction. Such muscles are called pennate muscles. The angle between the fascicles and the direction of pull along the tendon is called the pennation angle. This angle is not fixed but rather varies with the contraction of the muscle [11].

A measure for the strength of the muscle is the physiological cross-sectional area (PCA) of a muscle that takes into account the number of sarcomeres in parallel with the angle of pull of the muscle. The motor system can measure the actual muscle length and velocity through muscle spindles, which are integrated in the muscles and tendons. They are linked to the spinal cord by nerve fibers to give feedback to the motor system.

Physiology of Muscle Activation and Electromyography

The contraction of the muscle depicts a response to signals that originate from the so-called alpha-motor neurons sitting in the spinal cord or brain stem. The neurons produce an electrical impulse that travels along axons with an average velocity between 50 and 90 m/s to the neuromuscular junctions (motor endplates) sitting on top of the muscle fibers, usually near the middle or proximal to the middle. The axons branch before they reach the fibers, so that every alpha-motor neuron could be connected to a number of muscle fibers ranging from as low as 10 for eye muscles, to about 100 for muscles in the hand, and to approximately 2000 in leg muscles. The lower the number of innervated muscle fibers, the more fine-grained the control of the muscle activation could be performed by the nervous system [4, 11, 16]. However, every muscle fiber is only controlled by a single alpha-motor neuron. The alpha-motor neuron, together with the axon, motor endplate, and muscle fibers that belong to this motor neuron, is called a motor unit.

The action potential that is transmitted to the motor endplates initiates a biochemical process inside the junction to the muscle fiber, the synapse, and in the synaptic cleft between the synapse and the muscle fiber membrane (postsynaptic membrane). If the resulting depolarization at the postsynaptic membrane exceeds a certain threshold, a single muscle fiber action potential may be generated that travels along the muscle fiber with a velocity between 3 and 6 m/s to excite all sarcomeres [11, 16]. This leads to contraction of the sarcomeres and to a single twitch. To achieve a longer period of contraction, a series of action potentials needs to be generated by the motor neuron, that means, a motor unit action potential train. Since a motor unit may be able to perform only an all-or-nothing activation of the muscle fiber (the strength of a twitch cannot be modulated), the strength of contraction would depend on the number of recruited motor units, that means, the number of motor neurons that produce an action potential at the same time. Typically, weaker motor units, innervating less muscle fibers, are recruited first if progressively increasing force is required. In general, motor units fire in a random pattern and are not synchronized [16].

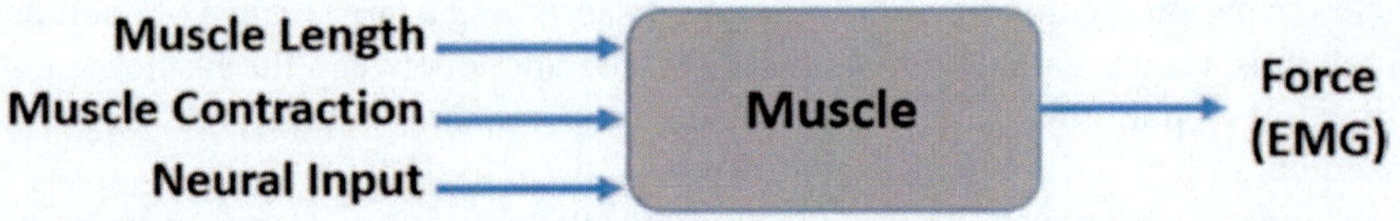

Fig. 1.26 Illustration of a basic muscle model

Muscles exhibit highly non-linear dynamic characteristics. The force from a muscle is dependent upon the muscle length and its contraction velocity. There are muscle models of varying complexities that depend on the overall behavior of the musculoskeletal system (Fig. 1.26).

Contribution of Muscle to Motion Control

Skeletal muscles are the main functional units of the muscular system. The basic contractile properties of a muscle, such as its contraction kinetics, force versus length relationship, and force versus velocity relationship, are also known to affect its mechanical output [16].

Although muscles are often viewed as motors that produce movement by shortening to perform mechanical work, they may serve a variety of other functions during movement. They may stabilize motion at joints, store elastic energy in connective tissues, in tendons or apodemes, and also absorb work as well as perform it. The steady locomotion over level ground often involves the use of muscles to produce force economically. Muscles facilitate elastic energy recovery to achieve minimal net- work output. When movement becomes non-steady or requires changes in grade, shifts in motor recruitment will reflect the changing need for muscles to perform or absorb work. The function of a muscle during movement may also depend on the biomechanical context.

Muscle as a Smart Material with Intrinsic Self-Stabilizing Properties

Not only can different muscles innervated by the same nerve exhibit different functions during movement, but different muscle segments within a single fascicle may also exhibit different mechanical output during a single contraction [11, 16].

In addition to generating force and producing or absorbing energy, muscles also play important intrinsic, self-stabilizing roles during movement due to their force versus velocity, force versus length, and viscoelastic properties. For example, when subjected to a higher force, the force output of skeletal muscle increases automatically to resist the imposed load. Similarly, if a muscle is suddenly unloaded, its rate of shortening increases and its force production decreases. This self-stabilization behavior results from the inverse force versus velocity relationship exhibited by all striated muscles.

Hence, the intrinsic force versus velocity and force versus length properties of striated muscle provide immediate impedance responses that help to stabilize movement in response to perturbations prior to the subsequent action of force-dependent and length-dependent reflexes that incur time delays and are known to modulate motor recruitment during ongoing tasks. These intrinsic musculoskeletal properties, including force versus length and force versus velocity behavior, can stabilize movement and simplify control. The viscoelastic properties of active muscle also provide self-stabilization during perturbations in load and contribute to active motor control.

Cerebral Blood Flow in the Human Locomotor System

Cerebral circulation is the blood flow in the brain. It is important for healthy brain function. Circulating blood supplies the brain with the oxygen and nutrients it needs to function properly [17, 18]. Blood delivers oxygen and glucose to the brain. CBF or perfusion is a measure of the rate of delivery of arterial blood to a capillary bed in the brain tissue.

The brain is unique with a high metabolic rate and its oxygen demand exceeds that of all organs except the heart. It is approximately 2% of body mass and receives 20% of the basal oxygen consumption and 15% of the resting cardiac output (700 ml.min^{-1} in the adult).

Blood supply to the brain originates from the dorsal aorta, provided by the common carotid arteries, which branch into internal carotid arteries, and the basilar artery, formed by the union of the two vertebral arteries, which are branch of the subclavian artery. The anastomoses between these two sets of vessels give rise to the Circle of Willis.

Autoregulation constitutes the ability of the cerebral circulation to maintain cerebral blood flow at a relatively constant level despite changes in cerebral perfusion pressure by altering cerebrovascular resistance (CVR).

Knowledge of the physiology and pathophysiology of the cerebral blood flow (CBF) is essential for the proper treatment of patients with major intracranial disease and for neuroanesthesia in particular [18, 19].

The brain, undertaking normal intellectual functioning, uses oxygen at a rate of approximately 35 mL/min/kg brain tissue, so for a 70 kg man with 1.5 kg brain, basal whole body oxygen consumption is 280 mL/min with brain oxygen consumption of 50 mL/min.

Fractal Dynamics in the Human Locomotor System

A fractal is a never-ending pattern that the laws of nature repeat at different scales. They are created by repeating a simple process over and over in an ongoing feedback loop. The human brain, with its exquisite complexity, can be seen as a fractal object, and fractal analysis can be successfully applied to analyze its wide physiopathological spectrum and to describe its self-similar patterns in both neuroanatomical architecture and neurophysiological time-series [21–23].

The human locomotor system integrates muscles, motor pattern generators, and the brain (motor cortex, cerebellum, and basal ganglia), as well as feedback from sense organs (visual, vestibular, and proprioceptive sensors), to produce a stable, highly consistent, and coordinated walking pattern not only in complex terrain but also when confronted with unexpected perturbations. In fact, the kinetics, kinematics, and muscular activity of gait appear to remain relatively constant from one stride to the next, even during unconstrained walking.

Recent research depicts the existence of some complex fluctuations in the gait pattern, even under constant environmental conditions. These fluctuations were usually considered as "noise" and something to be ignored and filtered out of any analysis. Therefore, quantitative studies of walking typically focused on the properties of each participant's average stride, ignoring the within-subject stride-to-stride fluctuations. However, it has been alleged that this noise actually conveys some important information. Fractal dynamics exhibit the apparent "noisy" variations in the stride interval of human walking. Analysis of the dynamic characteristics of these stride-to-stride fluctuations reveals a self-similar pattern; fluctuations at one time scale are statistically similar to those at multiple other time scales, at least over hundreds of steps, while healthy subjects walk at their normal rate. Statistical physics typically deals with phase transitions, fluctuations, and interactions that occur at the microscopic level during mobility. Knowledge of these fluctuations provides insight into the neural control of locomotion as well as its changes with aging and disease [21–23].

There are some walking patterns that may be observed. One such pattern involves walking that appears to be a periodic, regular process. However, reliable measurement would depict small fluctuations in the gait pattern, even under stationary conditions. A possible analytic interpretation of these stride-to-stride variations in the walking rhythm is that they simply represent uncorrelated (white) noise superimposed on a basically regular process. This is what one might expect a priori if one assumes that these subtle fluctuations are merely "noise."

Another way to interpret these fluctuations is that there are finite-range correlations: the current value is influenced by only the most recent stride intervals, but over the long term, fluctuations are random. Furthermore, a less intuitive possibility involves that the fluctuations in the stride interval exhibit long-range correlations, as seen in a wide class of scale-free phenomena. In this case, the stride interval at any instant would depend (at least in a statistical sense) on the interval at relatively remote times, and this dependence would decay in a scale-free (fractal-like) power-law fashion.

The magnitude of the stride-to-stride fluctuations and their changes over time during human locomotion may be useful in understanding the motor control of gait in quantifying pathologic and age-related alterations in the locomotor control system, and in augmenting objective measurement of mobility and functional status. Indeed, alterations in gait dynamics may help to determine disease severity and medication utility and to objectively document improvements in response to therapeutic interventions, above and beyond what can be gleaned from measures based on the average, typical stride [22, 23]. Upon these understandings, it may be asserted that gait dynamics has meaning that may be useful for providing insight into the neural control of locomotion and for enhancing functional assessment of aging, chronic disease, and their impact on mobility.

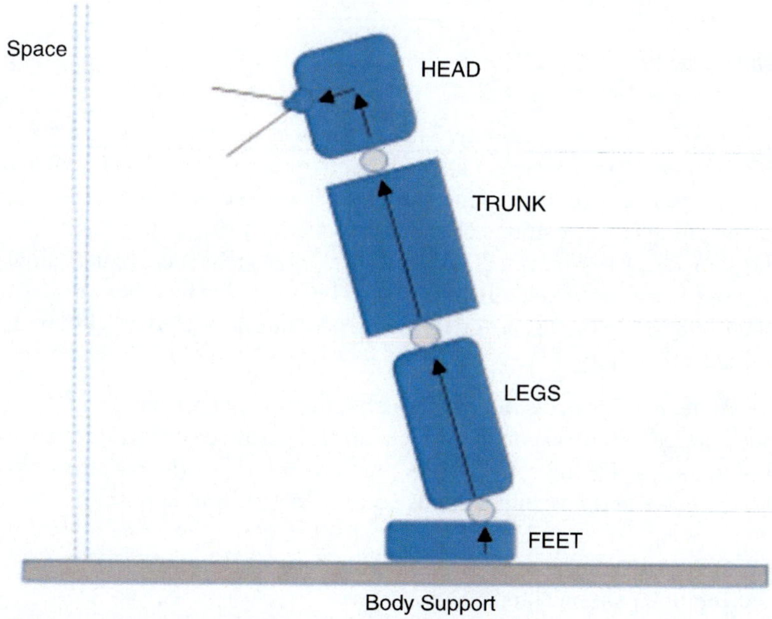

Fig. 1.27 Body on support in space

Perceptual and Feedback: Somatosensory System in the Human Locomotor System

The human locomotor system is structured in a hierarchy such that it consists of linked reference systems that are anchored in gravito-inertia space. This proposed reference system represents the internal reconstruction of the physics, the kinematic aspects of the biomechanical system.

The perceptual notion of space is derived mainly through vestibular sensors [2]. The vestibular canal signal that results from the head motion in space interacts with otolith and optokinetic signals. This interaction is then transferred to the feet by means of axial proprioceptive signals of the inter-segmental angles. This inter-segmental system attaches its space reference to the body's feet, legs (shank and thigh), trunk, and head by means of a set of different proprioceptive inputs, Fig. 1.27.

Both conscious perception and sensory control of action are derived from this resulting interactive system. The information pertaining to the conscious perception and sensory control of action (kinematics and kinetics of the inter-segmental system) is always available to the perceptual and sensorimotor systems because the interactions of the linked reference systems are continuously updated [2]. The kinematic signals and the related kinetic signals represent the global variables for the sensorimotor control. For sensorimotor control, these signals are fed as set point signals into the local proprioceptive feedback loops of the joints, together with voluntary commands.

Somatosensory System

Soma is the Greek word for body. The human somatosensory system serves three major functions. These include (1) exteroceptive (perceiving stimuli outside of our body), (2) interoceptive (perceiving stimuli inside of our body), and (3) proprioceptive (controlling body position and balance) functions [25, 26].

The somatosensory receptors are the sense organs that send signals along a sensory nerve to the spinal cord where they may be processed by other sensory neurons and then relayed to the brain for further processing activated by different stimuli such as the following:

1. Thermoreceptor: carries information about temperature changes.
2. Mechanoreceptors: responds to mechanical pressure or distortion such as lamellar corpuscles (Pacinian corpuscles), tactile corpuscles (Meissner's corpuscles), Merkel nerve endings, and bulbous corpuscles (Ruffini corpuscle).
3. Chemoreceptors: transduces a chemical substance to generate a biological signal, such as taste receptors, carotid bodies.
4. Nociceptors: pain receptor

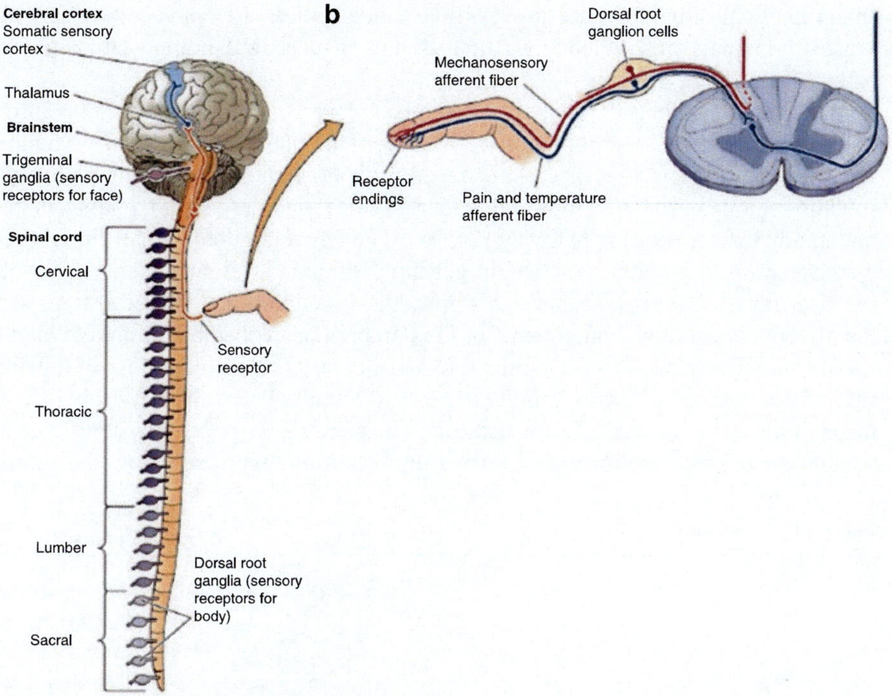

Fig. 1.28 Somatosensory information transmission

- External nociceptors found in the skin (cutaneous nociceptors), the corneas, and the mucosa
- Internal nociceptors found in muscles, joints, bladder, digestive tract
- The cell bodies of these neurons are located in either the dorsal root ganglia or the trigeminal ganglia. The trigeminal ganglia are specialized nerves for the face; the dorsal root ganglia are associated with the rest of the body.

Cutaneous sensory receptors may be classified as mechanoreceptors, thermoreceptors, and nociceptors.

Somatosensory Afferents Convey Information from the Skin Surface to Central Circuits

The sensation of touch is mainly mediated by mechanoreceptors, but there are a number of other processing channels within the somatosensory system for proprioception, pain, and temperature, Fig. 1.28.

Somatosensory information from the face is transmitted via the trigeminal nerve. Pain is conventionally viewed as a submodality of cutaneous sensation. Functional, anatomical, and imaging data suggest that pain impulses are conveyed by specific sensory channels or labeled lines that ascend in a central homeostatic afferent pathway.

A sensory channel is the sensory mechanism responsible for conveying the information needed for the recognition of a sensory modality. Each sensory channel includes one or more sets of sensory receptors, one or more ascending pathways, specific regions of the thalamus and cerebral cortex, and descending pathways that may modify the ascending pathway [24, 25]. The terms large-and small-fibred used were for want of a better pair. The large-fibred somatic afferent system deals with the discriminative-sensory aspects of somaesthesis and the small-fibred system with the affective-vegetative components of the perceptions evoked by all but the blandest of somatic stimuli. The discriminative-sensory systems include those of the dorsal and dorsolateral columns and the trigeminal lemniscal system that deal with the mechanoreceptive aspects of somaesthesis. The affective-vegetative systems convey somatosensory information from all the spinal columns, especially from the ventral

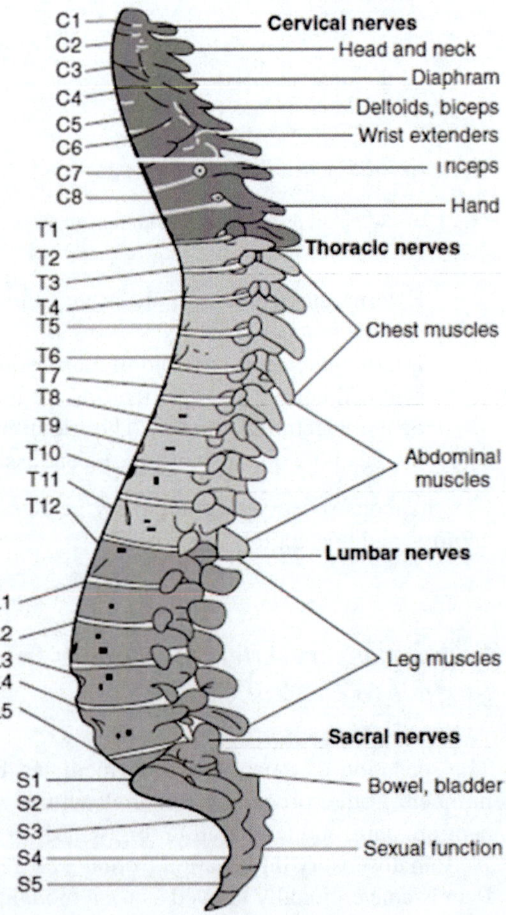

Fig. 1.29 Basic anatomy of the spinal cord

quadrants, and from the spinal trigeminothalamic tract, which not only converge on the relay nuclei of the discriminative-sensory systems but also have their own channels to frontal, lateral, and limbic cortical circuits to deal with affect and emotional reactions [29, 30].

The peripheral nerves of the upper extremity arise from the brachial plexus, formed by the union of the ventral branches of the C5–T1 segments, Fig. 1.29. The main peripheral nerves of the upper extremity are the axillary (C5, C6) and radial (C5–T1) nerves, arising from the posterior cord of the brachial plexus, the musculocutaneous nerve (C5–C7) from its lateral cord, the ulnar nerve (C7–T1) from the medial cord with a contribution from the posterior cord, and the median nerve (C5–T1) from the union of the remaining parts of the lateral and medial cords. Figure 1.29 illustrates the basic anatomy of the spinal cord [30–32].

The sensory branches are shown in Figs. 1.28 and 1.29, respectively. A lesion of a spinal nerve will show a strictly segmental distribution of the sensory deficit, whereas a **lesion** of a peripheral nerve (a mono-neuropathy) will show a typical sensory deficit distribution over the skin area innervated by that particular nerve. A mono-neuropathy is usually due to compression, affecting the myelinated fibers first [33].

The peripheral nerves of the lower extremity arise from the lumbosacral plexus. The main peripheral nerves for the lower limb are the lateral femoral cutaneous nerve (L2, L3), the femoral nerve (L2–L4) and the obturator nerve (L3, L4) from the lumbar plexus, and the superior (L4–S1) and inferior (L5–S2) gluteal nerves and the ischiadic nerve (L4–S3) from the sacral plexus.

Sensations are disparate pieces of information that are acquired and processed by the nervous system to enable the body and mind to adapt to the environment. The vestibular and proprioceptive senses are the least present in our conscious awareness because they operate at an unconscious level [32].

More severe damage may lead to radicular sensory and motor losses associated with abnormal reflexes.

Syndromes of root damage limited to individual cervical roots may be summarized as follows, Fig. 1.29:

1. C3, C4: pain in the neck and shoulder; rarely partial paresis of the diaphragm
2. C5: pain, potentially hypalgesia in dermatome C5; disorders in the innervation of the deltoid and biceps brachii muscles
3. C6: pain, potentially hypalgesia in dermatome C6; paresis of the biceps brachii and brachioradialis muscles; decreased biceps tendon reflex
4. C7: pain, potentially paraesthesias or hypalgesia in dermatome C7; paresis of the triceps brachii and teres muscles and possibly atrophy of the thenar muscles; decreased triceps tendon reflex
5. C8: pain, possibly paraesthesias and hypalgesia in dermatome C8; paresis and possibly atrophy of the hypothenar muscles; decreased triceps tendon reflex

Lumbar discs may also cause irritation to nerve roots, felt as pain and paraesthesias in the corresponding segments as lumbago or sciatica. More severe damage to the roots produces segmental sensory and motor deficits.

Syndromes of lesions limited to individual lumbar roots.
may be summarized as follows, Fig. 1.29:

1. L3: pain, possibly paraesthesias in dermatome L3; paresis of the quadriceps femoris muscle; decreased or missing quadriceps femoris tendon (patellar) reflex
2. L4: pain, possibly paraesthesias or hypalgesia in dermatome L4; paresis of the quadriceps femoris and tibialis anterior muscles; reduced patellar reflex
3. L5: pain, possibly paraesthesias or hypalgesia in dermatome L5; paresis of the extensor hallucis longus and extensor digitorum brevis muscles; absence of posterior tibial tendon reflex
4. S1: pain, possibly paraesthesias or hypalgesia in dermatome S1; paresis of the peroneal and triceps surae muscles; loss of triceps surae tendon (Achilles tendon) reflex

Significance of the Somatosensory System

The human somatosensory system is the system that subserves our sense of touch, which is so essential to our awareness of the world and of our own bodies. Touch could protect our bodies because many receptors are on the skin, which covers the whole body and detects harmful stimuli.

Somatosensory information from the skin, muscles, joint capsules, and viscera is conveyed by the dorsal root ganglion neurons innervating the limbs and trunk or by trigeminal sensory neurons that innervate cranial structures (the face, lips, oral cavity, conjunctiva, and dura mater). These sensory neurons perform major functions: (1) the transduction and encoding of stimuli into electrical signals and (2) the transmission of those signals to the central nervous system.

The somatosensory system has by far the largest number of receptor types of any of the primate sensory systems, including mechanoreceptors, chemoreceptors, nociceptors, and thermoreceptors [35, 36, 38].

The interoception was conceptually differentiated from the exteroception (sensory inputs activated from outside the body), proprioception (sensory inputs that relate to limb position), telereception (sensory input from a distance: vision and hearing), chemoreception (taste and smell), thermo-reception (temperature), and nociception (sensory inputs activated specifically by physically damaging or threatening stimuli). He categorized nociception and thermos-reception with the sense of touch as aspects of exteroception. A century later, Craig [25, 52] suggested to expand on the term interoception and included small-diameter sensory input from the whole body, not only from viscera, muscles, joints, and teeth but also from the skin, the largest organ of the body. In this conceptual framework, nociception and thermos-reception are aspects of interoception, not of exteroception, because they report aspects of the physiological condition of the body conveyed by small-diameter sensory fibers and the spinothalamic pathway to the interoceptive cortex [35].

Interoception is defined as "The sensory representation of the physiological condition of all tissues and organs of the body." Consequently, exteroception in this scheme is redefined to include only touch perception enabled by cutaneous low-threshold mechanoreceptors and the large-fiber system [32].

The sensation of touch is mainly mediated by mechanoreceptors, but there are a number of other processing channels within the somatosensory system for proprioception, pain, and temperature. The classic view of two independent channels for somatosensory information from the trunk and the extremities, i.e., the dorsal column-medial lemniscus system for tactile sensitivity and position sense and the anterolateral or spinothalamic system for pain and temperature sensitivity, has been modified through the discovery of additional spinal pathways for the transmission of sensory impulses to the brain and by new views on pain mechanisms. Somatosensory information from the face is transmitted via the trigeminal nerve [32, 34, 36].

Pain is conventionally viewed as a submodality of cutaneous sensation. Functional, anatomical, and imaging data suggest that pain impulses are conveyed by specific sensory channels or labeled lines that ascend in a central homeostatic afferent pathway.

The overarching purpose of the human somatosensory system is to inform the brain of the mechanical state of the body that it inhabits [36]. It shares this function with the vestibular system. But whereas the vestibular system operates in the low-dimensional space of head translations and rotations, the somatosensory system takes its input from almost the entire body. The main sources of information arise in part from the load-bearing structures represented by connective tissues, such as tendons and ligaments, in part from the motion-producing tissues, the muscles, and in part from the outer layers of body, that is the skin. As a result, unlike the vestibular system, which is sensitive to the movements of a rigid body, the cranium, the somatosensory system relates to mechanical domains that are in essence deformable bodies. This explains why, despite the fact that the two systems share the same overall task, they differ fundamentally [40, 41].

The vestibular system responds to body movement and changes in head position. This system is responsible for coordinating movements of the eye, head, and body through receptors in the inner ear [42, 45, 46]. It responds to the pull of gravity, registers where the body is in relationship to the earth, and is considered to be the bedrock of physical and emotional security. Vestibular input helps us maintain our balance by telling us whether we are at rest or in motion, how fast and in what direction we are moving, and by registering the movement of objects around us.

The vestibular inputs arise from small, easily identifiable organs in the inner ears; it is the low-dimensional description of the movements of a rigid body that is of interest. In contrast, the somatosensory system relates to what is essentially an infinite, dimensional solid (and liquid) domain and depends on the changes of its internal mechanical state to infer the properties of the objects that are being touched, such as their weight, the substance they are made of, or the existence and nature of the relative movement of the body in relation to external objects [43, 44]. In other words, it is a distributed system in the physical sense that its mechanical state is described by (tensor) fields rather than vector quantities. This basic fact is of course reflected in its general organization, where very large populations of specific detectors are found in all load-bearing and load-producing tissues. That is not to say that the somatosensory system is unique in its reliance on large populations of sensors. This is also true of all sensory systems, including vision, audition, and taste/olfaction, and of the vestibular system.

A somatosensory pathway will typically consist of three neurons: primary, secondary, and tertiary. In the periphery, the primary neuron is the sensory receptor that detects sensory stimuli like touch or temperature. The cell body of the primary neuron is housed in the dorsal root ganglion of a spinal nerve or, if sensation is in the head or neck, the ganglia of the trigeminal or cranial nerves [32]. The secondary neuron acts as a relay and is located in either the spinal cord or the brain stem. This neuron's ascending axons will cross, or decussate, to the opposite side of the spinal cord or brainstem and travel up the spinal cord to the brain, where most will terminate in either the thalamus or the cerebellum. In the case of touch, the tertiary neuron has its cell body in the thalamus and projects to the parietal lobe of the brain. In the case of the maintenance of posture, the tertiary neuron is located in the cerebellum.

Proprioception (Latin *proprius*, one's own) and it is through proprioceptive input that we know our own body, where it is, and how it is moving. Proprioceptors transmit information to the brain from the muscles, joints, and bones. Because there are so many joints, muscles, and bones in the human body, this system is almost as large as the tactile system (about 70% of the sensory receptors are located in the skin). It transmits continuous information to the brain.

The proprioceptive senses are generated as a result of our own actions that include the senses of position and movement of our limbs and trunk, the sense of effort, the sense of force, and the sense of heaviness. The receptors that enable proprioception are located in the skin, muscles, and joints. The information concerning the position of the limb and movement is not generated by individual receptors, but by population of afferents [27, 33, 35]. Afferent signals generated during a movement are processed to code for the end point position of a limb. The afferent input is referred to a central body map to determine the location of the limbs in space.

Receptors in skeletal muscle, joint capsules, and the skin enable us to have conscious awareness of the posture and movements of our own body, particularly the four limbs and the head. Although one can move parts of the body without sensory feedback from proprioceptors, the movements are often clumsy, poorly coordinated, and inadequately adapted to complex tasks, particularly if visual guidance is absent. Vibration produces sensations of limb displacement and movement, leading the subject to express astonishment at the unwilled nature of the sensations. This suggests that the will to move and the subsequent proprioceptive sensations are intimately linked. Proprioceptive sensations are mysterious because we are largely unaware of them [27, 36]. They are distinguishable from exteroceptors such as the eye and the ear in that they are not associated with specific, recognizable sensations.

If sensory receptors contribute to conscious sensation, their afferents must project to the cerebral cortex. It is now well established from studies on animals that afferents from muscle spindles and tendon organs, as well as those from skin and joint, project to the cortex. The pathway taken by skin and joint afferents is via the dorsal (posterior) columns, gracile and cuneate nuclei, medial lemniscus, and thalamus. Muscle afferents from the forelimb also project centrally via the dorsal columns [32, 35, 36]. However, for the hindlimbs, muscle afferents, unlike skin and joint afferents, leave the dorsal column in the upper lumbar region, synapse in Clarke's column, and project centrally in the dorsolateral funiculus as the dorsal spinocerebellar tract. Spindles provide information about length changes in muscles, and this is represented as changes in joint angles. Therefore, the vibration of one muscle group is able to generate a muscle-specific sensation. This stands to reason that the brain has access to information specific to individual muscles. While that is so, it does not preclude the possibility that input from the whole limb is used to calculate movement-related proprioceptive signals. The dynamic sensitivity of muscle spindles and skin stretch receptors provides the basis for our ability to detect small movements of our limbs [40].

Proprioceptive senses, particularly those of limb position and movement, deteriorate with age and are associated with an increased risk of falls in the elderly.

Both the vestibular and proprioceptive systems regulate our posture and muscle tone. They offer us information in relation to whether we are standing or sitting, slouching or squatting, bending or straightening, or stretching [37]. They assist in the negotiation of the distance from other people and objects. The vestibular and proprioceptive systems together excite such activities, including running, jumping, climbing, spinning, and swinging, etc.

Information Acquisition Through Proprioception

Robust locomotion relies on information from proprioceptors: sensory organs that communicate the position of body parts to the spinal cord and brain. Proprioceptive circuits in the spinal cord are known to regulate locomotion in challenging environments.

In mammals, information about body position in space and relative to the body itself (i.e., "proprioception") is conveyed by mechanosensory neurons found within muscles, tendons, and joints [36, 40].

Proprioceptive senses play significant roles in signaling body shape, body position and movement, and muscle force. Hence, proprioceptive senses are generated as a result of our own actions and/or behaviors. They include the senses of position and movement of our limbs and trunk, the sense of effort, the sense of force, and the sense of heaviness. In fact, the term *proprioceptor* has been restricted to receptors concerned with conscious sensations, and these include the senses of limb position and movement, the sense of tension or force, the sense of effort, and the sense of balance. Receptors involved in proprioception are located in the skin, muscles, and joints [39–41]. Information about limb position and movement is not generated by individual receptors, but by populations of afferents. Afferent signals generated during a movement are processed to code for the end point position of a limb. The afferent input is referred to a central body map to determine the location of the limbs in space.

However, like the proprioceptive system, the vestibular system contributes to a range of conscious sensations as well as the guidance of movement and posture. Proprioceptive sensations are mysterious because we are largely unaware of them. They are distinguishable from exteroceptors such as the eye and the ear in that they are not associated with specific, recognizable sensations. Yet, when we are not actually looking at our limbs, we are able to indicate with reasonable accuracy their positions and whether they are moving.

During limb movement and changes in position, the tissues around the relevant joints will be deformed, including the skin, muscles, tendons, fascia, joint capsules, and ligaments. All these tissues are innervated by mechanically sensitive receptors, and their density varies across muscles and regions of the body. Muscle spindles play a major role in kinesthesia, with some skin receptors providing additional information. Furthermore, the Golgi tendon organs contribute to proprioception, including the senses of force and heaviness. Joint receptors probably play only a minor role at most joints, acting as limit detectors. There is also some evidence of a contribution by joint receptors in the mid-range of movements at the finger joints [39, 41].

The afferents of sensory receptors project to the cerebral cortex in order to contribute to conscious sensations. In fact, animal studies have established that afferents from muscle spindles and tendon organs, as well as those from skin and joint, project to the cortex.

The pathway taken by skin and joint afferents is via the dorsal (posterior) columns, gracile and cuneate nuclei, medial lemniscus, and thalamus. Muscle afferents from the forelimb also project centrally via the dorsal columns. However, for the hindlimbs, muscle afferents, unlike skin and joint afferents, leave the dorsal column in the upper lumbar region, synapse in Clarke's column, and project centrally in the dorsolateral funiculus as the dorsal spinocerebellar tract.

Exteroception

Exteroception is the sense of direct interaction with the external world as it impacts on the body. The principal mode of exteroception is the sense of touch, which includes sensations of contact, pressure, stroking, motion, and vibration, and it is used to identify objects. Some touch involves an active motor component—stroking, tapping, grasping, or pressing, whereby a part of the body is moved against another surface or organism. The sensory and motor components of touch are intimately connected anatomically in the brain and are important in guiding behavior [40, 41].

Exteroception also includes the thermal senses of heat and cold. Thermal sensations are important controllers of behavior and homeostatic mechanisms needed to maintain the body temperature near 37 °C (98.6 °F). In fact, exteroception includes the sense of pain, or nociception, as a response to external events that damage or harm the body. Nociception is a prime motivator of actions necessary for survival, such as withdrawal or combat.

The third component of somatic sensation, **interoception,** is the sense of the function of the major organ systems of the body and its internal state. Although most of the events recorded by receptors in the viscera do not become conscious sensations, the information conveyed by these receptors is crucial for regulating autonomic functions, particularly in the cardiovascular, respiratory, digestive, and renal systems. Interoceptors are primarily chemoreceptors that monitor organ function through such indicators as blood gases and pH.

Information Acquisition Through Exteroception and Interoception

Exteroception, which is perceived as our sense and interaction with the external world, encompasses (according to most definitions) the primary sensory systems of vision, audition, olfaction, taste, and somatosensation.

Interoception refers to the perception of the physiological conditions of the body, including hunger, temperature, and heart rate. There is a growing appreciation that interoception is integral to higher-order cognition.

Interoception is described as the perception of the internal state of one's body; as such, signals including those relating to hunger, temperature, heart rate, and blood sugar levels are all interocep-tive in nature. These bodily signals are thought to be represented within the insula and anterior cingulate cortex (ACC), leading these structures to be collectively referred to as the interoceptive cortex.

In addition to the importance of interoception for physical health, recent research has suggested that interoception may play a role in higher-order cognition, such as in emotional memory [28].

The interoceptive system encompasses the ability of the nervous system to represent our own internal world.

Although the efferent role of the autonomic nervous system (ANS) in homeostasis has long been recognized, the afferent aspects of the ANS are increasingly recognized as equally important. Interoception is fundamental to the regulation of internal physiology, particularly as it is coordinated with contextually determined and adaptive behavioral processes. A cardinal, but often underappreciated, feature of interoception is its complex role in a myriad of integrative cognitive and affective processes in health and disease.

Interoception is a multidimensional construct, broadly encompassing the processing of afferent (sensory) information arising from internal organs, tissues, and cells of the body. This afference contributes to the regulation of homeostatic reflexes, and to the generation and regulation of cognitive and emotional behaviors.

Interoception can be encompassed by the broader construct of bodily afferents. The latter includes both visceral afferents and somatic afferents. We use the term visceral afferents to refer to the processing of internal sensory information derived from interoceptors that are located in the organs, tissues, and cells of the main cavities of the body (e.g., the viscera), as well as from olfactory and gustatory receptors, all being generally associated with the limbic system and the autonomic nervous system [45, 46]. We use the term somatic afferents to refer to the processing of sensory information (e.g., proprioceptive input and tactile sensitivity) derived from components of the somatic system (e.g., muscles, joints, skin). This distinction between somatic and visceral afferents does not imply complete independence. Indeed, in many cases, there is an integration of multiple modes of bodily or somatosensory information derived, for example, from metabolic changes in active muscle tissue. Hence, the term somato-visceral afferents is more appropriately applied to integrated, multimodal, or otherwise nonspecific internal sensory input from within the body. In these regards, the construct of interoception itself is more specifically aligned with that of visceral afferents, referring to the processing of sensory information from interoceptors that are located within the visceral organs and from interoceptors located elsewhere in the body that provide for local energy needs. Thus, in contrast to exteroceptors, interoceptors are tuned to sense internal events.

The so-called general visceral afferents (GVAs) that relay internal sensory information from interoceptors are carried by several cranial nerves, the most notable being the vagus nerve. These afferents carry information (e.g., pressor receptor activity) from the gut and the viscera more generally (i.e., organs and tissues located in the thoracic, abdominal, and pelvic cavities, as well as blood vessels and muscles). By comparison, special visceral afferents (SVAs) convey gustatory senses (i.e., taste) and olfaction (this means, smell and pheromonal senses). Although the SVAs detect environmental stimuli, they do so by virtue of those stimuli impinging on the bodily environment. Hence, they differ from exteroceptors, for example, in conveying information related to touch or audition [47]. Furthermore, the visceral senses have common central projections to cell groups in the thalamus that are distinct from those of somatic exteroceptors, and they link anatomically and functionally with a distinct set of central neural systems and processes. Moreover, they share biochemical markers in common with GVAs afferents and with autonomic. There are other classes of sensory systems, such as proprioceptors, that sense joint position, and vestibuloceptors, that sense body orientation in gravitational space. These might be considered interoceptors, as they are internal to the body. Yet, they are closely linked with somatic motor systems anatomically and functionally, and they have biochemical markers more in concert with somatic motor systems. Hence, they are sometimes considered within the unique class of proprioceptors or otherwise just included within the general class of exteroceptors.

Systems Analysis Approach in Human Locomotor System

A primary goal of the systems analysis approach is to simplify the immense complexity of the locomotor system (with its multi-body dynamics, biomechanics, multi-sensory integration, feedback and feed forward control, cognition, volition, etc.). The application of parameter identification on a given subject may offer insight into the control functions and the deficits, and may enable us to better understand the effects of therapy, and to design new therapeutic concepts. The systems approach is more convenient because the multi-segmented organism resulting from the system of linked kinematic references exhibits high complexity of biological systems and non-linear characteristics [2, 5].

The approach may provide a wealth of information on basic mechanisms, such as the postural reflexes. Noticeably, because of these features, it is essentially impossible to describe biological functions by means of mathematical formulations.

The systems analysis approach compares external stimuli and responses to make inferences on the mechanisms inside the multi-segmented linked references of the musculoskeletal system, thereby arriving at formal descriptions of the mechanisms. The application of psychophysics, that is, by measuring perception while controlling for action and cognition, becomes an essential methodology in the input-output relationships within the complex musculoskeletal system. Some internal representations of body kinematics can be accessed by psychophysics, that is, by measuring motion perception.

During dynamic behavior in space, many receptors in various sensor systems are activated in the body. Fortunately, the most relevant contributions come from a few sensors only. These include the proprioceptive, vestibular, and visual sensors.

The Vestibular System

Anatomically, this system consists of two parts: the macular organs and the semicircular canals. Their functions are those of a 3D accelerometer and a 3D gyrometer (angular speedometer), respectively. With only the biological accelerometers in the head, a distinction between head tilt and head linear acceleration would not be possible. In combination with the gyrometer, however, such a distinction is possible. This "canal-otolith interaction" represents a sensor fusion that yields from the two input organs three signals: i) head rotational velocity, ii) head angle with respect to the gravitational vector, and iii) head linear acceleration. Simulations of this sensor fusion predicted, in accordance with experimental findings, improvements of transmitted information. Improvements are required because signal transfer in the biological gyrometer from angular acceleration, the stimulus, to velocity, the signal in primary canal afferent nerve fibers, is not ideal. Mathematically, this transfer corresponds to an incomplete integration.

The Joint Angle Sense (Proprioception)

We know today that the position sense of a limb involves not only spindle receptors in the muscles but also receptors in the skin and in the capsules of the joints. When tested with psychophysical means, one finds that the brain combines these many signals into a rather accurate position sense over a broad dynamic range.

The technical equivalent of this sensor would be a goniometer (an angle measuring device, often simply realized in the form of a pivotable potentiometer). From the angular position signal, we may then derive, in addition, the corresponding velocity signal (by differentiation). Physiological and psychophysical evidence clearly indicates such an additional velocity signal. A simple way to demonstrate this is to activate, from the outside, primary spindle afferents by muscle vibration of an immobilized arm, for instance. The arm is then perceived as moving (but, paradoxically, not as displaced, which may indicate a functional segregation of central velocity and position pathways).

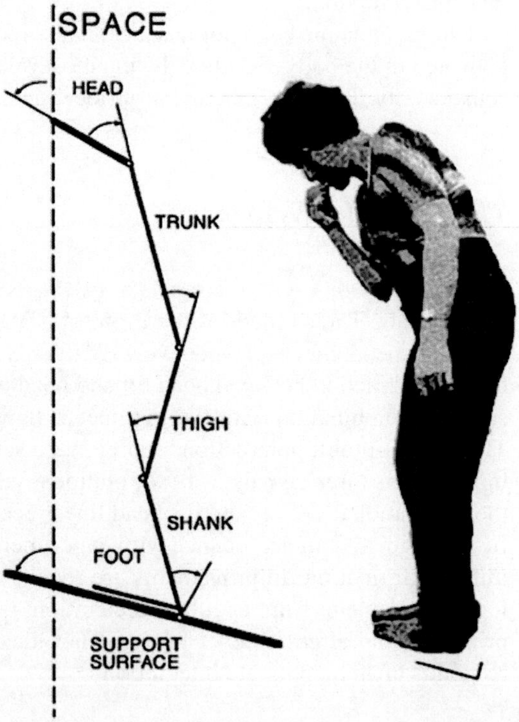

Fig. 1.30 Balancing of the body

The Visual Orientation Cues

Vision provides information about the relative motion (velocity) between body and the visual surroundings. Furthermore, vertical and horizontal items in the scene are used to establish a notion of the "visual vertical," that is, of the body angle with respect to the vertical elements in the scene (with an orientation parallel to that of trees and orthogonal to the horizon).

Corresponding technical sensors (cameras and image processing devices) which can extract those signals are nowadays available.

The Somatosensory Plantar Pressure Receptors

The sum of all reaction forces at the level of the foot soles and the support surface, Fig. 1.30, may be mathematically treated as acting at one point, called the Center of Pressure (COP; in posturography, COP shifts arising during body sway are measured with the help of a 2D force platform). Constituents of the COP are (i) the gravitational vector of the body's Center of Mass (COM), (ii) the active torque that is produced in the ankle joint, and (iii) the effects of external contact forces (such as a push against or a pull on the body) to the extent that they are transmitted to the feet [49, 50]. The receptors appear to be located in deep structures of the foot arch, whereas more superficial mechanoreceptors in the foot sole skin appear to be used to analyze the texture of the support surface, slip, etc.

Knowing the orientation of the head in space (e.g., through vestibular signals) and the angles between the body segments means that we also continuously know the orientation of the other segments in space, and here, during standing with firm foot-ground contact, also of the body support surface.

Data Acquisition in Human Locomotor System

During biped stance or locomotion, humans show remarkable skills in reactive balancing upon external disturbances. Balancing is often investigated in response to three external mechanical stimuli, Fig. 1.30. These include the body support surface tilt, the surface translation, and the contact force (e.g., a push or pull having impact on the body). It may be considered that the standing body is an inverted pendulum and is, therefore, inherently unstable. Gravitational torque arising with its angular excursions away from the ideal gravitational vertical tends to tip it over. Therefore, gravity will here be considered the fourth external stimulus.

Taken together, it stands to reason that the most relevant external stimulus types to be considered for posture control are the field force gravity, contact forces, support surface translational acceleration, and support surface rotation. With the human body's biomechanics modeled as a single inverted pendulum, the stimuli mechanically affect stance through torque disturbances (gravitational, external, inertial, and passive). For disturbance estimation and compensation, three sensors (vestibular, ankle angle, and ankle torque) appear to be instrumental.

These underlying physical stimuli are very essential toward the control of stance. Hence, the sensors and the physical variables they measure play such a significant role in the understanding of the modeling of the stance. It is essential to consider only the sensors that are instrumental for balancing: the vestibular, the joint angle, and the joint torque sensors, respectively. Visual and tactile orientation cues, as well as foot sole shear cues, remain unconsidered because they are not instrumental.

Since it is assumed that very large and fast body movements do not occur during stance, the field forces centrifugal and Coriolis forces are not considered at all. For purposes of simplicity, it may be assumed that all rotations occur at the ankle joint and that foot-support contact is fixed.

The main physical variables that arise with the four stimuli and are measured by the following three sensor systems include:

- The ankle angle proprioception (PROP): This yields the measurements of body-foot angle and the angular velocity, which are essentially ideal. They illustrate the unity sensor transfer function over the considered range of body sway amplitudes and velocities, and frequencies.
- The vestibular system (VEST): This yields broad-band pass measures of the vertical body-space angle and the angular velocity. Furthermore, it provides a measure of the translational head-space acceleration. Therefore, one may actually distinguish between a vestibular rotation sensor and a vestibular translation sensor.
- The ankle torque sensor (TORQUE): This measure of the torque may be derived from the Golgi tendon organs of the ankle muscles (located at the muscle-tendon junctions) or from the COP sensing pressure receptors deep in the foot soles.

Studies in psychophysics have shown that humans use the vestibular signals for perceiving not only head-space motion but also trunk-space motion (e.g., of motion of the COM in the torso) or foot-space motion (or even foot support-space motion, given fixed haptic foot-support contact). This is achieved through inter-link sensory coordinate transformations using, for example, neck proprioception during head-trunk movements.

Data Acquisition: The Vestibular System

The vestibular system supports four physiological functions of spatially oriented behavior: perceptual orientation, body stabilization, retinal image stabilization (mainly via compensatory eye movements, the vestibulo-ocular reflex, VOR), and autonomic nervous functions that include reactive blood redistribution upon changes in body orientation.

The vestibular organs are located as a symmetric pair in the right and left inner ear. Encapsulated by bone, they encode field forces such as gravito-inertial forces. Each organ comprises two receptor systems; (1) In the otolith transducer, the shear force between a heavy membrane and its support bends the sensory hair cells, yielding a neural signal of gravito-inertial acceleration, and (2) The semicircular canal transducers comprise fluid filled circular tubes, one for each of the three orthogonal rotational planes. Upon angular acceleration, fluid inertia exerts pressure on hair cells embedded in a gelatinous membrane (cupula) that obstructs fluid rotation around the canal. This leads to a neural signal during head rotation, but not translation. Due to its viscous-elastic properties, the responds to angular accelerations at frequencies above 0.1 Hz with a 90-degree phase lag, i.e., with a velocity signal. (Thus, the transducer can be viewed as an angular velocity meter with a decay time constant; canal time constant $T = 5$ s.)

Vertigo and disequilibrium arise most often from unilateral canal lesions. Underlying is that vestibular coding of rotation is a bilaterally distributed push-pull system that modulates activity about a resting level. Consequently, pathological firing reduction on one side leads to an imbalance of the push-pull system, resulting in a self-rotation illusion.

The ensuing attempt to compensate for the apparent rotation destabilizes posture and gaze and induces nausea. Fortunately, the symptoms occur only transiently. Symmetry of "vestibular tone" tends to become re-established within days and weeks by adaptive mechanisms [48, 50, 51].

Chronic bilateral vestibular loss hardly leads to clinical complaints.

However, closer scrutiny may reveal blurred vision during fast self-motion as well as postural instability when standing or walking on soft ground in the absence of a visual or tactile space reference. The impairments may be quantified by electro-oculography and posturography, respectively. They witness the importance of the vestibular system for human gaze and body stabilization.

Each of the two vestibular transducers comes with a problem. The canal transducer problem arises from the integrations of the input signal (mathematical angular acceleration-to-velocity and velocity-to-position integrations). This processing emphasizes low-frequency components of noise.

Data Acquisition: The Joint Angle Sensor

Muscle lengthening during a rotation of a joint leads to a firing increase of sensory endings in muscle receptors called muscle spindles. These spindles make a major contribution to the joint angle and movement sense ("muscle proprioception," "kinesthesia"). Since the muscle is elastic, however, the sense has to take into account, in addition to muscle length, also force, both in active and passive conditions. The force information appears to stem from transducers in the tendons. With active movements, there also appears to be a contribution from central sources. Furthermore, transducers in the skin and in the joint capsules participate.

The joint angle sensor makes a major contribution to reactive balancing. This is witnessed by the lean responses that result when spindles of ankle muscles are selectively activated through tendon vibration. Loss of large-diameter sensory nerve fibers stemming from the spindles dramatically degrades postural skills, although the effect may be obscured by visual control or mentally commanded postures.

Data Acquisition: The Force/Torque Sensor

Mainly, two transducer types are important, one being the aforementioned GTOs in the tendons. At the other end of the causal chain that links the force produced by the ankle muscles to the ground reaction force are pressure and tension transducers spatially distributed in the foot and its sole. From both transducers, a measure of ankle torque may be derived.

It is often held that humans use force cues to maintain the COP within the base of the foot support (a popular notion also in robotics). A "tonic excursion limiter" was identified in vestibular-able and vestibular-loss humans, but vestibular-able subjects appear to use vestibular cues for it. A likely reason why they do not prioritize force cues is that these may become less reliable than vestibular body-space cues when standing or walking on compliant support surfaces.

Data Acquisition: The Psychophysical Evidence for Sensor Concept

The concept of a vestibular sensor that arises from canal-otolith fusion is based on congruent evidence from psychophysics and.

VOR studies. Two intuitive psychophysical observations on canal-otolith interaction are briefly described. First, the self-motion percept during passive whole-body rotation in the earth-horizontal plane shows the canals' decay time constant, whereas it shows static (gravitational) sensitivity in the vertical planes. Second, unusual rapid and lasting horizontal translation, e.g., during aircraft catapult launch on a carrier ship, leads to a body-space tilt illusion, which is a typical feature of the complementary filter fusion method. Also, the concept of a joint angle sensor holds.

In summary, canal and otolith transducer signals are fused in the brain, yielding a vestibular sensor that provides 3D information of the head angular velocity, attitude, and translational acceleration in space. Furthermore, sensors of joint angle and angular velocity and of joint torque can be posited. The brain obviously makes an effort to derive from many transducer signals explicit measures of the relevant kinematic and kinetic variables (the same that are used in physics and engineering sciences). This makes it easy to implement human-like technical sensors in humanoids for balancing. The vestibular sensor may be mimicked, for example, by combining gyros and accelerometers, the ankle angle sensor by a goniometer, and the ankle torque sensor by force sensors in the "muscle tendons" and/or by compression load cells under the feet.

The Meta Level Concept

The psychophysics of human perception of passive self-motion provides information not only on how humans fuse transducer signals to obtain physical variable measures (see above, sensor concept) but also on how these measures are further used (in neuroscience, these aspects are often combined in the term "multisensory integration"). Psychophysics provides here information that, to date, is not directly available from single neuron recordings in animals nor from other electrophysiological or imaging data (psychophysics contributes to make the "black box" system identification approach a "gray box" approach). The psychophysical function principles are applied to reactive sensorimotor control with the argument that, there, basically the same variables are processed as in spatial orientation. This is a plausibility assumption that draws on a mostly valid action-perception congruency.

It has been suggested that a neck proprioceptive body-head signal is used to transfer the vestibular space reference from the head to the trunk (to the COM) and further proprioceptive signals to transfer it to the foot (for the foot-space estimate).

Another important finding was that conscious self-motion perception combines different sensor signals to reconstruct external stimuli. During passive body rotation on a platform, for example, subjects relate their sensation of self-motion not to the vestibular signal in the head but to the physical stimulus, i.e., to the platform-space motion (and physically correct consider their body motion a consequence of the platform motion).

They are not aware of the underlying sensory transformations of the space reference from the vestibular organs in the head via the body to the feet and their support.

Postural Reflex Concept

The basic neural feedback loop from a sensor via the spinal cord back to a muscle may clinically be tested using muscle stretch (muscle stretch reflex), for example, in the form of the Achilles tendon tapping reflex. The reflex tends to be weakened with impaired nerve conduction or spinal transmission. An "exaggerated" reflex response may result from lesions of higher control centers ("release from inhibition" effect). The reflex is involuntary and stereotype and occasionally referred to as a "short latency reflex" (latency -20–80 ms). It may represent the early part of a more complex response, where the later parts, involving ascending loops in the spinal cord and brain, are called median-latency and long-latency reflexes. These arise in a context-dependent way. The stretch reflex uses the joint angle sensor and stabilizes a given desired segment (link) position. A local version of it is the neck-spinal (or cervico-spinal) reflex. These proprioceptive intersegmental reflexes together with vestibulo-spinal reflexes represent the basis of the involuntary and stereotypical postural stabilizations that one may observe in newborn babies, brain-injured human adults, and so-called reduced animal preparations (in which higher brain functions are eliminated).

In this classical postural reflex concept, reflex chains are assumed to explain more complex (multi-segmental) stabilizations, such as the righting of a body from a horizontal to a vertical posture after falling down. It is suggested that all movements build in some way on reflexes that serve as a kind of motor primitives.

In intact and mature individuals, postural reflexes do not stand out anymore. They appear to be hidden (integrated) in more complex mechanisms that took over during childhood sensorimotor development with training and maturation of neural pathways and centers. But they may re-emerge in the adult upon large brain lesions.

Posture-Movement Problem

In traditional reflex physiology, it was often assumed that the postural reflexes stabilize postures that, per definition, are static. It was believed that the reflexes would have to be suppressed during voluntary movements, so as not to hinder them. A modern view would be that not only static postures but also movements require stabilization against unforeseen or foreseen external disturbances. Another concern was that the sensory inflow arising upon voluntary movement interferes with the sensory inflow that evokes the postural reflexes. These issues are still debated to date.

There were several suggestions of how to solve the problems, some being still influential to date. One suggestion was that, instead of reflex inhibition, reflex interactions might do the job. The example considered was an interaction of the vestibulo-spinal and cervico-spinal reflexes.

Each of them tries to stabilize the body when evoked in isolation (vestibular: whole body rotation; neck: trunk-only rotation). But when combined during head rotation, they are opposite in sign and tend to cancel each other. Therefore, so it was held, head movements can occur unhindered and furthermore do not endanger body stability. The flaw of this hypothesis is that the reflexes would still hinder active body movements. The idea of vestibular-proprioceptive interaction, however, is still present in the meta-level concept.

In primates and humans, meta-level concept is commonly applied in models of the visuo-oculomotor system, where eye muscle proprioception appears to play no considerable role and the eyeball mechanics and motor control are relatively simple (e.g., gravity can remain unconsidered). Concerning the human skeleton-motor system, evidence is still sparse and speculative, at least as concerns the efference copy of the motor command signal. In engineering, this is often used as a constituent of

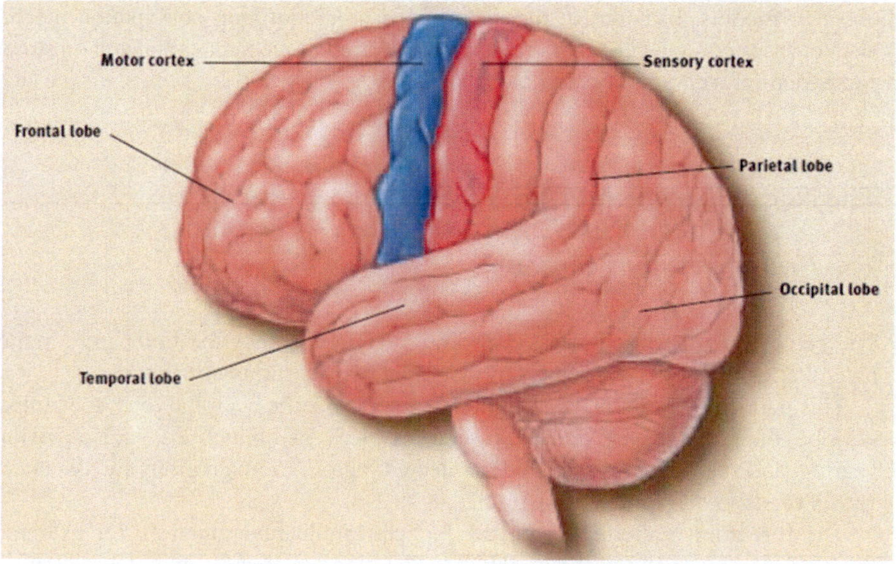

Fig. 1.31 The brain: cerebral cortex

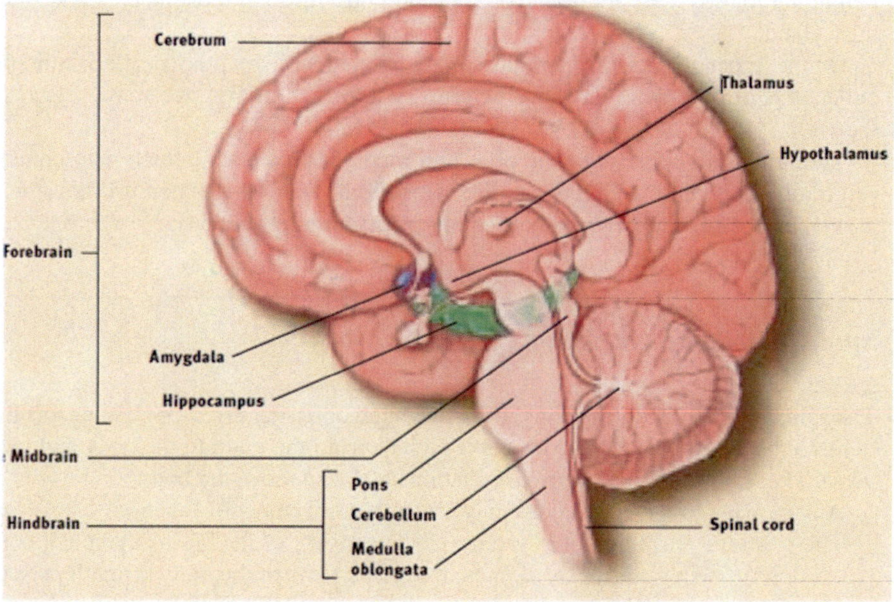

Fig. 1.32 The brain major internal structures

observer models. In contrast, the disturbance prediction and anticipation mechanism of the meta-level concept draws on the desired movement signal and on stored contextual information of external disturbances.

The Nervous System (CNS, PNS)

The Brain (Cerebral Cortex)

The cerebral cortex, as illustrated in Fig. 1.31, constitutes the outer gray matter layer that completely covers the surface of the two cerebral hemispheres.

This part of the brain is divided into four sections: the occipital lobe, the temporal lobe, the parietal lobe, and the frontal lobe [53, 54]. Functions, such as vision, hearing, and speech, are distributed in selected regions. Some regions are associated with more than one function, Fig. 1.32.

The forebrain is credited with the highest intellectual function, for example, thinking, planning, and problem-solving. The hippocampus is involved in memory [45, 51, 53, 54]. The thalamus serves as a relay station for almost all of the information coming into the brain. Neurons in the hypothalamus serve as relay stations for internal regulatory systems by monitoring information coming in from the autonomic nervous system and commanding the body through those nerves and pituitary gland.

On the upper surface of the midbrain are two pairs of small hills called colliculi, collections of cells that relay specific sensory information from the sense organs to the brain.

The hindbrain consists of the pons and medulla oblongata, which help control respiration and heart rhythms, and the cerebellum, which helps control movement as well as cognitive processes that require precise timing [53, 54].

Spinal Cord and Nerves

The spinal cord constitutes a long, tube-like band of tissue, Fig. 1.33. It connects the brain to the lower back. It carries nerve signals from the brain to the body and vice versa. These nerve signals help in sensations and also move the body.

Any damage to the spinal cord may affect the movement of function.

The mature central nervous system (CNS) consists of the brain and spinal cord [53, 54]. The brain sends nerve signals to specific parts of the body through peripheral nerves, known as the peripheral nervous system (PNS). Peripheral nerves in the cervical region serve the neck and arms; those in the thoracic region serve the trunk; those in the lumbar region serve the legs; and those in the sacral region serve the bowels and bladder. The PNS consists of the somatic nervous system that connects

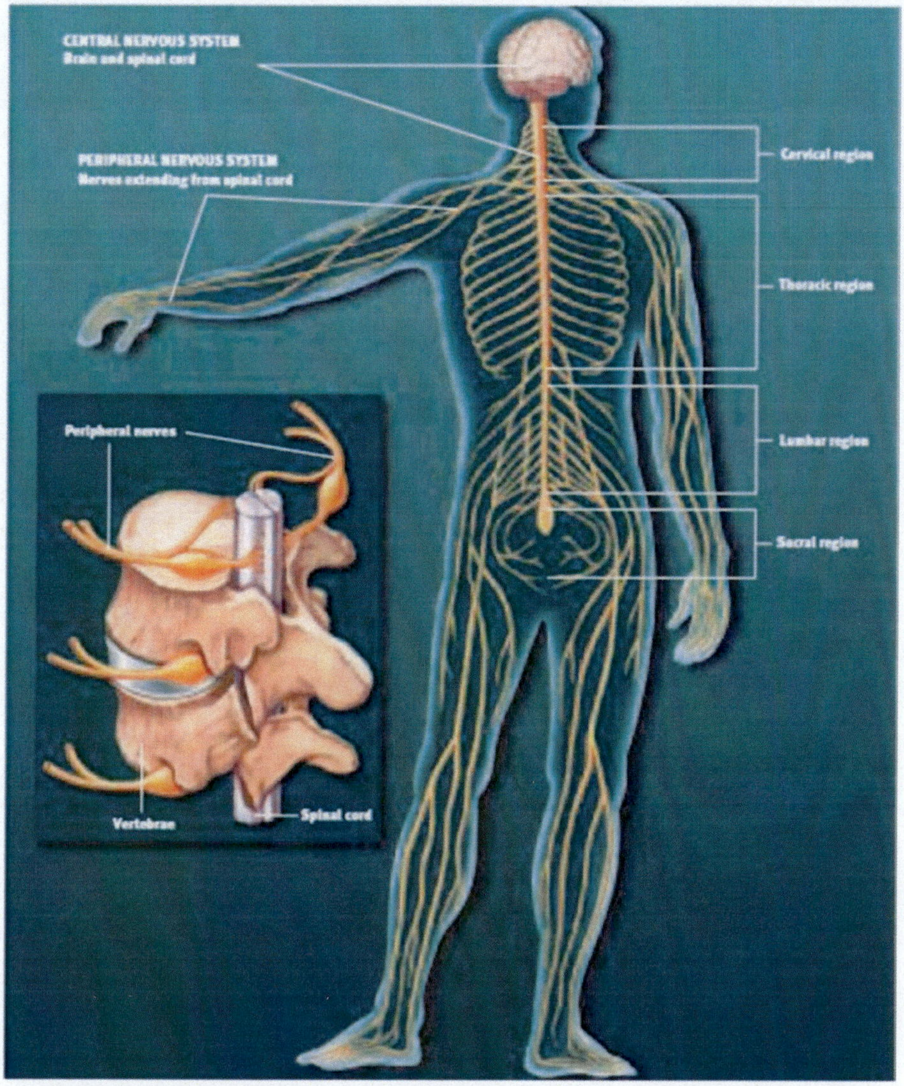

Fig. 1.33 The spinal cord and nerves

voluntary skeletal muscles with cells specialized to respond to sensations such as touch and pain. The autonomic nervous system is made of neurons connecting the CNS with internal organs [53, 54]. It is divided into the sympathetic nervous system, which mobilizes energy and resources during times of stress and arousal, and the parasympathetic nervous system, which conserves energy and resources during relaxed states.

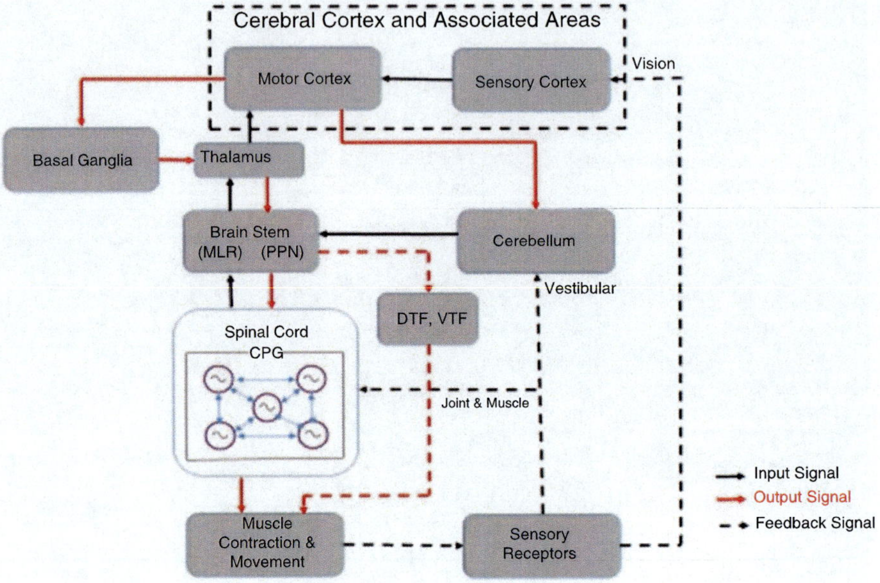

Fig. 1.34 Principle of the motor system of the human body

The Motor System of the Human Body

This section seeks to describe the process of intention in the brain or reflex that elucidates the flow of neural information in the human body, resulting in locomotion. Figure 1.34 illustrates the principle underlying the motor system of the human body.

The motor system of the human body is responsible for converting neural signals into physical energy. A thought or intention initiates a motion. Furthermore, both conscious brain activity and inputs from the sensor system of the human body may be able to initiate movements. Reflexes cause the conversion of physical energy into neural signals, which in turn stimulate muscles without going through the brain.

According to neurophysiology, movements may be viewed in three categories depending on the influence of voluntary control. The categories include:

Reflex responses are the simplest form of motor behavior. Examples include the withdrawal of the hand from a hot object, the knee jerk, or swallowing. Reflexes are rapid, stereotyped responses and can be performed without any voluntary control, although they can be modified with conscious effort.

Rhythmic motor patterns are typically initiated and terminated voluntarily, but in-between no conscious effort to maintain repetitive movement is necessary, although it can adapt to certain circumstances. Some examples of this type of movement include walking, running, or swimming.

Voluntary movements are the most complex movements that include activities like playing an instrument and driving a car. Those movements are goal-directed and can be improved with practice. The better those movements have been learned, the less conscious effort they require.

According to the different levels of voluntary influence on a movement, the human motor system can be divided into three levels of motor control, as illustrated in Fig. 1.34, the spinal cord, the brain stem, and the motor cortex.

The motor areas of the cerebral cortex are responsible for complex voluntary movements, passing control signals to the lower levels of the hierarchy, the brain stem, and the spinal cord. The rhythmic movement patterns and reflex responses are generated, which can be modulated from the higher levels through the parallel pathways. The cerebellum is responsible for correcting the current movement to resemble the desired movement.

These levels are hierarchically organized and in parallel with the spinal cord being the lowest level. The reflexes and the rhythmic motor patterns have local circuits, which are optimized for quick responses.

The spinal cord: Neurophysiology reports that human locomotion may be generated by a rhythm-generating neuronal network in the spinal cord. This rhythm-generating neuronal network is known as the central pattern generator (CPG). The locomotion is evoked by stimulus input from the mesencephalic locomotor region (MLR) within the brainstem. However, the adaptive locomotion may not be generated solely by the rhythmic signals produced by the CPG, but also the information from the afferent proprioception is equally essential [48, 51]. Hence, the afferent proprioceptive information may be necessary to be mutually coordinated in the spinal cord toward the construction of appropriate coordination of the intra- and inter-limb.

Furthermore, the spinocerebellar neurons depict that the sensory feedback signals from proprioceptors in the muscles and joints may be integrated in the spinal circuitry in order to encode global parameters of the limb movement, such as the orientation and length of the axis connecting the most proximal joint and distal position of a limb (limb axis). This implies that the global parameters describing limb kinematics, along with individual local proprioceptive inputs, are actually utilized as important sensory inputs for locomotion [46]. In addition, muscle activation pattern is also suggested to be generated primarily based on global kinematic parameters.

The central pattern generators (CPG) have been generally hypothesized to be an oscillator that generates the orientation and length of the corresponding limb axis, and the spinal circuitry of interneurons somehow generate muscle activation patterns based on the output signal of the CPG.

The brain stem consists of the medulla oblongata, pons, and midbrain, which connect the cerebrum and cerebellum with the spinal cord. The medulla oblongata is the most caudal part of the brain stem. Its ventral surface rests upon the basilar portion of the occipital bone, while its dorsal surface is, in large part, covered by the cerebellum.

The pons interposes between the midbrain and the medulla oblongata and is characterized by the huge ventral swelling. It is covered by the cerebellum dorsally. The midbrain is the shortest brain stem segment. It connects the pons and cerebellum with the cerebrum. The ventral surface of the midbrain is a pair of longitudinal columns of nerve fibers, the cerebral peduncles [45, 47, 48].

The internal structure of the brain stem comprises the cranial nerve nuclei, non-cranial nerve nuclei, as ascending and descending pathways, and reticular formation. Almost all the cranial nerve nuclei are located in the brain stem except the nuclei of the olfactory and optic nerves; they are evaginations of the brain itself.

The non-cranial nerve nuclei involve (1) the nucleus gracile and nucleus cuneate, (2) inferior olivary nuclear complex, (3) superior olivary nucleus, (4) pontine nuclei, (5) nucleus ceruleus, (6) inferior colliculus, (7) superior colliculus, (8) red nucleus, (9) substantia nigra, and (10) pretectal area.

The long ascending pathways consist of (i) medial lemniscus, (ii) spinothalamic lemniscus, (iii) lateral lemniscus, (iv) trigeminal lemniscus, and (v) medial longitudinal fasciculus.

The axons of the spinal and pontine nuclei of the trigeminal nerve successively cross the median plane to form the trigeminal lemniscus, which conducts the tactile, pressure, pain, and thermal impulses, ascending with the medial lemniscus to the thalamus and terminating in the ventral posterolateral nucleus of the thalamus [47, 48].

The long descending pathways consist of i) the pyramidal tract and ii) tectospinal tract.

The pyramidal tract includes the corticospinal tract and the corticonuclear tract. Both continue downwards from the cerebral peduncle, entering the basilar part of the pons. They become dispersed into numerous smaller bundles from a compact collection of fibers, separated from one another by the pontine nuclei and transverse fibers of the pons. The corticonuclear tract terminates successively in the motor nuclei of the cranial nerve located in the brain stem. The corticospinal tract descends through the whole length of the pons and enters the pyramids of the medulla oblongata, where they converge into compact tracts again. In the lower part of the pyramid, about three-quarters of the nerve fibers cross the median plane and continue down the spinal cord in the lateral funiculus as the lateral corticospinal tract. The uncrossed fibers retain their ventromedial position and descend in the anterior funiculus of the spinal cord as the anterior corticospinal tract [44, 45, 54].

The tectospinal tract arises from the tectum of the midbrain, descends on the ventral to the medial longitudinal fasciculus, passing through the pons and medulla oblongata to end in the anterior gray column of the spinal cord.

The reticular formation of the brain stem has long been recognized that outside the more conspicuous fiber bundles and nuclei of the brain stem, there is an extensive field of intermingled gray and white matter collectively termed the reticular formation. According to the traditional view, the reticular formation forms the central core throughout the brain stem; it appears in the upper cervical cord, lying laterally between posterior and anterior horns; in the medulla oblongata, it lies among the olivary nucleus, inferior cerebellar peduncle, floor of the fourth ventricle, and medial lemniscus; in the pons and midbrain, it fills most of the tegmentum, embedding various nuclei and tracts.

The fibers contributing to the reticular formation interlace each other to form a widespread network. The neurons contributing to the reticular formation can make up nuclei, such as the rapheal nuclei throughout the brain stem, the lateral reticular nucleus and gigantocellular reticular nucleus of the medulla oblongata, the rostral and caudal pontine reticular nuclei of the pons, and the mesencephalic reticular nucleus, etc.

The afferent projections of the reticular formation are (1) the direct corticoreticular fibers and the collaterals of the corticospinal tract, (2) the spinoreticular fibers and the collaterals of the long ascending tracts, such as the lemnisci, the vagal pathway, and the optic pathway via the tectoreticular fibers, and (3) the fibers deriving directly or indirectly from the diencephalon to the corpus striatum [46–48].

The efferent connections of the reticular formation are (1) the reticulospinal fibers to the spinal autonomic and locomotor control centers, (2) by short descending pathways to sensory and locomotor nuclei of the cranial nerves, and (3) ascending reticular fibers to the nonspecific nuclei of the diencephalon and through to the corpus striatum and to the cerebral cortex, including most regions of neocortex and many areas of the limbic system.

The reticular formation is an important integration center for the vital activity. Its major functions may sum up as follows:

1. Somatic motor control: The reticular formation exerts controlling influences on activity patterns of the skeletal muscles ranging from relatively simple reflex loops, coarse to fine locomotion and to complicated patterns associated with emotional expression, involving the complexities of speech, gestures, and fluctuations of facial expression.
2. Activation of the behavioral arousal is essential for the maintenance of the conscious state of the cerebral cortex. In human, lesion of the brain stem, which results in damages to the reticular formation, often produces disturbances of consciousness that range from fleeting unconsciousness to deep and sustained coma.
3. Visceromotor control: The cardiovascular readjustments and the activity of nonstriated muscles and many glandular cells in the thoracicoabdominal viscera are, in most cases, under the controlling influences of postganglionic autonomic neurons. Their preganglionic neuronal somata and dendrite are either directly, or through the intermediary of local interneurons, partially controlled by reticulobulbar and descending reticulospinal fibers. Based mainly on the physiological investigations, a series of vital functional centers such as the "cardiovascular controlling center," "respiratory center," "vomiting center," etc. have been established.

The brain stem is the next level of control and is subdivided into two parts, notably the medial systems, which are mainly responsible for the control of the body posture through the integration of information from the eyes and the balance system, and the lateral systems that are connected to the distal muscles of the limbs, such as the arms and hands, in order to control goal-directed movements [44–46].

The neuronal mechanisms that control the body posture play a very significant role toward the generation of robust locomotion. Basically, the body orientation and force equilibrium are controlled by tone in the limb muscles, and this tone is regulated by the pedunculopontine tegmental nucleus (PPN) in the brainstem through the dorsal tegmental field (DTF) and ventral tegmental field (VTF). This tone system may utilize the global kinematic parameters for control. The combination of the DTF/VTF encodes the degree of stiffness of the limb axis, and the spinal circuitry of the interneurons generates the muscle activation patterns according to its activity.

Both the pedunculopontine tegmental nucleus (PPN) and the mesencephalic locomotor region (MLR) receive inputs from the basal ganglia to integrate the posture and the locomotion and to initiate and terminate locomotion in a coordinated manner. The cerebellum also seems to play a crucial role in this process, as this region is where multi-modal sensory information from the vestibular organ (VO), proprioceptors, and exteroceptors is integrated. Therefore, integrated control of the posture and locomotion in the basal ganglia-brainstem-cerebello-spinal system is envisaged as providing the basis of adaptive locomotion.

The cerebellum is the second-largest portion of the brain and occupies the inferior and posterior aspects of the cranial cavity. Specifically, it is posterior to the medulla and pons and inferior to the occipital lobes of the cerebrum. It is separated from the cerebrum by the transverse fissure and by an extension of the cranial dura mater called the tentorium cerebelli, which supports the occipital lobes of the cerebral hemispheres.

The cerebellum has two so-called cerebellar hemispheres joined by a sagittal, transversely narrow median vermis [42, 48, 54].

The cerebellar surface is called the cortex, which is separated into numerous curved, transverse fissures, giving it a laminated appearance and separating its folia. Consists of gray matter in a series of slender, parallel ridges called **folia**. Beneath the gray matter are white matter tracts that resemble the branches of a tree. Deep within the white matter are masses of gray matter, which are called the cerebellar nuclei or central nuclei.

The cerebellum is attached to the brain stem by three paired bundles of fibers (tracts) called inferior, middle, and superior cerebellar peduncles.

1. **Lobes of the cerebellum**: The cerebellum can be primarily divided into a flocculonodular lobe and the corpus cerebelli, the latter having anterior and posterior lobe: (1) The flocculonodular lobe is predominantly vestibular in its connections and constitutes the oldest part of the cerebellum, so it is called the vestibulocerebellum or archicerebellum; (2) The anterior lobe and the rostral part of the inferior vermis are predominantly spinocerebellar in its connections and is phylogenetically the next part to appear, so it is also called the spinocerebellum or

paleocerebellum; (3) The posterior lobe is predominantly corticopontocerebellar in its connections and constitutes the pontocerebellum or neocerebellum [48, 54].

2. **Cerebellar cortex:** The cerebellar cortex is uniformly structured in all parts, and three layers are evident in histologic sections. From the surface to the white matter of the folium, these are the molecular layer, the Purkinje cell layer, and the granular layer. There are two types of afferent fibers in the cortex. Mossy fibers terminate in synaptic contact with cells of the granular layer, through which they affect Purkinje cells, whereas climbing fibers enter the molecular layer and wind around the dendrites of Purkinje cells. The only fibers leaving the cortex are axons of Purkinje cells; these fibers terminate in central nuclei of the cerebellum, with the exception of some fibers from the cortex of the flocculonodular lobe that proceed to the brain stem.
3. **Central nuclei of cerebellum:** Three gray masses are embedded in the cerebellar white core on each side. Most lateral and largest is the nucleus dentatus, medial to which are the smaller nuclei emboliformis and globosus, the most medial being the nucleus fastigial from which most fibers run directly to the brain stem through the inferior cerebellar peduncle, ending in the vestibular nuclei of both sides and in the reticular formation of the medulla oblongata. The dentate nucleus is the largest one and lies most laterally, which receives the fibers from the cerebellar cortex.

Important Functions of the Cerebellum

Functionally, the cerebellum is an area of the brain concerned with coordinating subconscious contractions of skeletal muscles (the cerebellar peduncles are the fiber tracts that channel information into and out of the cerebellum). The cerebellum constantly receives input signals from proprioceptors in muscles, tendons, and joints, receptors for equilibrium, and visual receptors of the eyes. Such input permits the cerebellum to collect information on the physical status of the body with regard to posture, equilibrium, and all movements at joints. In addition, when other motor areas of the brain, such as the motor cortex of the cerebrum and basal nuclei, send signals to skeletal muscles, they also send a duplicate set of signals to the cerebellum. The cerebellum compares this input information regarding the actual status of the body with the intended movement determined by motor areas of the brain (cerebrum and basal ganglia). If the intent of these motor areas is not being attained by the skeletal muscles, the cerebellum detects the variation and sends feedback signals to the motor areas, either to stimulate or inhibit the activity of the skeletal muscles. This interaction produces smooth, coordinated movements of our body's skeletal muscles [46–48].

The cerebellum also functions in maintaining equilibrium and controlling posture, for example, receptors for equilibrium send nerve impulses to the cerebellum, informing it of body position. When the direction of movement changes, the cerebellum sends corrective signals to the motor cortex of the cerebrum. The motor cortex then sends signals over motor tracts to somatic motor neurons to skeletal muscles to reposition the body.

Another function of the cerebellum is related to predicting the future position of a body part during a particular movement. Just before moving a part of the body to reach its intended position, the cerebellum sends signals over motor tracts to somatic motor neurons to the skeletal muscles to slow the moving part and stop it at a specific point. This function of the cerebellum is used in actions such as walking.

There is some evidence that the cerebellum may play a role in a person's emotional development, modulating sensations of anger and pleasure and allowing normal emotional expression and interpretation.

The Motor Cortex

It is the highest level of control with the highest layer of abstraction. It is responsible for coordinating and planning complex movements. To perform this, it is connected to the cerebellum and also to the basal ganglia. During a voluntary movement, the cerebellum compares the actual movement through responses from the sensory systems to the desired movement and corrects the movement if necessary.

The output to the cerebral cortex is excitatory, initiating movements [47, 49, 50]. The basal ganglia, on the other hand, works inhibitory on the cerebral cortex, suppressing certain movements to allow others to be performed. The basal ganglia and the cerebellum are connected to the motor cortex via the thalamus, a relay station.

The hierarchical structure of the control system ensures that simpler movements can be performed without conscious effort. But through the additional parallel neural pathways, it is possible for higher levels to modulate lower levels to adapt the movement to special circumstances, for example, changing the stride length or stepping over an obstacle.

In addition to the neural commands from the higher levels, every level is fed with sensory information that is needed for appropriate control. There is a permanent flow of information about the position and orientation of the body limbs, the degree of muscle contraction, and the information about events in the environment through the skin or eyes.

All pathways from the different control levels are connected in complex networks of interneurons in the spinal cord. They are ultimately converging into common pathways that lead to the motor neurons, which innervate the muscles. Those motor neurons are situated in the spinal cord and connect to the skeletal muscles by axons (nerve fibers).

The signals (action potentials) sent over the axons to the muscles lead to contraction of muscle fibers, shortening the muscle. Since every muscle is connected to the human skeleton at least at two points spanning one or more joints, the shortening of

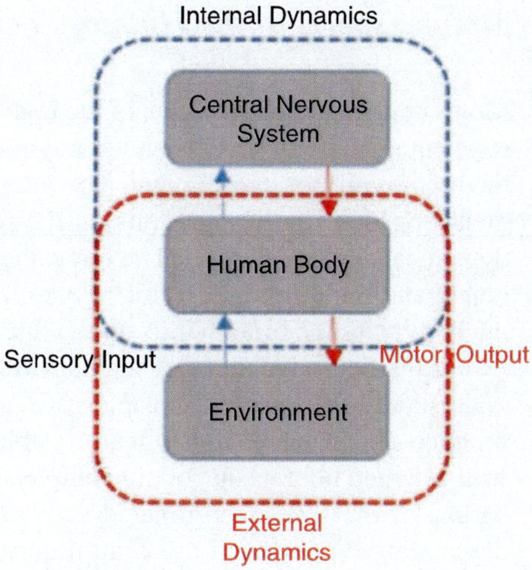

Fig. 1.35 Illustrating the brain-body-environment system

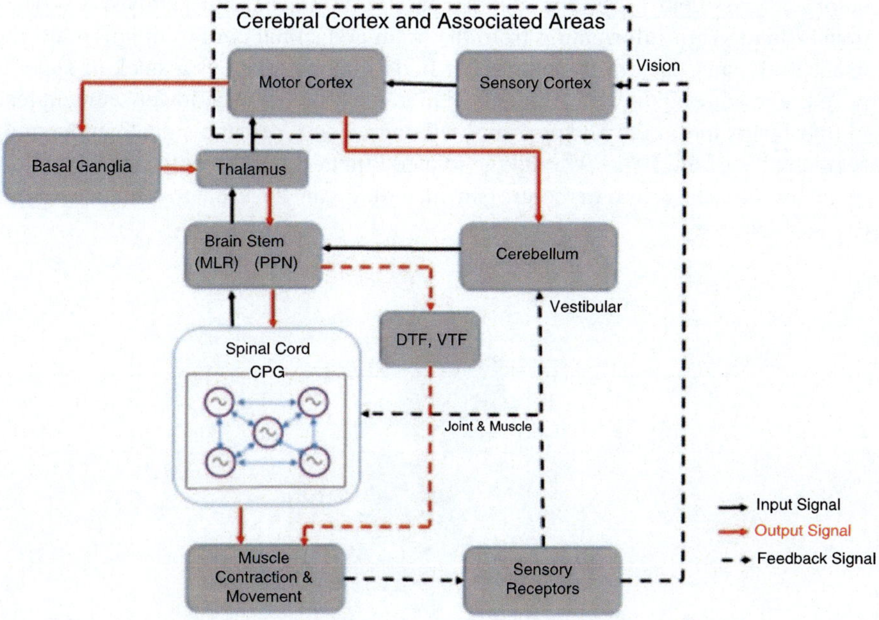

Fig. 1.36 Concept of neuromechanics in the locomotor system

the muscle creates a torque in those joints. If the torque is large enough, a motion is performed.

Motor Aspects of the Human Locomotor System

Motor control system consists of the body, the central nervous system, and the environment. The body includes a large number of sensors and actuators and connects the brain with the environment. The interaction between the body and the environment imposes some constraints on the redundant degrees of freedom in the total dynamical system [2, 5, 6]. The body and the environment constitute the controlled object and build up the external dynamics to the brain, whereas the connection of the body with the brain builds up the internal dynamics, respectively, Fig. 1.35.

For the neurodynamics of the human locomotor system, it becomes essential to understand how muscles, sense organs, and the central nervous system interact to produce coordinated, dynamically stable movement under steady conditions, as well as when humans negotiate complex terrain and experience unexpected perturbations, Fig. 1.36. Coordinated movement emerges from the interplay among descending output from the central nervous system, sensory input from the body and environment, muscle dynamics, and the emergent dynamics of the whole musculoskeletal system [2, 5, 51]. Fundamentally, motor control involves a series of transformations of information from the brain and spinal cord to the muscles, then to the body, and back to the brain. The body may be taken as a stack of superimposed segments, again with a kinematic hierarchy, in that each lower segment tends to transfer its motion to all superimposed ones. The kinematic signals in the multisegmented system, Fig. 1.37, and the related kinetic signals would represent global variables for sensorimotor control in that they can be used to produce a given

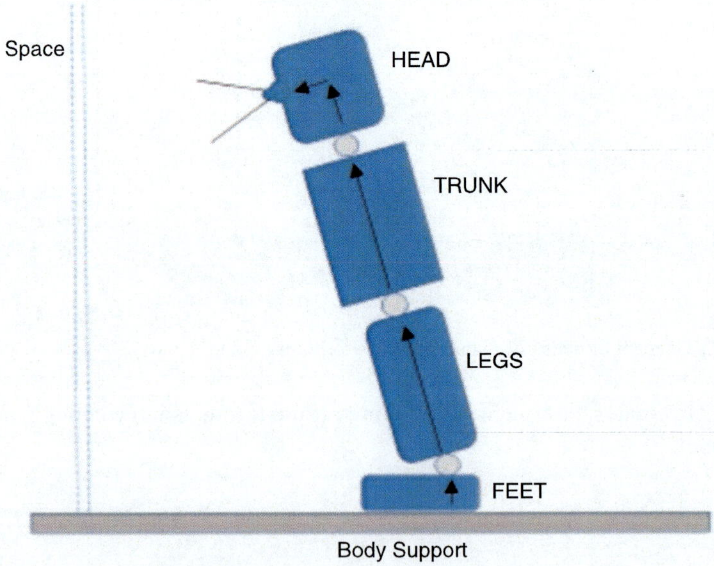

Fig. 1.37 Multisegmented musculoskeletal system

reaction in one or the other inter-segmental joint, for example, a head-in-space movement in the neck, hip, or leg joints.

Through the fusion of multiple sensor sources, the internal reconstruction of the physical links between the references may be established [5].

The control of both the perceptual and the motor aspects of the spatial behavior takes place in one common kinematic reference system. This envisaged reference system represents a rather faithful internal reconstruction of physics (that is, of the kinematic aspects of mechanics). The system is hierarchically structured, consisting of linked references that are anchored in gravito-inertial space. The perceptual notion of space is derived mainly through vestibular sensors. The internal kinematic reference system represents a reconstruction of physics, with a gravitational anchoring of head on trunk on legs (shank and thigh) on feet support in space.

Bipedal locomotion involves the body movement carried out by cyclical and dynamical interactions of the legs with the ground.

During the bipedal locomotion in the real world, the body will receive unpredictable forces, depending on various factors, e.g., changes of the ground conditions, changes of the wind conditions, etc. Perturbations to the environment caused by these external forces depend on the condition of the system itself, that is, the dynamical properties of the body. For example, the body posture will be directly affected by the external forces if the body stiffness is high and will be less affected if it is low [5, 49, 50].

Muscle tone is one of the important parameters to determine the body properties such as the stiffness and viscosity. Recent physiological studies have revealed that the appropriate setting of the muscle tone before and during movements is essentially important for carrying out various movements, including locomotion.

References

1. Hausdirff, J. M. (2005). Gait variability: Methods, modeling and meaning. *Journal of Neuro Engineering and Rehabilitation, 2*, 19. https://doi.org/10.1186/1743-0003-2-19
2. Mergner, T. (2002). The Matryoshka dolls principle in human dynamic behavior in space: A theory of linked references for multisensory perception and control of action. *Cahiers de Psychologie Cognitive/Current Psychology of Cognition, 21*(2–3), 129–212.
3. Karduna, A. R. (2003). Introduction to Biomechanical Analysis. In Carol Oaths (Eds.), *Kinesiology: Mechanics and pathomechanics of human motion*. Lippincott Williams and Wilkens, 1st edition 2003, 2nd edition 2009.
4. Fleischer, C. (2007). *Controlling exoskeletons with EMG signals and a biomechanical body model*. PhD Dissertation, Fakultaet IV: Elektrotechnik und Informatik, Technische Universitaet Berlin.
5. Sarkodie-Gyan, T., Huiying, Y., Bogale, M., Hernandez, N. V., & Pirela-Cruz, M. (2017). Application of multiple sensor data fusion for the analysis of human dynamic behavior inspace: Assessment and evaluation of mobility-related functional impairments. *Journal of Biomedical Science and Engineering, 10*, 182–203.

6. La Scaleia, V., et al. (2018). *Early manifestation of arm-leg coordination during stepping on a surface in human neonates* (Experimental brain research). Springer-Verlag GmbH Germany.
7. Pandis, P. (2013). *Musculoskeletal biomechanics of the shoulder in functional activities*. PhD Dissertation, Department of Bioengineering, Imperial College London.
8. Catarina de Sousa e Silva. (2020). *Biomechanical study of the shoulder joint complex and associated injuries*. MS Degree in Biomedical Engineering, Universidade do Porto.
9. Eschweiler, J., Li, J., Quack, V., Rath, B., Baroncono, A., Hildebrand, F., & Migliorini, F. (2022). *Anatomy, biomechanics, and loads of the wrist joint*. MDPI Life. https://mdpi.com/journal/life
10. Harris, M. D. (2013). *The geometry and biomechanics of normal and pathomorphologic human hips*. PhD Dissertation, The University of Utah.
11. Uhlmann, K. (1996). *Lehrbuch der Anatomie des Bewegungsapparates*. UTB für Wissenschaft.
12. Gordon, A. M., Huxley, A. F., & Julian, F. J. (1966). The variation in isometric tension with sarcomere length in vertebrate muscle fibres. *The Journal of Physiology, 184*(1), 170–192.
13. Delp, S. L., Loan, J. P., Hoy, M. G., Zajac, F. E., Topp, E. L., & Rosen, J. M. (1990). An interactive graphics-based model of the lower extremity to study orthopaedic surgical procedures. *IEEE Transactions on Biomedical Engineering, 37*(8), 757–767.
14. Buchanan, T. S., Lloyd, D. G., Manal, K., & Besier, T. F. (2004). Neuromusculoskeletal modeling: Estimation of muscle forces and joint moments and movements from measurements of neural command. *Journal of Applied Biomechanics, 20*, 367–395.
15. Zajac, F. E. (1989). Muscle and tendon: Properties, models, scaling, and application to biomechanics and motor control. *Critical Reviews in Biomedical Engineering, 17*(4), 359–411.
16. Basmajian, J. V., & De Luca, C. J. (1985). *Muscles alive: Their functions revealed by Electromyogr*. Williams & Wilkins.
17. Kandel, E. R., Schwartz, J. H., Jessell, T. M., et al. (1995). *Essentials of neural science & behavior*. Appleton & Lange.
18. Menon, D. K., & Veenith, T. (2012, November). *The cerebral circulation*. https://doi.org/10.13140/2.1.1186.2405.
19. Sokoloff, L., Field, J., Magoun, H., Hall, V., et al. (1960). The metabolism of the central nervous system in vivo. In *Handbook of physiology. Section I, neurophysiology* (pp. 1843–1864). American Physiological Society.
20. Hamill, J., Knutzen, K. M., & Derrick, T. R. (2015). *Biomechanical basis of human movement* (4th ed.). Copyright: 2015, 2009, 2003, 1995 Lippincott Williams & Wilkens, a Wolters Kluwer business.
21. Hausdorff, J. M., Ashkenazy, Y., Peng, C. K., Ivanov, P. C., Stanley, H. E., & Goldberg, A. L. (2001). When human walking becomes random walking: Fractal analysis and modeling of gait rhythm fluctuations. *Elsevier Physica A, 302*, 138–147.
22. Hausdorff, J. M., Purdon, P. L., Peng, C.-K., Ladin, Z., Wei, J. Y., & Goldberg, A. L. (1996). *Fractal dynamics of human gait: Stability of long-range correlations in stride interval fluctuations*. American Physiological Society.
23. Hausdorff, J. M. (2007, August). Gait dynamics, fractals and falls: Finding meaning in the stride-to-stride fluctuations of human walking. *Human Movement Science, 26*(4), 555–589. NIH Public Access.
24. Murphy, J., Brewer, R., Catmur, C., & Bird, G. (2017). Interoception and psychopathology: A developmental neuroscience perspective. *Developmental Cognitive Neuroscience, 23*, 45–56. https://doi.org/10.1016/j.dcn.2016.12.006
25. Craig, A. D. (2002). How do you feel? Interoception: The sense of the physiological condition of the body. *Nature Reviews. Neuroscience, 3*(8), 655–666.
26. Damasio, A., Damasio, H., & Tranel, D. (2012). *Persistence of feelings and sentience after bilateral damage of the insula*. Cereb, Cortex bhs077.
27. Feinstein, J. S., Khalsa, S. S., Salomons, T. V., Prkachin, K. M., Frey-Law, L. A., Lee, J. E., Tranel, D., & Rudrauf, D. (2016). Preserved emotional awareness of pain in a patient with extensive bilateral damage to the insula, anterior cingulate, and amygdala. *Brain Structure & Function, 221*(3), 1499–1511.

28. Pollatos, O., & Schandry, R. (2008). Emotional processing and emotional memory are modulated by interoceptive awareness. *Cognition and Emotion, 22*(2), 272–287.
29. Werner, N. S., Jung, K., Duschek, S., & Schandry, R. (2009). Enhanced cardiac perception is associated with benefits in decision-making. *Psychophysiology, 46*(6), 1123–1129.
30. Wang, L., Ma, L., Yang, J., & Wu, J. (2021). Human somatosensory processing and artificial Somatosensation. *Review Article, AAAS, Cyborg and Bionic Systems, 2021*, Article ID 9843259.
31. Vallbo, A. B., & Johansson, R. S. (1984). Properties of cutaneous mechanoreceptors in the human hand related to touch sensation. *Human Neurobiology, 3*(1), 3–14.
32. Hayward, V. (2018). Chapter 3: A brief overview of the human somatosensory system. In *Sorbonne Universites, Universite Pierre et Marie Curie*. Institut des Systemes Intelligents et de Robotique.
33. Hayward, V. (2011). Is there a plenhaptic function? *Philosophical Transactions of the Royal Society of London. Series B, Biological Sciences, 366*(1581), 3115–3122.
34. Terekhov, A. V., & Hayward, V. (2015). The brain uses extrasomatic information to estimate limb displacement. *Philosophical Transactions of the Royal Society of London. Series B, Biological Sciences, 282*, 166.
35. Dietz, V. (2002). Proprioception and locomotor disorders. *Nature Reviews. Neuroscience, 3*, 781–790. https://doi.org/10.1038/nrn939
36. Proske, U., & Gandevia, S. C. (2012). "The proprioceptive senses: Their roles in signaling body shape, body position and movement, and muscle force," Department of Physiology, Monash University, Victoria, Australia, and Neuroscience Research Australia and University of New South Wales, Sydney. *Australia Physiological Reviews, 92*(4), 1651–1697.
37. Adrian, E. D., & Umrath, K. (1929). The impulse discharge from the pacinian corpuscle. *The Journal of Physiology, 68*, 139–154.
38. Grigg, P. (1994). Peripheral neural mechanisms in proprioception. *Journal of Sport Rehabilitation, 3*, 2–17.
39. Banks, R. W., Hulliger, M., Saed, H. H., & Stacey, M. J. (2009). A comparative analysis of the encapsulated end-organs of mammalian skeletal muscles and of their sensory nerve endings. *Journal of Anatomy, 214*, 859–887.
40. Clark, F. J., Grigg, P., & Chapin, J. W. (1989). The contribution of articular receptors to proprioception with the fingers in humans. *Journal of Neurophysiology, 61*, 186–193.
41. Ferrell, W. R., Gandevia, S. C., & McCloskey, D. I. (1987). The role of joint receptors in human kinaesthesia when intramuscular receptors cannot contribute. *The Journal of Physiology, 386*, 63–71.
42. Landgren, S., & Silfvenius, H. (1969). Projection to cerebral cortex of group I muscle afferents from the cat's hind limb. *The Journal of Physiology, 200*, 353–372.
43. McIntyre, A. K. (1974). Central actions of impulses in muscle afferent fibres. In C. C. Hunt (Ed.), *Handbook of sensory physiology: Muscle receptors* (pp. 235–288). Springer.
44. McIntyre, A. K., Proske, U., & Rawson, J. A. (1984). Cortical projection of afferent information from tendon organs in the cat. *The Journal of Physiology, 354*, 395–406.
45. Oscarsson, O., & Rosen, I. (1963). Projection to cerebral cortex of large muscle-spindle afferents in forelimb nerves of the cat. *The Journal of Physiology, 169*, 924–945.
46. Bosco, G., & Poppele, R. E. (2001). Proprioception from a spinocerebellar perspective. *Physiological Reviews, 81*, 539–568.
47. Landgren, S., Silfvenius, H., & Nucleus, Z. (1971). The medullary relay in the projection path to the cerebral cortex of group I muscle afferents from the cat's hind limb. *The Journal of Physiology, 218*, 551–571.
48. Rosen, I. (1969). Afferent connexions to group I activated cells in the main cuneate nucleus of the cat. *The Journal of Physiology, 205*, 209–236.
49. Mergner, T. (2010). A neurological view on reactive human stance control. *Elsevier Journal Annual Reviews in Control, 34*, 177–198.

50. Mergner, T., Maurer, C., & Schweigart, G. (2007). *Sensorimotor control of human dynamic behavior in space implemented into a hominoid robot.* Neurocenter, University of Freiburg, Germany. A research Project sponsored by the German Research Foundations (DFG) Me 715/5-3, 715/6-1.
51. Nishikawa, K., Biewener, A. A., Aerts, P., Ahn, A. N., Chiel, H. J., Daley, M. A., Daniel, T. L., Full, R. J., Hale, M. E., Hedrick, T. L., Kristopher Lappin, A., Richard Nichols, T., Quinn, R. D., Satterlie, R. A., & Szymik, B. (2006). "Neuromechanics: An integrative approach for understanding motor control," From the symposium "Biomechanics and Neuromuscular Control" presented at the annual meeting of the Society for Integrative and Comparative Biology, January 4–8, 2006, at Orlando, Florida. *Integrative and Comparative Biology, 47*(1), 16–54. https://doi.org/10.1093/ich/icm024
52. Craig, A. (2003, June). A new view of pain as a homeostatic emotion. *Trends in Neurosciences, 26*(6), 303–307.
53. Sarkodie-Gyan, T. (2005). *Neurorehabilitation devices: Engineering design, measurement and control.* McGraw-Hill Publishing.
54. Brain Facts: A Primer on the Brain and nervous system; The Society for Neuroscience, 4th edition, January 2002. ISBN-13: 978-0916-110000.

Suggested Reading

van der Kooij, H., Koopman, B., & van der Helm, F. C. T. (2008). *Human motion control.* Delft University/Twente University.

Chapter 2
Significance in the Understanding of the Human Locomotor System

Introduction

The locomotor system is an organ system that gives humans the ability to move using the muscular and skeletal systems. Its primary functions include supporting the body, allowing motion, and protecting vital organs. The bones of the skeletal system provide stability to the body.

The locomotor system is an essential component of human health. In addition to providing the body with structure and the means for movement, the musculoskeletal system acts as endocrine system, stimulated by exercise, interacting through biochemical signaling with other organs in the body [1].

As people age, they face new challenges in maintaining musculoskeletal health because the interactions between muscle and bone are complicated by the ageing process. As the population ages, the conditions of sarcopenia, osteoporosis, and arthritis will increase, as will hip and knee replacements.

Musculoskeletal disorders are the leading cause of disability worldwide, with enormous impact on quality of life and longevity. They encompass a broad range of conditions that impair normal activity due to injury, pain, or disease of the bones, joints, ligaments, muscles, and tendons. Many of these disorders are associated with or made more complex by the ageing process [1].

The main significance of the human locomotor system is to better understand the underlying and fundamental sciences in order to be able to improve the care and treatment of these disorders [3].

© The Author(s), under exclusive license to Springer Nature Switzerland AG 2023

T. Sarkodie-Gyan, H. Yu, *The Human Locomotor System*,
https://doi.org/10.1007/978-3-031-32781-0_2

The Human Locomotor System

The locomotor system includes bones, muscles, tendons, ligaments, and soft tissues that provide our body with movement, stability, shape, and support. It is an organ system that gives humans the ability to move using their muscular and skeletal systems. The muscular system includes all types of muscles in the body. In particular, skeletal muscles are the ones that act on the body joints to produce movements. The muscular system contains the tendons, which attach the muscles to the bones.

Figure 2.1 illustrates the human locomotor system as a feedback control system in which a (position) or set point is generated within the central nervous system (CNS). This set point is compared with the actual position of the limb, and the CNS will send a neural signal to the muscles. The muscles exert forces on the bones of the skeletal system, which will start moving if it is not constrained by the environment [12]. Notably, the interaction between the body and environment imposes some constraints on the redundant degrees of freedom in the total dynamical system. The movements are detected by the sensors, which are all over the body, in the muscles, joints, and skin, as well as the visual and vestibular systems. The sensory receptors in the muscles, joints, and in the skin provide the brain with essential information about the position and motion of the body and limbs. Using these sensory signals, the actual position of the skeletal system may be reconstructed.

The coordinated and/or adaptive mobility behaviors inherent in this locomotor system are a result of the intelligent sensory-motor functions inherent in the musculoskeletal system. The interactions within the musculoskeletal system exhibit the tendency to regulate of its internal environment while maintaining a constant and stable condition. This regulatory process is known as homeostasis. It constitutes the attempt of the body to maintain a stable and constant internal environment through constant monitoring and adjustments of the human physiological system [6, 7]. The skeletal and muscular systems work hand-in-hand in the process of homeostatic regulation in a combined system otherwise known as the muscular-skeletal system (or the musculoskeletal system).

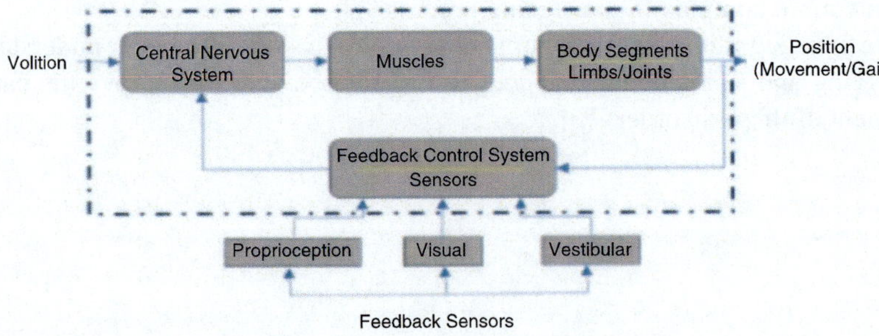

Fig. 2.1 The human locomotor system as a homeostatic regulation system

The Skeletal System

As the structural framework for the human body, the skeletal system consists mainly of the over 206 bones but also includes cartilages, ligaments, and other connective tissues that stabilize and interconnect them. The bones work in conjunction with the muscular system to aid in posture and locomotion. Many bones of the skeleton function as levers, which change the magnitude and direction of forces generated by the skeletal muscles. Protection is a pivotal role occupied by the skeletal system, as many vital organs are encased within the skeletal cavities (cranial, and spinal or dorsal), and bones form much of the structural basis for other body cavities (examples include the thoracic and pelvic cavities). The skeletal system also serves as an important mineral reserve [7]. The skeletal system provides the calcium needed for all muscular contractions.

In summary, the skeletal system is that body system composed of bones and cartilage and performs critical functions for the human body. These critical functions include supporting the body; facilitating movement; protecting internal organs; producing blood cells; and storing and releasing minerals and fat.

The Muscular System

The muscular system is one of the most versatile systems in the body. It contains the heart, which constantly pumps blood through the body. It is also responsible for involuntary (e.g., digestion, heartbeat, breathing) and voluntary (e.g., walking, picking up objects) actions. The muscles also help to protect organs in the cavities of the body. Furthermore, it exhibits functions including movement, support, protection, heat generation, and blood circulation. The muscular system includes all types of muscles in the body [7]. In particular, skeletal muscles are the ones that act on the body joints to produce movements. The muscular system contains the tendons, which attach the muscles to the bones.

A predominant characteristic of the skeletal muscle tissue is its contractility, and nearly all movement in the body is the result of muscle contractions [9]. Four functions of muscle contraction constitute movement, posture, joint stability, and heat production. Three types of muscle are skeletal, smooth, and cardiac.

Figure 2.2 illustrates a conceptual scheme of homeostatic regulation of the human locomotor system. However, the functions of such coordinated/adaptive behaviors may be disturbed in the human due to impairment, disorder, injury, or disability (intractable medical conditions).

Gaining insight into the dynamic behavior of the locomotor system in space may significantly lead towards ameliorating human health problems.

Ameliorating human health concerns may include the restoration of mobility following injury to the brain or spinal cord that may lead to the possible design of

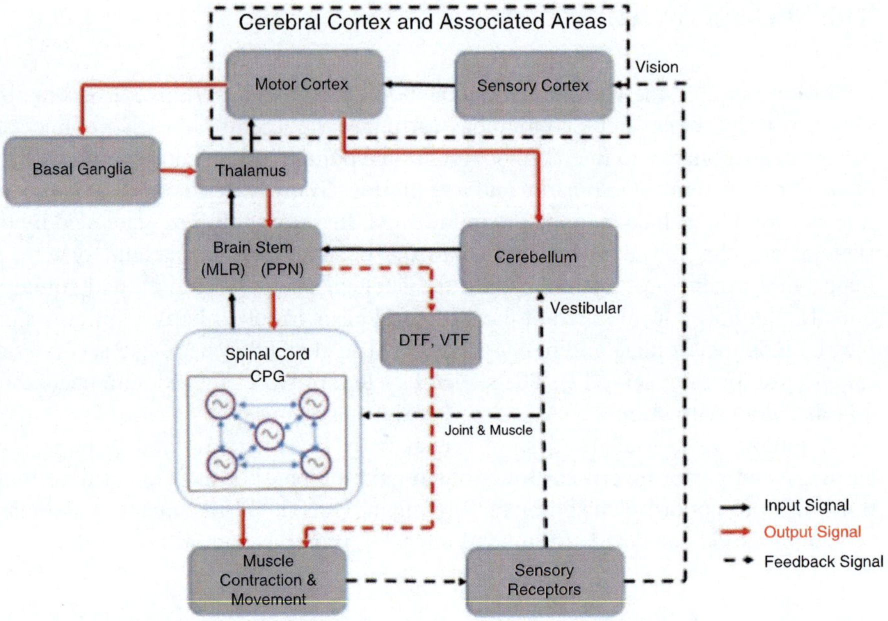

Fig. 2.2 Conceptual scheme of homeostasis regulation

orthotics and prosthetics, as well as the design, actuation, and control of wheelchairs and other human-aided assistive devices.

Furthermore, it may be borne in mind that motor control involves a series of transformations of information. This information is transmitted from the brain and spinal cord to the muscles, to the body, and then back to the brain, Fig. 2.2. The control problem revolves around the specific transfer functions that describe each transformation. The transfer functions depend on the rules of organization and operation that determine the dynamic behavior of each subsystem (that is, central processing, force generation, emergent dynamics, and sensory processing). The dynamics of human locomotion possess content and meaning [29]. These attributive features may obviously be very useful in providing insight into the neural control of the locomotion, neuro-musculoskeletal system.

Homeostatic regulation involves three mechanisms: (1) the receptor that receives information about some change in the environment (the feedback sensors consisting of proprioception, visual, and vestibular systems); (2) the control center or integration center that receives and processes information from the receptor (the central nervous system); and (3) the effector that responds to the commands of the control center by either opposing or enhancing the stimulus (the muscles as actuators). This is an ongoing process that continually works to restore and maintain homeostasis.

The central nervous system exhibits a significant role as the organ through which the body as a whole is controlled. Our embodiment is an essential dimension of our existence, of our capacities for perception and action, for language and emotion.

The healthy functioning of the central nervous system plays a dominant role in the regulations in our bodies, our capacities for autonomous agency, our conceptions of ourselves and our relationships with others, and, in fact, in our abilities to lead fulfilling lives. It has evolved such that neurons assemble into the clear "grey matter" nuclei, such as the thalamus or the basal ganglia, which can be located deep within the central nervous system. This efficient organization characterizes the basis of a broad regional specialization of function and thus appears to elucidate communication. Nerve fibers entering or leaving these nuclei may also be identified as "white matter" bundles or tracts, which establish key routes of communication between the nuclei of the central nervous system [1]. Covering the whole surface of the brain is a convoluted, layered sheet of grey matter, which has a thickness of two to three millimeters in man, but with a surface area of several hundred square centimeters. This is the cerebral cortex, one of the most important parts of the central nervous system that is responsible for perception of sensations, voluntary movement, learning, speech, cognition, and emotional control.

The cerebral cortex is broadly divided into specialized regions (sensorimotor, visual, auditory, and olfactory) as well as those for high-level perceptual analysis of faces, places, bodies, and visually presented words. Many cortical regions have multiple integrating and analytic properties, and so, despite many popular accounts, regions cannot be simply ascribed to a single function. Furthermore, many brain nuclei control exhibit less obvious, but equally important functions, for example, hormonal and autonomic functions such as cardiovascular control. Much experience from human, such as the perception of pain, has sensory, emotional, and autonomic components (entailing bodily responses that are separate from the brain) and do not involve a single cortical area but rather involve a spatial and temporal pattern of activity in multiple brain regions that are also activated in other contexts, such as fear. The complexities of cortical circuitry are immense.

The nervous system is in charge of controlling other body parts by sending an electrochemical signal to the brain whenever there is any change from optimum levels. The brain then corresponds by sending the required stimulus to the respective body organ. The circulatory system, made up of arteries and veins, maintains this balance through the circulation of blood to all other body parts. Through capillaries, an exchange of nutrients in each cell is initiated. This system also enhances the transportation of toxic excretory wastes and other liquid material to the excretory organs. Through lymphatic vessels, the lymphatic system is involved in homeostasis by collecting excess tissue fluids and taking them back to the veins. The endocrine system, the main regulator of this process of homeostasis, on the other hand, contains hormones (chemical messengers) that keep circulating throughout the blood stream and act on the respective target organs. As much as the endocrine system cannot work without coordinating with the nervous system, its effects are long lasting even though they are slow.

Skeletal muscle fibers contain numerous nucleus components in each cell, and the cell nuclei are found beneath the plasma membrane of the skin. Both the skeletal and muscular systems work together in the process of performing basic and essential functions necessary for life that include protection (of the brain and other internal organs), supportive services of maintaining an upright posture, blood cell formation in a process called hematopoiesis, and storage of minerals and fats, among other minor duties like leverage that involves magnifying movement or force speed. The musculoskeletal system consists of five major tissues that include bones, tendons, ligaments, cartilage, and skeletal muscles. The ligaments are the tissues on which one bone is attached to another bone. Cartilages are protective and gel-like substances that line intervertebral discs and joints, while tendons usually attach muscles to bones. Each of these tissues has four connective blocks that build tissues. They include fibroblasts, collagen, elastic fibers, and proteoglycans. Fibroblasts are the mother of all these other connective tissues as it is the one that produces them. Collagens are the principal proteins that are long and thin and are organized into various intertwining fibers to form strong ones that do not stretch. Elastic fibers are found in the walls of arteries, whereas proteoglycans are ground substances normally called matrix in which the other connective tissues reside.

On account of homeostatic regulation, the human is capable of exhibiting adaptive behaviors even in diverse and complex environments. Examples of such adaptive behaviors include locomotive behaviors like swimming, flying, and walking and manipulative behaviors such as reaching, capturing, grasping by using hands and arms, etc. These adaptive behaviors involve the intelligent sensory-motor functions that are of utmost significance and the most essential and indispensable ones for the survival of the human.

There are several physiological systems that play significant roles in homeostasis in the musculoskeletal system. These systems include the nervous system, the muscular system, the skeletal system, the respiratory system, the cardiovascular system, the digestive system, etc.

The Nervous System

The nervous system maintains homeostasis by controlling and regulating the other parts of the body. A deviation from a normal set point acts as a stimulus to a receptor, which sends nerve impulses to a regulating center in the brain. The brain directs an effector to act in such a way that an adaptive response takes place. The adaptive response returns the body to a state of normalcy, and the receptor, the regulating center, and the effector temporarily cease their activities. Regulating centers are located in the central nervous system, consisting of the brain and spinal cord. The hypothalamus is a portion of that and is particularly concerned with homeostasis. It influences the action of the medulla oblongata, a lower part of the brain, the autonomic nervous system, and the pituitary gland.

The nervous system has two major portions: the central nervous system (CNS) and the peripheral nervous system (PNS). The peripheral nervous system consists of the cranial and spinal nerves. The autonomic nervous system is a part of the peripheral nervous system and contains motor neurons that control internal organs. It operates at the subconscious level and has two divisions: the sympathetic and parasympathetic systems. In general, the sympathetic system that brings about those results we associate with emergency situations, often called fight or flight reactions, and the parasympathetic system that produces those effects necessary to our everyday existence.

The Cardiovascular System

The cardiovascular system, in addition to needing to maintain itself within certain levels, plays a role in the maintenance of other body systems by transporting hormones (heart secretes ANP and BNP) and nutrients (oxygen, EPO to bones, etc.), taking away waste products, providing all living body cells with a fresh supply of oxygen, and removing carbon dioxide. Homeostasis is disturbed if the cardiovascular or lymphatic systems are not functioning correctly. The skin, bones, muscles, lungs, digestive tract, nerves, endocrine, lymphatic, urinary, and reproductive systems all use the cardiovascular system as their "road" or "highway" as far as distribution of things that take place in the body. There are many risk factors for an unhealthy cardiovascular system. Some diseases associated are typically labeled "uncontrollable" or "controllable." The main uncontrollable risk factors are age, gender, and a family history of heart disease, especially at an early age.

The Respiratory System

The respiratory system works in conjunction with the cardiovascular system to provide oxygen to the cells within every body system for cellular metabolism. The respiratory system also removes carbon dioxide. Since carbon dioxide is mainly transported in the plasma as bicarbonate ions that act as a chemical buffer, the respiratory system also helps maintain proper blood pH levels, a fact that is very important for homeostasis. As a result of hyperventilation, carbon dioxide is decreased in the blood levels. This causes the pH of the body fluids to increase. If the acid levels rise above 7.45, the result is respiratory alkalosis. On the other hand, too much carbon dioxide causes pH to fall below 7.35, which results in respiratory acidosis. The respiratory system also helps the lymphatic system by trapping pathogens and protecting the deeper tissues within. An increase in the thoracic space would provide abdominal pressure through the contraction of the respiratory muscles. This may assist in defecation. And the lungs are the gateway to our breath of life.

The Digestive System

Without a regular supply of energy and nutrients from the digestive system, all body systems would soon suffer. The digestive system absorbs organic substances, vitamins, ions, and water that are needed all over the body. In the skin, the digestive tract provides lipids for storage in the subcutaneous layer. Food undergoes three types of processes in the body: digestion, absorption, and elimination. If one of these is not working, there may arise problems that would be extremely noticeable. Mechanics of digestion may include chemical digestion, movements, ingestion, absorption, and elimination. In order to maintain a healthy and efficient digestive system, it may be necessary to observe the components involved. If these components are disturbed, the digestive health may be compromised.

Disabilities, Disorders, Impairments

Unfortunately, there is the emergence of several intractable medical conditions, including injuries, infections, or diseases, as a result of damage to the brain and spinal cord. Many of these types of damages lead to serious disorders that affect the memory, cognition, movement, and consciousness.

A very high percentage of individuals worldwide suffer injury to the peripheral neuromuscular circuitry that can cause different degrees of limb function loss and even ulceration and articular deformity. The peripheral nerve system is responsible for relaying electrochemical messages between the brain-spinal cord unit and the muscles, joints, and skin. Therefore, any injury suffered to the peripheral neuromuscular circuitry would cause different degrees of limb function loss and even ulceration and articular deformity. This type of injury, thus, becomes a serious health problem leading to life-long disabilities. The most serious of such damages include traumatic brain injury (TBI) that accounts as an impact to the skull; stroke that involves damage resulting from the interruption of the blood supply to the brain; and a range of neurodegenerative disorders, such as Parkinson's disease or Alzheimer's disease in which populations of brain cells undergo degenerative change. When the brain is damaged through illness or injury that causes a fall in its functions, there can be devastating personal consequences.

Neuromuscular Disorders

Neuromuscular disorders constitute the diseases of the neurons that control the voluntary muscles. Therefore, these disorders affect a part of the nervous system and the muscles. The nerves that facilitate the communication of sensory information between the brain and the voluntary muscles are affected by these disorders. The peripheral nervous system is mainly affected due to these disorders. Neuromuscular disorders are mostly genetic disorders. They can arise due to new mutations in our genes as well. They show symptoms such as muscle weakness, movement issues, balance problems, droopy eyelids, troubled swallowing, double vision, and troubled breathing. Some neuromuscular disorders are autoimmune diseases. Some examples of neuromuscular disorders include amyotrophic lateral sclerosis, muscular dystrophy, diabetic neuropathy, toxic neuropathy, myasthenia gravis, small fiber neuropathy, spinal muscular atrophy, etc. They can pose a direct or indirect impact on an individual, leading to a loss of functional capacity. However, most neuromuscular disorders are treatable and can be improved to increase mobility and lengthen life if diagnosed at an early stage.

Musculoskeletal Disorders

Musculoskeletal disorders constitute diseases that affect muscles, bones, and joints. They are one of the most common work-related diseases. The development of musculoskeletal disorders increases with age. Musculoskeletal disorders show symptoms such as pain, redness, swelling, muscle weakness, etc. However, people of all ages can be affected by musculoskeletal disorders. Other than age, occupation, activity level, lifestyle, and family history are the risk factors for these disorders [19–21]. Furthermore, significant mental health decline and deteriorated functioning are highly associated with musculoskeletal disorders. Tendinities, carpal tunnel syndrome, osteoarthritis, rheumatoid arthritis (RA), fibromyalgia, and bone fractures are several musculoskeletal disorders. The pain and discomfort associated with these disorders affect the day-to-day activities of the affected people.

Osteoporosis: It is the weakening of bones that occurs when the body loses bone tissue that is not adequately replaced. As people age, particularly women, they are more likely to develop osteoporosis and be at higher risk of breaking bones.

Arthritis: It is common and can be debilitating. Osteoarthritis results from the deterioration of the cartilage that coats and cushions bones, enabling joints to operate smoothly. Loss of cartilage results in pain, swelling, and movement problems. Rheumatoid arthritis is one of several forms of arthritis caused by inflammation of the joints and other tissues. Often, they are caused by autoimmune disorders in which the immune system attacks its host body.

Bone and joint: Problems can develop in association with other conditions, such as diabetes, chronic kidney disease, or genetic disorders.

Cancer: Several types of cancer can originate in bone tissue. In addition, multiple myeloma, a cancer of blood plasma, causes abnormalities in the bone marrow and other bone tissues.

Bone fractures: These are a common injury subject to a broad range of research, from surgical techniques to compounds to promote healing to issues of nutrition and rehabilitation.

It is extremely essential to enhance our understanding of neuro-and pathophysiology in order to gain more insight into capabilities and enable the improvement in the quality of diagnosis, prognosis, and follow-up for individual patients suffering from neurological disorders, degenerative diseases, and injuries.

It may be extremely essential to illuminate our basic understanding regarding some common musculoskeletal disorders [19, 21]. Some of these common disorders may include, but not limited to, the following:

- Carpal tunnel syndrome.
- Tendonitis.
- Muscle/tendon strain.
- Ligament sprain.
- Tension neck syndrome.
- Thoracic outlet compression.
- Rotator cuff tendonitis.
- Epicondylitis.
- Radial tunnel syndrome.
- Digital Neuritis.

Carpal Tunnel Syndrome

Understanding the wrist may assist in better understanding carpal tunnel syndrome.

Carpal tunnel syndrome is a common nerve disorder that may interfere with hand strength and sensation, causing a decrease in function. It constitutes a common neurological disorder that occurs when the median nerve, which runs from the forearm into the palm of the hand, becomes pressed or squeezed at the wrist [31, 32]. Figure 2.3 illustrates the carpal tunnel. It is essentially a pinched nerve in the wrist.

The median nerve is one of the main nerves of the hand. It controls feeling in the palm side of the thumb, index, middle, and ring fingers. Along with the nine tendons that bend the fingers, the median nerve travels from the forearm into the hand through the narrow carpal tunnel [33, 34].

Carpal tunnel syndrome occurs when the tissues surrounding the tendons in the wrist swell and put pressure on the median nerve. These tissues are called the synovium. The synovium lubricates the tendons and makes it easier to move the fingers.

This swelling of the synovium narrows the small space of the carpal tunnel and, over time, crowds the nerve. This can result in hand pain, numbness, tingling, and weakness.

Tendinitis

Tendinitis is a condition caused by the tearing of tendon fibers and subsequent inflammation in the tendon. Tendons are the strong connective tissue that connects muscle to bone.

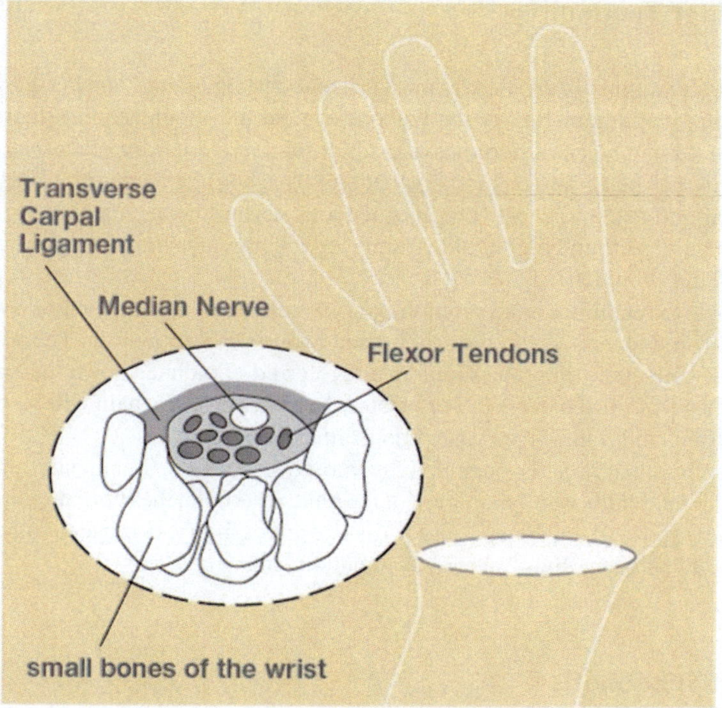

Fig. 2.3 The carpal tunnel

Tendinitis is inflammation of the thick, fibrous cords that attach muscle to bone. These cords are called tendons. The condition causes pain and tenderness just outside a joint. Tendinitis can occur in any tendon. Tendons that commonly become inflamed include:

- Tendons of the hand.
- Tendons of the upper arm that affect the shoulder.
- Tendons of the forearm at the elbow.
- The tendon of the quadriceps muscle group at the knee.
- The Achilles tendon at the ankle.

Some of the main symptoms of tendonitis include:

- Pain and tenderness in the affected tendon, which is often worse when you move it.
- Swelling.
- A grating sensation as the tendon moves.
- A lump on the tendon.
- Weakness in the affected area.
- Decreased range of motion.

Shoulder Tendonitis

The shoulder joint is encircled by muscles, tendons, ligaments, and bursae. Shoulder tendonitis is inflammation of the tendons around the shoulder joint (including the rotator cuff and biceps tendon) as a result of injury, overuse, or other factors.

There may be certain risk factors that provoke shoulder tendonitis. These include (1) wear and tear and degeneration of tendons during the ageing process, (2) overuse of tendons around the shoulder joint, and (3) previous trauma or surgical operations to the shoulder.

Shoulder tendinitis is a common overuse injury in sports (such as swimming, baseball, and tennis) where the arm is used in an overhead motion. The pain, which is usually felt at the tip of the shoulder and referred or radiated down the arm, occurs when the arm is lifted overhead or twisted. In extreme cases, pain will be present all of the time, and it may even wake one from a deep sleep.

The shoulder joint is encircled by muscles, tendons, ligaments, and bursae. Shoulder tendonitis is inflammation of the tendons around the shoulder joint (including the rotator cuff and biceps tendon) as a result of injury, overuse, or other factors, such as diabetes mellitus, thyroid dysfunction, etc.

Wrist Tendonotis

Wrist tendonitis, Fig. 2.4, involves pain in the wrist and forearm with repetitive activity [35, 36]. The inflamed tendon is tender to the touch, and there may arise swelling around the inflamed tendon.

Fig. 2.4 Wrist tendonitis

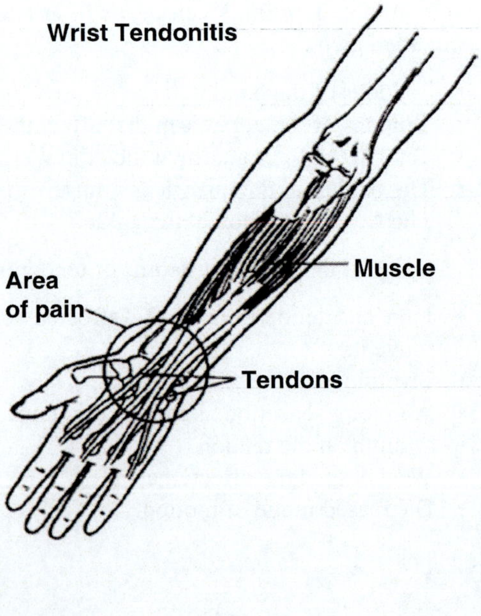

Biceps Tendonitis

The tendon tendinitis of the proximal biceps and the associated tenosynovitis may be characterized by pain at the front of the shoulder and the upper arm that may be caused by inflammation of the biceps tendons sheath or strain of the upper biceps tendon. When the lining becomes inflamed, the tendon is unable to glide smoothly in its covering (sheath). The biceps tendon is one of the anchor points of the biceps muscle, which is important for bending the elbow and rotating the wrist. It also plays a role in shoulder function.

Tibialis Posterior Tendinopathy

Tibialis posterior is a muscle in the calf. It has a tendon that runs around the inside of the ankle and attaches in the instep of the foot. Figure 2.5 illustrates the tibialis posterior tendinopathy, which functions like a brace to hold up the arch of the foot [38, 39]. Tibialis posterior tendinopathy is an injury or problem with the tibialis posterior tendon.

The main cause of tendinopathy is the wear and tear of the tendon over a period of time. The main symptom is pain, usually around the inside of the ankle and into the instep of the foot, where it attaches. There may also occur some stiffness. The pain is usually worse first thing in the morning and may interfere with day-to-day life.

Quadriceps Tendonitis

Alignment or overuse problems of the knee structures can lead to strain, irritation, and/or injury of the quadriceps muscle and tendon. This produces pain, weakness, and swelling of the knee joint. These problems can affect people of all ages, but the

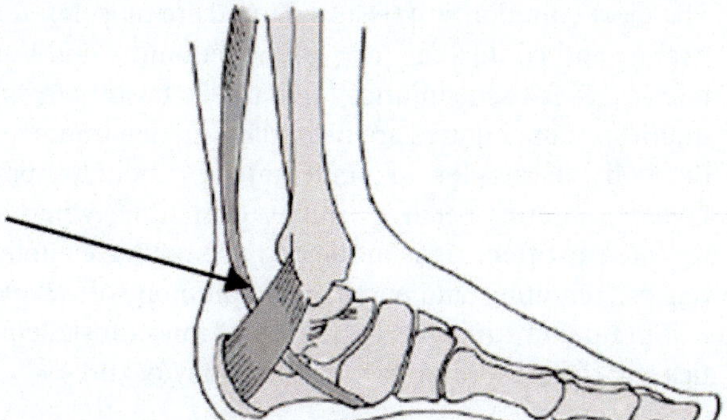

Fig. 2.5 Tibialis posterior tendinopathy

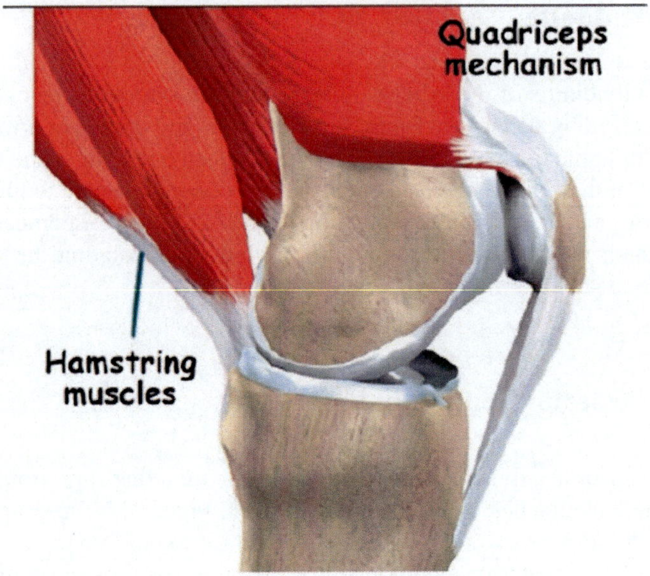

Fig. 2.6 Mechanisms of quadriceps

majority of patients with overuse injuries of the knee (and specifically quadriceps tendonitis) are involved in soccer, volleyball, or running activities [40–42].

The patella (kneecap) is the moveable bone on the front of the knee. This unique bone is wrapped inside a tendon that connects the large muscles on the front of the thigh, the quadriceps muscles, to the lower leg bone.

The large quadriceps muscle ends in a tendon that inserts into the tibial tubercle, a bony bump at the top of the *tibia* (shin bone) just below the patella. The tendon together with the patella is called the quadriceps mechanism, Fig. 2.6.

Sprain-Strain

The most common soft tissues injured are muscles, tendons, and ligaments. Sprains, strains, and contusions, as well as tendinitis and bursitis, are common soft-tissue injuries. Soft-tissue injuries fall into two basic categories: acute injuries and overuse injuries. Acute injuries are caused by sudden trauma, such as a fall, twist, or blow to the body. Examples of acute injuries include sprains, strains, and contusions. Overuse injuries occur gradually over time; when an athletic or other activity is repeated so often, areas of the body do not have enough time to heal between occurrences. Tendinitis and bursitis are common soft-tissue overuse injuries.

Sprain and strain are two types of musculoskeletal injuries that exhibit similarities and differences in the way they may occur.

A sprain is an injury to a ligament, which is the thick connective tissue that holds two or more bones in place at a joint. When a sprain occurs, one or more ligaments are overstretched, partially torn, or fully torn [18, 19, 21].

A strain is an injury to a muscle or tendon. A tendon is the strong, fiber-like tissue that connects a muscle to a bone and helps their movement. In a strain, a muscle or tendon is overstretched, partially torn, or fully torn.

A sprain may cause moderate pain, swelling, bruising, and difficulty moving the affected joint. In sharp contrast to sprains, strains cause severe pain. They may also cause muscle spasms or cramps and additional swelling.

Low back injuries, including muscular strains and/or ligament sprains, are exceedingly common in the general population. These injuries may be the result of mechanical stresses and/or functional demands placed on the low back by everyday activities or may be related to an acute injury. Symptoms are believed to be caused by a partial stretching or tearing of the soft tissues (muscles, fascia, ligaments, facet joint capsule, etc.).

A cervical musculoligamentous injury (sprain/strain) may cause neck pain due to a partial stretching or tearing of the soft tissues (muscles, fascia, ligaments, tendons, etc.). Non-specific upper extremity complaints such as stiffness, muscle fatigue, and paresthesias may also be reported [23, 26]. Although injury to the neck can result in a fracture or neurologic impairment, by definition, a diagnosis of cervical sprain or strain excludes a fracture.

Radial Tunnel Syndrome

The radial nerve constitutes the principal nerve in the upper extremity. It subserves the extensors of the arms and fingers and the sensory nerves of the extensor surface of the arm. It exhibits a long and winding pathway that makes it vulnerable to injury. Radial neuropathies are commonly a consequence of acute traumatic injury and are only rarely caused by entrapment in the absence of such an injury. Since it serves the sensory and motor nerves of the dorsal arm, it derives innervation from most-to-all of the spinal roots that participate in the brachial plexus and all three derived trunks.

Radial tunnel syndrome involves compression of the radial nerve in the proximal forearm. In the region of the proximal forearm, the radial nerve splits into the posterior interosseous nerve branch (the main trunk) and the sensory branch of the radial nerve (the minor trunk) in the proximal forearm [26]. Compression can occur either before or after this split off of the sensory branch of the radial nerve has occurred. Multiple sites of potential entrapment of the radial nerve include: the origin of the extensor carpi radialis brevis origin; the fibrous bands overlying the radial head; the radial recurrent arterial fan; and the arcades of Frohse, at the entrance to the supinator muscle. The condition has multiple causes, including: space-occupying lesions, such as tumors; local edema or inflammation; overuse of the hand and wrist through repetitive movements, causing the nerve to be compressed; blunt trauma to the proximal forearm with secondary bleeding; and idiopathic onset. The condition can occur at any age but is generally seen in younger individuals, most commonly in women aged 30–50. Early radial tunnel symptoms can mimic early lateral epicondylitis, occasionally causing these two entities to be confused.

The sensory branches of the radial nerve supply sensation to the dorsum of the arm, forearm, and hand. There are three major sensory branches: (1) the posterior brachial cutaneous nerve, (2) the posterior antebrachial cutaneous nerve, and (3) the superficial branch of the radial nerve. There is great variation in the sensory area innervated by the radial nerve, primarily because of overlapping contributions from other nerves.

The main key towards the confirmation of a radial mononeuropathy is through electrodiagnosis and the localization of the area of the injury. This may involve radial sensory nerve conduction, radial motor nerve conduction, including segments in the proximal arm, and needle electromyography (EMG) study of relevant muscles.

Degenerative Disc Disease

The human spine, Fig. 2.7, consists of 24 bones or vertebrae in the cervical (neck) spine, the thoracic (chest) spine, and the lumbar (lower back) spine, including the sacral bones. The vertebrae are connected by several joints, which enable actions involving bending, twisting, and carrying loads [43]. The main joint between two vertebrae is called an intervertebral disc. The disc consists of two parts: a tough and fibrous outer layer (annulus fibrosis) and a soft, gelatinous center (nucleus pulposus). These two parts work in conjunction to enable the spine to move and also to provide shock absorption.

Degenerative disc disease (spondylosis) constitutes a general term for the description of changes that could occur along any area of the spine (cervical, thoracic, or lumbar) with respect to ageing, and these changes are most common in the lumbar area.

In fact, this is actually not a disease. Rather, it constitutes a condition in which the discs degenerate, or lose their flexibility and ability to cushion the spine. Since the discs do not exhibit a good blood supply, they are unable to repair themselves once injured [44].

Degenerative disc disease is a spinal condition caused by the breakdown of the intervertebral discs. With the process of ageing, the spine begins to exhibit signs of wear and tear in which the discs dry out and shrink. These age-related changes may lead to arthritis, disc herniation, or spinal stenosis, which may put pressure on the spinal cord and nerves and may cause back pain.

The understanding of degenerative disc may be based on the basic understanding of the functions of the spine. Notably, the spine consists of a column of bones called vertebrae. Between each vertebra is a gel-filled disc that acts like a shock absorber that keeps the vertebrae from rubbing together. Discs have a tough outer wall,

Fig. 2.7 The human spine

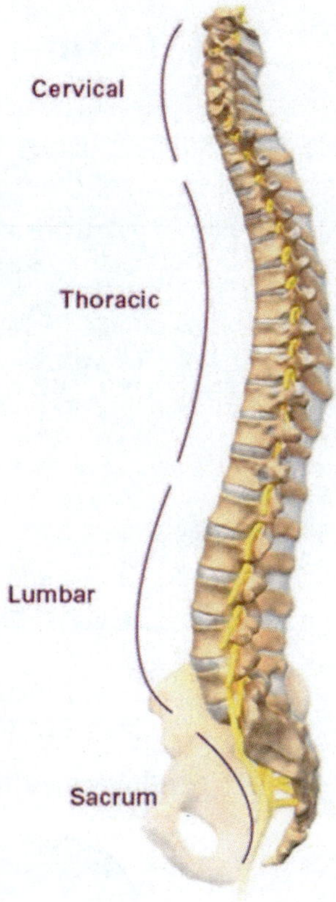

annulus, and a soft center, nucleus. Discs are made up of about 80% water, but slowly lose water with increasing age, and hence, the ability to act as shock absorbers.

Degenerative disc disease typically begins with a decrease in the water content of the nucleus pulposus and can lead to tears in the annulus fibrosis. Figure 2.8 illustrates the normal disc, whilst Fig. 2.9 illustrates the degenerative disc. In Fig. 2.9, the degenerative disc disease causes the discs to dry out and shrink.

With advanced degenerative disc disease, the loss of disc height may lead to segmental instability that would result in disc slippage (degenerative spondylolisthesis) or asymmetric disc height loss [43–45]. This will cause a side-to-side curvature of the spine (degenerative scoliosis). These advanced degenerative changes that affect the discs, joints, and surrounding soft tissues may further result in the narrowing of the spinal canal (degenerative stenosis). This degenerative stenosis may exert increased pressure on the spinal cord and spinal nerves that pass through the spinal canal.

Emerging medical technologies need to recognize the pivotal point of research into the use of information technology and other engineering disciplines to improve

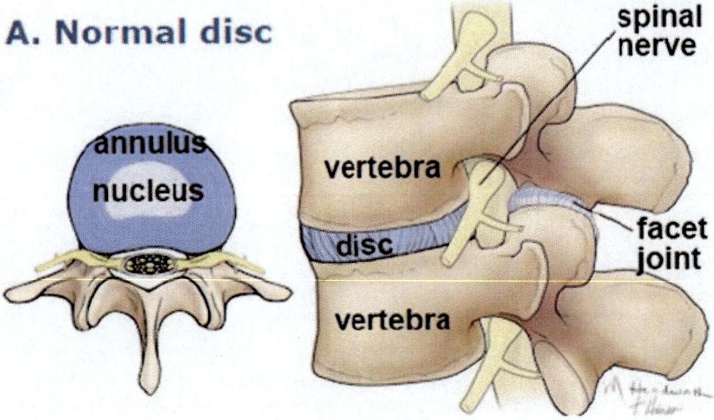

Fig. 2.8 Normal disc

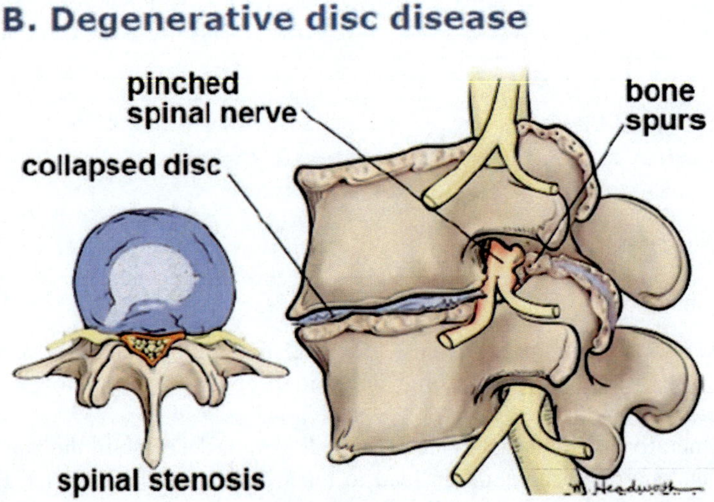

Fig. 2.9 Degenerative disc

the quality of life and wealth-creating abilities of the elderly and disabled, as well as the effectiveness and efficiency of care of that section of the population [1, 8, 9].

Through the application of engineering, mathematics and computational techniques, neurosciences, and social and cognitive sciences, it may be possible to gain insight into ageing and disabling diseases, coupled with technologies for maintaining independence and reasonable quality of life, and also gain understanding of the response to progressive degenerative diseases and age-related cognitive decline [10, 11, 14].

One way to gain insight into the musculoskeletal system involves the ability to mathematically model it as a complex system comprising the skeleton, skeletal muscles, ligaments, tendons, joints, cartilage, and other connective tissues. The system involves a human body system that provides the body with movement, stability, shape, and support. This system is mechanically very complex, and hence, the applied computational models may be highly simplified to enable some level of reasonable efficiency. The system is typically assumed to be a rigid-body system in order to be able to take advantage of the standard methods of multibody dynamics for modeling [6–8]. Additionally, the model must exhibit some reasonable representation of the muscle geometry, including the recruitment pattern of the muscles. These characteristics constitute complicated, dynamic problems. The muscles consist of soft tissues that are so much connected/linked to each other and to the bones, ligaments, and other anatomical elements in a very complicated manner. A good model of these complicated geometries may serve well for biomechanical modeling [9, 10]. Furthermore, the muscles are activated by the central nervous system (CNS) by mechanisms that are not well understood toward the modeling of the system in the required detailed. Therefore, the modeling of these mechanisms is based on assumptions, typically some kind of optimality condition. The fundamental problem is that there are more muscles than necessary to drive the degrees of freedom of the system, which implies that there are infinitely many muscle recruitment patterns that are acceptable from a dynamical point of view. This problem is often referred to as the redundancy problem of the muscle recruitment.

Musculoskeletal models may be divided into two groups: (1) the forward dynamics and (2) the inverse dynamics models, respectively. The forward dynamics methodology computes the motion based on a predicted muscular activation. While this is attractive in view of the detailed modeling of various physical phenomena, it is a very computationally intensive control problem and requires a costly optimization to make the model perform a specific task [18, 19, 21]. On the other hand, the inverse dynamics methodology computes the muscle activation based on a specified task, for example, a known motion. This puts many restrictions on the model, but it is computationally much more efficient. This efficiency may be exploited to build more complex models comprising more muscles, that means, a finer level of details of the mechanical model of the body.

Furthermore, gaining insight into the brain would enable further technological initiatives that would involve interventions in the brain, the organ that furnishes us with the capacities that underpin our existence, personal and social [15, 18, 21]. Some of the interventions may include invasive physical intrusions into the brain, while others rely on methods that interact with the brain from outside, typically by exposing the brain to electromagnetic fields, in order to provide treatments of further neurological and psychiatric disorders, which at present lack safe and effective treatments.

References

1. The Nuffield Council on Bioethics. (2013). *Novel neurotechnologies: Intervening in the brain*. Nuffield Council on Bioethics.
2. Lebedev, M. A., & Nicolelis, M. A. L. (2017). Brain-machine interfaces: From basic science to neuroprostheses and neurorehabilitation. *Physiological Reviews, 97*, 767–837.
3. Sarkodie-Gyan, T., Huiying, Y., Alaqtash, M., Bogale, M. A., Moody, J., & Brower, R. (2013). Application of fuzzy sets for assisting the physician's model of functional impairments in human locomotion. *Journal of Intelligent & Fuzzy Systems, 1*(1). https://doi.org/10.3233/IFS-120704. IOS Press. ISSN: 1064-1246.
4. Sarkodie-Gyan, T., Alaqtash, M., Huiying, Y., Spier, E., Abdelgawad, A., & Brower, R. (2011). Measurement of functional impairments in human locomotion using pattern analysis. *Elsevier Journal of Measurement, 44*, 181–191.
5. Hesse, S., Uhlenbrock, D., & Sarkodie-Gyan, T. (1999). Pattern of severely disabled Hemiparetic subjects on a new controlled gait trainer as compared to assisted treadmill walking with partial body weight support. *Clinical Rehabilitation, 13*, 401–410, 0269-2155(99) CR2710A.
6. Mergner, T. (2002). The Matryoshka dolls principle in human dynamic behavior in space: A theory of linked references for multisensory perception and control of action. *Cahiers de Psychologie Cognitive/Current Psychology of Cognition, 21*(2–3), 129–212.
7. Yin, J., & Yuan, Q. (2015). Structural homeostasis in the nervous system: A balancing act for wiring plasticity and stability. *Frontiers in Cellular Neuroscience, 8*, Article 439.
8. Francis Heylighten and Cliff Joslyn. (2001). Cybernetics and second-order cybernetics. In R. A. Meyers (Ed.), *Encyclopedia of physical science & technology* (3rd ed.). Academic.
9. Elting, J. W., van der Hoeven, J. H., Renken, R. J., et al. (2003). *Clinical neuroengineering: Engineering principles applied to clinical neuroscience*. NeuroImaging Center Groningen Group Leader: Prof. Dr. Ir. N. M. Maurits.
10. Kriegeskorte, N., Mur, M., & Bandetini, P. (2008, November). Representational similarity analysis-connecting the branches of systems neuroscience. *Frontiers in Systems Neuroscience, 2*, Article 4.
11. Newell, A. F. (1996). Technology and the disabled. *Technology, Innovation and Society, 12*(1), 21–23.
12. van der Kooij, H., Koopman, B., & van der Helm, F. C. T. (2008). *Human motion control*. Delft University/Twente University.
13. Grossman, M. (2002). Frontotemporal dementia: A review. *Journal of International Neuropsychological Society, 8*, 566–583.
14. Bak, T., & Hodges, J. R. (1999). Cognition, language and behavior in Motor Neurone disease: Evidence of frontotemporal dysfunction. *Dementia and Geriatric Cognitive Disorders, 10*, 29–32.
15. Sarkodie-Gyan, T., Huiying, Y., Bogale, M., Hernandez, N. V., & Pirela-Cruz, M. (2017). Application of multiple sensor data fusion for the analysis of human dynamic behavior inspace: Assessment and evaluation of mobility-related functional impairments. *Journal of Biomedical Science and Engineering, 10*, 182–203.
16. Brun, A. (1987). Frontal lobe degeneration of the non-Alzheimer type: I. Neuropathology. *Archives of Gerontology and Geriatrics, 6*, 193–208.
17. Elfgren, C., Passant, U., & Risberg, J. (1993). Neuropsychological findings in frontal lobe dementia. *Dementia, 4*, 214–219.
18. Malouin, F., Richards, C. L., Jackson, P. L., Dumas, F., & Doyon, J. (2003). Brain activations during motor imagery oflocomotor-related tasks: A PET study. *Human Brain Mapping, 19*, 47–62. [PubMed: 12731103].
19. Adams, R., & Parsons, O. (2003). *Neuropsychology for clinical practice: Etiology, assessment, and treatment of common neurologic disorders*. American Psychological Association.

20. Liu-Ambrose, T., Pang, M. Y., & Eng, J. J. (2007). Executive function is independently associated with performances of balance and mobility in community-dwelling older adults after mild stroke: Implications for falls prevention. *Cerebrovascular Diseases, 23*, 203–210. [PubMed: 17143004].
21. Woodruff-Pak, D. S., & Papka, M. (1999). Theories of neuropsychology and aging. In V. L. Bengston & K. W. Schaie (Eds.), *Handbook of theories of aging*. Springer.
22. Schmidt, R. A., & TDL. (1999). *Motor control and learning. Human kinetics*. Human Kinetics.
23. Dubost, V., Kressig, R. W., Gonthier, R., et al. (2006). Between dual-task related changes in stride velocity and stride time variability in healthy older adults. *Human Movement Science, 25*, 372–382. [PubMed: 16714067].
24. Braak, H., & Braak, E. (1998). Evolution of neuronal changes in the course of Alzheimer's disease. *Journal of Neural Transmission Supplementum, 53*, 127–140.
25. Ernst, J., Krapp, S., Schuster, T., Forstl, H., Kurz, A., & Diehl-Schmid, J. (2007). Car driving ability of patients with frontotemporal lobar degeneration and Alzheimer's disease. *Nervenarzt, 81*(1), 79–85.
26. Wang, L. H., & Weiss, M. D. (2013). Anatomical, clinical, and electrodiagnostic features of radial neuropathies. *Physical Medicine and Rehabilitation Clinics of North America, 24*, 33–47.
27. Alaqtash, M., Huiying, Y., Brower, R., Abdelgawad, A., Spier, E., & Sarkodie-Gyan, T. (2011). Application of wearable sensors for human gait analysis using fuzzy computational algorithm. *Elsevier Journal of Engineering Applications of Artificial Intelligence, 24*(6), 1018–1025.
28. Huiying, Y., Alaqtash, M., Spier, E., & Sarkodie-Gyan, T. (2010). Analysis and diagnosis of muscle activity during gait cycle using fuzzy rule-based reasoning. *Elsevier Journal of Measurement, 43*, 1106–1114.
29. Nishikawa, K., Biewener, A. A., et al. (2007). Neuromechanics: An integrative approach for understanding motor control. In *Symposium "Biomechanics and neuromuscular control" Society for Integrative and Comparative Biology, January 4–8, 2006*, at Orlando, Florida.
30. Hesse, S., Uhlenbrock, D., & Sarkodie-Gyan, T. (1999). Gait pattern of severely disabled Hemiparetic subjects on a new controlled gait trainer as compared to assisted treadmill walking with partial body weight support. *Clinical Rehabilitation, 13*, 401–410, 0269-2155(99) CR2710A.
31. Ibrahim, I., Khan, W. S., Goddard, N., & Smithham, P. (2012). Carpal tunnel syndrome: A review of the recent literature. *The Open Orthopaedics Journal, 6*(Suppl i: MS), 69–76.
32. Carpal Tunnel Syndrome: Orthoinfo Basics, American Academy of Orthopedic Surgeons (AAOS).
33. Carpal Tunnel Syndrome, Information for patients, Berkshire Healthcare, NHS Foundation Trust.
34. Historical Review: The Mayo Clinic and Carpal Tunnel Syndrome.
35. Robert Giffin, J., & Stantsh, W. D. (1993, September). Overuse tendonitis and rehabilitation. *Canadian Family Physician Medicin de familie canadien, 39*, 1762–1769.
36. The Sports Medicine Patient Advisor: Wrist Tendonitis.
37. Wrist Tendinopathy, International Tennis Federation.
38. Beeson, P. (2014). Posterior tibialis tendinopathy. *Journal of the American Podiatric Medical Association, 104*(5), 455–467.
39. Premkumar, A., Perry, M. B., Dwer, A. J., Gerber, L. H., et al. (2002). Sonography and MR imaging of posterior Tibial tendinopathy. *American Journal of Roentgenology, 178*(1), 223–232.
40. Nance, E. P., Jr., & Kaye, J. J. (1982). Injuries of the quadriceps mechanism. *Radiology, 142*(2), 301–307.
41. Advanced Physical Therapy: Orthopedics, Spine, Sports Medicine " A patient's guide to Quadriceps Tendonitis".
42. The Sports Medicine Patient Advisor "Quadriceps contusion (Thigh Bruise) and Strain".

43. Schizas, G. K., & Kosmopoulos, V. (2010). Disc degeneration: Current surgical options. *European Cells and Materials, 20*, 306–315.
44. Kun, H. S., Pang Hung, W., & Jang, I. I.-T. (2020). Lumbar Degenerative Disease Part 1: Anatomy and pathophysiology of intervertebral discogenic pain and radiofrequency ablation of basivertebral and sinuvertertebral nerve treatment for chronic review of literature. *International Journal of Molecular Sciences, 21*, 1483.
45. Chester, J., Donnally, I. I. I., & Varacation, M. (2018). *Lumbar degenerative disk disease*. Penn Highlands Healthcare System, ResearchGate.

Chapter 3
Challenges and Concerns to Society: The Human Locomotor System

The central nervous system controls all body activities, ranging from heart rate and sexual function to emotion, learning, and memory. It influences the response to disease of the immune system and to determine, in part, how well people respond to medical treatments. Ultimately, it shapes our thoughts, hopes, dreams, and imagination [1]. It may be concluded that the central nervous system is what makes us human. However, there are several illnesses and injuries that result in damages to the brain and its functions. These damaging conditions may lead to serious disorders that affect memory, cognition, movement, or consciousness or cause conditions such as chronic pain. Several disorders of the brain and the nervous system result in more hospitalizations than any other disease group, including heart disease and cancer.

Musculoskeletal conditions are prevalent and their impact is pervasive. They are the most common cause of severe long-term pain, functional limitations, physical disability, and social and economic implications, and they affect a vast majority of people around the world [31]. They significantly affect the psychosocial status of affected people as well as their families and caregivers. They are also highly associated with significant mental health decline and deteriorated functioning [1–3]. Because of population increases and ageing, the number of people with musculoskeletal conditions is rapidly increasing. Musculoskeletal conditions are the main cause of disability among older age groups. Moreover, the pain and physical disability brought about by musculoskeletal conditions affect social functioning and mental health, further diminishing the quality of life of the patient.

Current medical interventions remain marginal and unable to offer prevention and effective treatment for most chronic musculoskeletal conditions because their etiology and pathogenesis are unknown [1]. This may be attributed to our marginal understanding of the functions of the locomotor system in both healthy and unhealthy states. This may be manifested by the "absence" of the theory of etiology and pathogenesis of prevalent chronic musculoskeletal conditions, which already dominate the list of major causes of disability. These include:

© The Author(s), under exclusive license to Springer Nature Switzerland AG 2023

T. Sarkodie-Gyan, H. Yu, *The Human Locomotor System*,
https://doi.org/10.1007/978-3-031-32781-0_3

- Degenerative conditions in the spine and joints (osteoarthritis, degenerative disk disease, etc.)
- Skeletal deformities (scoliosis, kyphosis, etc.)
- Many types of muscle pathology (muscular dystrophy, myofascial trigger points, fibromyalgia, etc.)
- Back and neck pain and other musculoskeletal pain with an indeterminate etiology.

New technology initiatives have helped to increase the years of life expectancy so that the average age at death for both men and women has now become increasingly high. In fact, the increase in life expectancy may be partly due to medical advances. However, treatments for disorders like Alzheimer's have not advanced at the same rate as those for other common diseases. The understanding of the central nervous system (the brain and spinal cord) as the control center for the human body may constitute a paramount focus in the human locomotor system. The nervous system has a special status in human life that distinguishes it from other organs [1–3]. Its healthy functioning plays a central role in the operation of our bodies, our capacities for autonomous agency, our conceptions of ourselves and our relationships with others, and thus in our abilities to lead fulfilling lives. Hence, the brain plays an irreducible role in the maintenance and performance of the body, including controlled movements and autonomic functions, as well as thoughts, emotions, memories, and behavior.

Illness or injury that results in damage to the brain and its functions may lead to serious disorders that affect memory, cognition, movement, or consciousness or cause conditions such as chronic pain. The brain exhibits limited capacity to repair damaged tissue, although new functional connections may be formed. Therefore, there arises the need for novel initiatives to address some of the distressing and disabling effects of brain damage by intervening in the functions of the brain itself.

Neuroscience has made some significant discoveries in areas, including, but not limited to, genetics, neural plasticity, molecular neuropharmacology, and imaging, among other techniques [4–7]. The ability to understand these techniques may directly lead to the improvement of life and also improve the efficiency and effectiveness of services provided to people with disabilities, neurological diseases, and injuries in order to improve the quality of care.

The understanding of genetics has provided new insights into underlying disease mechanisms and is beginning to suggest new treatments. With the mapping of the human genome, neuroscience may be able to make more rapid progress in identifying genes that either contribute to human neurological disease or that directly cause disease. Mapping animal genomes may support the search for genes that regulate and control many complex behaviors [7–9]. The identification of key disease genes may help to underlie several neurodegenerative disorders, including Alzheimer's disease, Huntington's disease, Parkinson's disease, and amyotrophic lateral sclerosis.

Research thrusts into the molecular bases of neural plasticity may reveal how learning and memory occur and how declines might be reversed. It may also lead to new approaches toward the treatment of chronic pain.

Gaining insights into the mechanisms of molecular neuropharmacology may provide insight into the mechanisms of addiction [8, 9]. These advances may further lead to new treatments for depression and obsessive-compulsive disorder.

Revolutionary imaging techniques that include magnetic resonance imaging and positron emission tomography may, to some extent, be able to reveal the brain systems that underlie attention, memory, and emotions and may assist in indicating dynamic changes that occur in schizophrenia [1, 10].

The insight and understanding of how and why neurons die, as well as the discovery of stem cells, which divide and form new neurons, may offer good contributions in clinical applications. It would dramatically improve the outlook for reversing the effects of injury, both in the brain and spinal cord.

Knowledge and insight into new principles and molecules responsible for guiding nervous system development may lead to a better understanding of certain disorders of childhood. Together with the discovery of stem cells, these advances may point toward novel strategies for helping the brain or spinal cord regain functions lost to diseases.

Musculoskeletal conditions are the leading contributors to disability worldwide. The diseases consist of countless conditions that affect the human locomotor system. The human locomotor system includes the muscles, bones, joints, tendons, and ligaments and has a growing impact worldwide. Those conditions are typically characterized by pain (often persistent) and limitations in mobility, dexterity, and overall level of functioning, reducing the ability of the human to work (low back pain), with low back pain being the single leading cause of disability worldwide [1, 2]. The impact of these conditions is measurable (according to the World Health Organization, WHO) using disability-adjusted life years (DALYs), which combine the years lived with disability (YLDs) and the years of life lost (YLLs) through premature death [13].

The WHO defines the following parameters as measures of disability:

Disability-Adjusted Life Year, DALY

This is a measure of overall disease burden.

$$\mathbf{DALY = YLL + YLD}$$

Where **YLL**: Years of Life Lost.
And **YLD**: Years Lived with Disability.
And

$$\mathrm{YLL = N \times L}$$

Where N = number of deaths, and.
L is the standard life expectancy at age of death in years.

The DALY Index is a summary measure that combines time lost through premature death (YLLs) and time lived in states of less-than-optimal health, loosely referred to as "disability," or years lived with disability (YLDs). One DALY can be thought of as one lost year of "healthy" life. DALYs for a specific cause are calculated as the sum of the YLLs from that cause and the YLDs for people living in states of less than good health resulting from a specific cause.

The challenges of musculoskeletal conditions on the society include such disorders as osteoarthritis, rheumatoid arthritis, osteoporosis, and low back pain.

Low back pain is the main contributor to the overall burden of musculoskeletal conditions [11, 12, 72, 76]. It affects nearly everyone at some point in time and about high percentage of the population at any given point. Cultural factors greatly influence the prevalence and prognosis of low back pain. Other contributors to the overall burden of musculoskeletal conditions include fractures (fragility fractures, traumatic fractures), osteoarthritis, other injuries (muscles, joints, traumatic brain injuries), neck pain, amputations, and rheumatoid arthritis. Musculoskeletal conditions significantly limit mobility and dexterity, leading to lower levels of well-being and a reduced ability to participate in society [71, 74, 78].

Human mobility may be characterized as coordinated and adaptive behavior that emerges from the interplay among descending output from the central nervous system, sensory input from the body and environment, muscle dynamics, and the emergent dynamics of the whole body. This coordinated and adaptive behavior in mobility may be the result of the intelligent sensory-motor functions. The functions of the central nervous system (brain and spinal cord) are subject to damage caused by injury, infection, or disease [1, 2]. Many of these types of damages lead to serious disorders that affect memory, cognition, movement, or consciousness and cause conditions such as chronic pain, neurological disorders, and some of the most intractable medical conditions. Some of these intractable health conditions include:

Stroke

Stoke constitutes a cardiovascular disease that affects the supply of blood to the brain. The interruption in blood flow to the affected area of the brain starves the cells of valuable nutrients and oxygen, which in a matter of minutes may be able to kill the cells. After the death of these cells, the human body parts controlled by them become unable to function [10]. In fact, the overwhelming effects of strokes may often be permanent because dead brain cells may normally not be replaced.

The effects of stroke depend upon the type of stroke, the area of the brain affected, and the extent of the brain injury. Brain injuries sustained from stroke can affect the senses of the patient, motor activity, speech, and the ability to comprehend speech. Paralysis or weakness on one side of the body is also common.

The effects of loss of senses, motor activity, speech, and the ability to understand speech may be devastating for patients. These losses may culminate and cause victims to feel as if they are not whole people.

Traumatic Brain Injury

Traumatic brain injury (TBI) is one of the leading causes of death and disability in children and young adults around the world. TBI is defined as the result of the application of either external physical force or rapid acceleration/deceleration forces that disrupt brain function, as manifested by immediately apparent impairments in cognitive and/or physical function. It is important to note that it is the application of such forces to the brain, rather than to the head per se, that produces a TBI. In other words, not all head injuries produce brain injuries, and some brain injuries (particularly those resulting from acceleration/deceleration forces) may occur without apparent head injury.

A closed head injury arises when the head suddenly and viciously hits an object, but the object does not pierce the skull. A penetrating injury transpires when an object punctures the skull and enters the brain tissue. Besides, for distinguishing brain injuries as open or closed, there exist three separate processes that cause injury to the brain: bruising, tearing, and swelling.

Bruising, or a contusion, can occur when the brain slams into the front or back of the skull. This action can cause blood vessels to rupture, which releases uncontrollable amounts of blood. The flowing of blood can be very bad due to the fact that the skull is very hard. The accumulation of blood can cause pressure against the delicate brain, which may interrupt some functions of the brain.

Tearing can occur when the brain hits the front and back of the skull, and the energy actually tears apart some of the brain. The tearing or cutting of brain tissue can be a very serious problem. The portions of the brain that are being torn can be the wires that make the brain work. If the connections are lost, certain functions controlled by the brain become lost.

Swelling occurs in the brain in much the same way as if any other body parts were injured. For example, when a hammer smashes the finger, the finger will bleed, and the body will realize that the finger is hurt and send agents to help heal the finger. The difference between the finger and the brain is that the finger has room to swell; the skull encapsulates the brain and it has no room to swell. When the brain does swell, it puts pressure on the brain, which can distort the structure and even cause important portions of the brain to lose their ability to control breathing or heart rate.

Traumatic brain injuries can cause several types of disabilities. The disabilities ensuing from a TBI depend upon the severity of the injury, location of the injury, age, and general health of the patient. Some of the more common disabilities include cognition, sensory processing, communication, and behavioral or mental health.

Traumatic brain injury (and also mild TBI and concussion) can have a serious and long-term impact on the cognitive, physical, and psychological functions of a person. In fact, many people suffer from balance and instability problems after (m)TBI.

There is a very large target population associated with (mild) traumatic brain injury. These include, but are not limited to, football, boxing, auto/motor vehicle/transportation crashes, etc.

Spinal Cord Injuries

The spinal cord is protected by the spinal column and acts as the major link between the brain and the rest of the body. When an injury is inflicted on the spinal cord, there can be significant physiological consequences. The vertebrae can still be broken or dislocated and cause traumatic injury to the spinal cord. The injuries to the spinal cord can occur at any level. Depending on the segment of the cord that is injured, and the severity of the injury, certain body functions are compromised.

Spinal cord injuries usually result when there is dramatic trauma inflicted to the spinal column [7, 10, 16]. These traumas can arise from numerous accidents, including catastrophic falls, being thrown from a horse or through the windshield of an automobile, or any other kind of trauma that causes the vertebrae in the back or neck to be crushed or compressed. These types of accidents can cause permanent damage at the cervical level of the spinal cord and below. Paralysis of most of the body, called quadriplegia, is a likely result of a spinal cord injury. Paralysis of the lower trunk and lower extremities, called paraplegia, is also possible and often the result of automobile accidents [14–17].

Spinal cord injuries are usually termed complete or incomplete, in which complete spinal cord injuries being the worst. The class of incomplete spinal cord injuries determines how many motor functions will remain. A complete spinal cord injury results in no communication whatsoever between the body and the brain below the injury and is termed class A in the American Spinal Injury Association (ASIA) Impairment Scale. Incomplete injuries are usually a more promising condition. Class B means that there is sensory, but not motor, function preserved below the level of injury. Class C constitutes that motor function is preserved, but more than half of the key muscles below the injury are not strong enough to move against gravity. Class D states that motor function is preserved and at least half of key muscle groups are able to move against gravity [16–18]. Class E is termed as having normal motor and sensory functions.

Spinal cord injuries affect each patient differently, and the complications caused depend upon which area of the spinal cord was injured and how bad it was. Spinal cord injuries can cause patients to have trouble with a number of different complications, including, but not limited to, breathing, pneumonia, irregular heartbeat, low blood pressure, blood clots, spasm, autonomic dysreflexia, pressure sores, pain, bladder and bowel problems, reproductive and sexual function, and finally being able to ambulate and move certain portions of the body.

Parkinson's Disease

Parkinson's disease occurs when the nerve cells, or neurons, in the substantia nigra portion of the brain die or become impaired. The neurons in this portion of the brain typically produce a brain chemical known as dopamine. This chemical messenger is responsible for relaying signals between the substantia nigra and the corpus striatum, which produces smooth, purposeful muscle activity [2, 23]. The loss of dopamine causes erratic firing of the nerve cells in the striatum, leaving the patient unable to direct or control their movements in a normal manner. Parkinson's patients lose about 80% of their dopamine-producing cells, and the cause of the loss of these cells is not known or understood.

Parkinson's disease is a neurological disorder that falls into a group of conditions called motor system disorders. The four key symptoms are tremors or trembling in hands, arms, legs, jaw, and face; rigidity or stiffness of the limbs and trunk; bradykinesia or slowness of movement; and postural instability or impaired balance and coordination. As these conditions become more severe, patients tend to have trouble walking, talking, or completing simple tasks [81]. Parkinson's disease is usually termed chronic and/or progressive. By chronic, it means that the disease will persist over a long period of time. Progressive means that the symptoms tend to grow worse over time. The disease is also neither contagious nor usually inherited, meaning it does not pass from one family member or generation to the next.

Amyotrophic Lateral Sclerosis

Amyotrophic lateral sclerosis, or (ALS), sometimes called Lou Gehrig's disease, is a rapidly progressive, invariably fatal neurological disease that attacks the neurons responsible for controlling voluntary muscles. The disease belongs to a group of disorders known as motor neuron diseases, which are characterized by the gradual degeneration and death of motor neurons.

The motor neurons are nerve cells located in the brain, brainstem, and spinal cord that serve as controlling units and vital communication links between the nervous system and the voluntary muscles of the body. Messages from motor neurons in the brain are transmitted to motor neurons in the spinal cord and, from thereafter, to particular muscles. In ALS, both the neurons in the brain and spinal cord degenerate or die, ceasing to send messages to the muscles. Unable to function, the muscles gradually weaken, waste away, and twitch. Eventually, the ability of the brain to start and control voluntary movement is lost.

Alzheimer's Disease

Alzheimer's disease (AD) is an irreversible neural degenerative disease that attacks the brain, causing disruptions in memory, cognition, personality, and other functions that eventually lead to death from complete brain failure. Patients with Alzheimer's disease experience difficulties performing daily tasks independently. Studies have found that low levels of physical and mental activity significantly increase the risk of Alzheimer's disease [35].

Frontotemporal Dementia (FTD)

Frontotemporal dementia (FTD) primarily affects the frontal and anterior temporal lobes of the brain. These areas control the human "executive functions" like reasoning, personality, social behavior, movement, speech, language, and some aspects of memory [21, 23, 34, 36].

Osteoarthritis (OA) is the most common form of arthritis. It is a potentially devastating joint disease. Its pathophysiology is marked by a gradual degenerative process accompanied by low-grade inflammation, and although there is a strong correlation between age and OA risk, the abnormal changes that occur in the articular cartilage of people with OA differ notably from the typical changes associated with joint ageing in several important ways.

Risk factors for OA are multiple and span a variety of risk domains, such as lifestyle issues (e.g., obesity and engagement in manual labor), genetic predisposition, sex and ethnicity (risk is higher in women and African Americans), and comorbidities. Clinical outcomes for people with OA typically involve pain, limitations of daily living activities, and an overall diminution of quality of life (QOL). It is characterized by a loss of joint cartilage, which leads to pain and loss of function primarily in the knees and hips and affects 9.6% of men and 18% of women aged above 60 years.

The need to evaluate the degree of this burden, as well as to determine treatment approaches and measure their success, requires instruments for measuring QOL.

Rheumatoid arthritis is an inflammatory condition that usually affects multiple joints. It affects 0.3–1.0% of the general population and is more prevalent among women and in developed countries. Persistent inflammation leads to joint destruction, but the disease can be controlled with drugs.

Osteoporosis is characterized by low bone mass, and micro-architectural deterioration is a major risk factor for fractures of the hip, vertebrae, and distal forearm. Hip fracture is the most detrimental fracture, being associated with 20% mortality and 50% permanent loss of function.

Neurodegeneration

Peripheral nerve injuries are a serious health problem that result from damage to the nerves located outside of the brain and spinal cord, often resulting in the loss of motor or sensory function or both functions. This is the progressive loss of structural integrity and functional capacity in individual nerve cells associated with specific disorders. It causes weakness, numbness and pain, usually in the hands and feet [22, 24]. It can also affect other areas and body functions, including digestion, urination and circulation. Parkinson's disease, Alzheimer's disease, multiple sclerosis, and Huntington's disease are examples of neurodegenerative disorders in which specific populations of neurons undergo such change.

Peripheral neuropathy, or diabetic sensory neuropathy, is an anatomically diffuse process primarily affecting sensory and autonomic fibers. People with diabetes have mild to severe forms of nervous system damage, causing peripheral neuropathy with impaired sensation or pain involving the feet and/or hands. Foot deformity and callus formation resulting from focal areas of high pressure, carpal tunnel syndrome, poor glucose control leading to impaired wound healing, and other nerve problems are some of the complicated conditions.

Long nerves are affected first, with symptoms typically beginning insidiously in the toes and then advancing proximally. This leads to loss of protective sensation (LOPS), whereby a person is unable to feel minor trauma from mechanical, thermal, or chemical sources. When foot lesions are present (foot ulceration), the reduction in autonomic nerve functions may also inhibit wound healing, which leads to amputations. A substantial percentage of people with diabetes develop at least one foot ulcer in their lifetime.

Lumbar Degenerative Disk Disease

The sequelae of disk degeneration are among the leading causes of functional incapacity in both men and women and belong to a common source of chronic disability in the years of working. Disk degeneration constitutes structural disruption and cell-mediated changes in composition. Mechanical, traumatic, nutritional, and genetic factors seem to play a role in the cascade of disk degeneration, albeit to a variable degree in different individuals. The presence of degenerative change is by no means an indicator of symptoms, and there is a very high prevalence in asymptomatic individuals. The etiology of pain as the symptom of degenerative disease is complex and appears to be a combination of mechanical deformation and the presence of inflammatory mediators.

Scoliosis

The lateral curvature of the spine (scoliosis), as shown in Fig. 3.1, exhibits several varieties, an unknown cause, and an unpredictable course [63, 66].

The curve is usually "S" or "C" shaped over three dimensions. Mild scoliosis does not typically cause problems, but more severe cases can affect breathing and movement. Pain is usually present in adults and can worsen with age. There is a serious need for innovative thinking and research to enable our understanding of scoliosis [64, 65].

There exists also the secondary scoliosis due to neuropathic and myopathic conditions, which may lead to a loss of muscular support for the spinal column. Hence, the spinal column is pulled in abnormal directions. Some conditions that may cause secondary scoliosis include muscular dystrophy, spinal muscular atrophy, poliomyelitis, cerebral palsy, spinal cord trauma, and myotonia [62, 65–70].

Muscular Dystrophy

Duchenne muscular dystrophy and myotonic dystrophy are genetic, progressive muscle diseases. These muscular dystrophies, which are currently incurable, cause muscle wasting or muscle weakness and decrease the quality of life of patients.

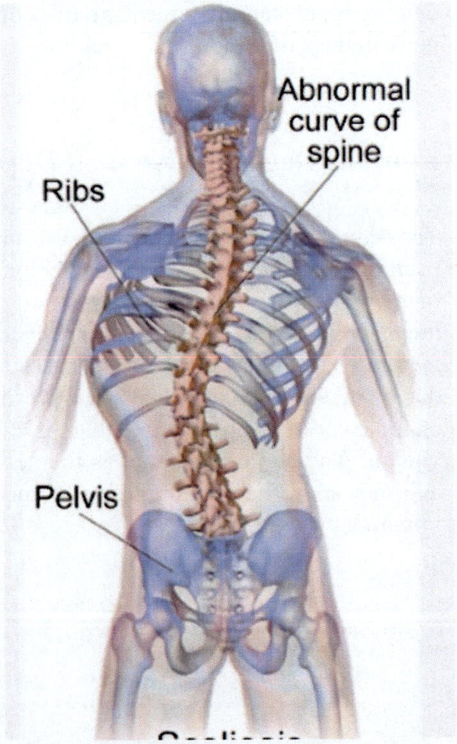

Fig. 3.1 Illustration of scoliosis

In addition to muscular impairments, cognitive impairments are also reported in both Duchenne muscular dystrophy and myotonic dystrophy [68, 69, 72]. Cognitive impairments in each type of muscular dystrophy are different and closely related to psychosocial variables and the quality of life of the patients.

Muscular dystrophy is a genetic, progressive disease of the muscles with several clinical forms, all of which have an early onset and are incurable with current medical technology [69, 70]. These diseases severely decrease motor functions and make it difficult to live an independent social life or engage in an occupation.

Muscular dystrophy causes not only physical impairments but also cognitive impairments.

Low Back Pain

Low back pain is a leading cause of disability. It occurs in similar proportions in all cultures and interferes with quality of life and work performance. There are a few cases of back pain that may be due to specific causes, while most cases are non-specific. Acute back pain is the most common presentation and is usually self-limiting, lasting less than 3 months regardless of treatment [73–75]. Chronic back pain is a more difficult problem, which often has a strong psychological overlay (work dissatisfaction, boredom), and a generous compensation system contribute to it. Among the diagnoses offered for chronic pain is fibromyalgia, an urban condition that does not differ materially from other instances of widespread chronic pain.

Low back pain is neither a disease nor a diagnostic entity of any sort. The term refers to pain of variable duration in an area of the anatomy afflicted so often that it has become a paradigm of responses to external and internal stimuli. The incidence and prevalence of low back pain are roughly the same the world over. This type of pain ranks high (often first) as a cause of disability and inability to work, as an interference with the quality of life, etc.

Myofascial Pain Syndrome

The myofascial pain syndrome is defined by the presence of myofascial trigger points in a palpable taut band that produces pain and tenderness in the muscle or its fascia. It is a common cause of musculoskeletal pain that is associated, in the majority of cases, with other pathologies and increases the comorbidities of these primary entities [76–80]. Understanding the pathophysiology and the pharmacological and non-pharmacological treatments provides better tools to the clinician in order to improve the quality of life of the patients who suffer pain.

Military Action

Blast-related trauma is receiving a lot of attention due to the increase in the use of improvised explosive devices (IEDs) in Iraq and Afghanistan [28]. These injuries may be due to the overpressure of the blast itself (primary blast injury), or due to objects put into motion by the blast overpressure (secondary blast injury), or by people being put into motion by the blast (tertiary blast injury). Blast-related injuries account for 88% of military personnel treated at an echelon II medical unit in Iraq [29, 30, 32]. About 47% of these injuries involved the head. A full 97% of the injuries to one Marine unit in Iraq were due to explosions (65% IEDs, 32% mines). Of these injured, 53% involved the head or neck. Most (82%) returned to duty following an average of 3 light duty days, which ranged from 0 days to 30 days of light duty [25, 27, 32].

As more powerful explosives are used in warfare, more soldiers are rendered dazed or unconscious by the explosion that may cause no external injury [26]. Retrograde or anterograde amnesia is commonly present upon regaining consciousness, along with severe headaches, tinnitus, hypersensitivity to noise, and tremors. This effect was observed and reported during the Great War and again during World War II.

Mission Essential Fitness (Military)

Current research has suggested that non-battle injuries (NBIs) have become a major cause of morbidity and mortality during combat operations [25]. There arises a significant shift from the influence of infectious diseases to non-battle ones, with special reference to the first and second world wars and the Korean War. Musculoskeletal and connective tissue conditions made up the second-leading category of hospitalizations in Operations Desert Shield and Storm (14%) and the fourth-leading category in Operation Joint Endeavor (10%), respectively. The number of non-battle deaths from unintentional trauma during the Operations Desert Shield and Storm exceeded the number of battle-related deaths. Even though the impact of NBIs during military operations is well recognized, the epidemiology of these injuries is poorly understood [26–29]. During past military operations, analyses to describe injury incidence, types, severity, causes, and treatment outcomes were conducted at the completion of the operations when copies of the medical records were centralized for review [26, 27]. Lessons learned from these retrospective analyses led to major advancements in medical evaluation, treatment, and rehabilitation that have greatly benefited injured service members. This retrospective approach to injury surveillance did not, however, allow identification of injury problems early in the deployments, when changes in practice and policy could have lowered the injury risk for deployed soldiers.

The combined categories of injuries and musculoskeletal disorders in the US Army accounted for over 30% of all hospitalizations in 1992. The ten leading types of injury for the army that accounted for 41% of all injury hospitalizations included ankle fractures, intracranial injury, fracture of face bones, sprains and strains on the knee, dislocation of the knee, fracture of radius/ulna, fracture of one or more fingers, and open wounds of fingers [28–30]. Injuries and musculoskeletal disorders have a greater impact on the health and readiness of the United States military than any other category of medical complaint. Soldiers require high levels of strength, power, speed, and agility to carry out their mission in today's modern combat environment. Physical training and physical fitness are required to accomplish military missions, and many military occupations routinely require a higher level of physical exertion and fitness [31–33]. There is the urgent need for fitness and the requisite physical training to maintain mission readiness, the burden and impact of training injuries, and the protective effects of fitness in preventing subsequent injuries in a complex and dynamic matrix of competing requirements [25–27]. There is a need for a novel technology that will offer neurophysiological measures towards the understanding of the functioning of the nervous system within operationally relevant settings and may extend the capabilities to develop cognitive state classification algorithms that may be used to assess soldier-system performance.

Physiological Functions in Human During Spaceflight

Humans depend on a sophisticated sensory-motor system to sustain proper balance control [23, 79]. This system is calibrated to terrestrial gravity and relies on the application of gravitational forces to provide a frame of reference in order to supply a necessary data stream to the central nervous systems (CNS) located in the brain. Key motion sensors include subtle organs located in the vestibular system (inner ear) that function as ultra-sensitive accelerometers for motion and direction [37, 38]. The somatosensory system is that system that includes tension, pressure, and motion receptors located in the skin, muscles, and joints that assist in the spatial awareness and perception through proprioception, and the visual and auditory senses complete the spatial orientation system [44–47, 52].

However, in the absence of gravity, signals from the central vestibular system, peripheral pressure receptors, and visual sense become misleading, to such a point that immediate perceptual confusion and subsequent disorientation usually occur [39–41, 52, 53]. Many astronauts suddenly feel as if they are upside-down or may even have difficulty sensing the location of their own arms and legs. This disorientation is described as space adaptation syndrome (SAS) and is widely recognized as the main cause of space motion sickness (SMS) [41–44].

The exposure to microgravity and the space environment during short- and long-duration space missions has important medical and health implications in astronauts [47–51]. These include neuro-vestibular problems, involving space motion sickness and disorientation during the flight, and impaired balance and neuromuscular coordination after landing, cardiovascular and fluid-related problems of orthostatic hypotension immediately following spaceflight, the possibility of altered cardiac susceptibility to ventricular arrhythmias, and reduced cardiac muscle mass and diminished cardiac function, muscle-related problems of atrophy, involving loss of muscle mass, strength and endurance, decrease in the bone mineral density, circadian rhythm-related problems, involving sleep and performance, and immune-related problems, involving infections and immunodeficiency [39, 40, 45, 54–56]. Sensory-motor control deconditioning is one of the most critical effects to consider as it leads to deconditioning of posture and gait controls during long range missions [57–61].

Intervening in the Brain

There are approximately 100 billion neurons in the brain, residing in a complex network of non-neuronal cells called glia. While there are ten times more glia cells than neurons, it is the neurons that are responsible for sensing changes in the environment, communicating these changes to other neurons, and commanding the responses of the body to them [81–84]. Neurons are.

specialized in complex ways to fulfill particular functions: the number, length, and pattern of extensions (axons and dendrites) that extend from their cell bodies; the connections they make with other neurons; the neurotransmitters that they release to pass on information; and the surface channels and receptors, selectively sensitive to particular chemicals [84–88]. These are just some of the features that distinguish individual neurons. Glia appear more uniform, though they too contribute to information processing in individual ways [1, 89, 90]. The three main types of glia are astrocytes, which regulate the chemical environment between neurons; oligodendrocytes, which provide electrical insulation; and microglia, the resident immune cells of the brain, which clean up debris and react to disruptions in brain homeostasis such as those caused by brain damage.

There may be some symptoms that imply a brain tumor; these include:

- Headaches.
- Seizures (convulsion or fits).
- Nausea and vomiting.
- Changes to eyesight.
- Drowsiness/lethargy.
- Changes in personality.
- Language problems/slurring speech.
- Co-ordination problems.
- Dizziness.
- Difficulty swallowing.
- Problems with smell, hearing, and sight.
- Inability to gaze upwards.
- Changes in facial expression.

Symptoms that relate to spinal cord tumors include:

- Pain.
- Numbness.
- Weakness in either the arms or legs.
- Loss of bladder or bowel control.

There are four categories of technology initiatives to intervene in the brain. They include (1) transcranial brain stimulation (TBS), (2) deep brain stimulation, DBS, (3) brain-computer Interface, BCIs, and (4) neural stem cell therapy, NSCT, respectively.

None of these technology initiatives has provided any cure for neurological or mental health disorders, but they could ameliorate symptoms or fulfil assistive roles in ways that help improve the quality of life of patients [105–109].

The technology initiative involving transcranial brain stimulation (TBS) depicts the methodology of stimulating the brain either by inducing an electrical field using a magnetic coil placed against the head (transcranial magnetic stimulation (TMS)) or by applying weak electrical currents via electrodes on the scalp (transcranial direct current stimulation (TDCS) and transcranial alternating current stimulation (TACS)). However, their therapeutic applications are still being explored, where the most established application is in treating drug-resistant depression. The exact mechanisms by which TBS achieves its therapeutic effects are still being researched.

The initiative involving deep brain stimulation (DBS) depicts a process that alters the functioning of brain cells and neural networks by using electrical currents, but in this case, the stimulation is delivered by electrodes implanted deep in the brain. The applications of this methodology in therapeutic interventions include the treatment of movement disorders, such as those associated with Parkinson's disease, and of neuropathic pain. There is also considerable research activity exploring its use to treat a wide range of psychiatric disorders. The exact mechanisms by which DBS achieves its therapeutic effects are still unknown.

The technology initiative involving the brain-computer interface, or BCI, uses electrodes (either implanted in the brain or resting on the scalp) to record the brain signals of the user, which are then translated into commands to operate computer-controlled devices. By actively producing brain signals, users may be able to control these devices. BCIs could in principle assist users to communicate, control prostheses or wheelchairs, support rehabilitation, or facilitate the detection of consciousness. This initiative exhibits potential application for those with paralysis. However, the therapeutic application of BCIs is still confined to research contexts, in which non-invasive techniques are the most prevalent.

The technology initiative involving the neural stem cell therapy, or NSCT, involves the injection of stem cells into the brain in order to repair damage caused by acute events such as stroke or neurodegenerative conditions such as Alzheimer's disease. However, this initiative may still be at an early phase of development. The precise ways in which stem cell grafts may assist in repairing lost brain tissue are not known, but these could include direct replacement of lost cells or stimulating repair by the brain itself.

Nonetheless, our understanding of the brain has improved dramatically in recent years, in part because of the development of fMRI scanners, which produce the familiar striking images of brain activity and provide suggestive hints concerning the basis of neurological disorders.

Some of the technology initiatives involve interventions in the brain, some requiring invasive physical intrusions into the brain, others relying on methods of interacting with the brain from outside, typically by exposing the brain to electromagnetic fields. Furthermore, some of these initiatives are still at an early experimental stage (such as the introduction of human neural stem cells into the brain), while others build on existing techniques (such as deep brain stimulation) to provide treatments for further neurological and psychiatric disorders, which at present lack safe and effective treatments.

In fact, these technological initiatives involve interventions in the brain, the organ that furnishes us with the capacities that underpin our existence, personal and social; interventions here are liable to affect our lives in the most intimate and fundamental ways. So, while the aim of these interventions is to meet the need for neurological and psychiatric therapies, there is always the risk of damaging side effects. And despite recent advances in neuroscience, this risk has to be taken seriously because of the limited knowledge of the ways in which these new technologies affect the brain.

There exist serious uncertainties about the long-term and unintended effects of these technological initiatives of intervening in the brain. There may be a lack of alternative treatments for some neurological disorders. Besides, many initiatives only address conditions that impair the decision-making capacities of the patients [1]. The lack of clear evidence of risks and benefits of some interventional techniques also presents challenges to responsible clinical decision-making.

Data concerning brain function and neurological health collected by devices such as those delivering DBS or BCIs may be sensitive and stigmatizing.

These technology initiatives come with very hard problems that need to be addressed. These hard problems may include demand for much more basic science and technology. These may include supporting frail people and people with dementia, aphasia following a stroke, memory loss within the community, and transportation and care for those with failing senses and slowing reaction times. The unmet and/or unfulfilled needs and wants of elderly and disabled people may lead directly to the need for new basic and strategic science and technology, which may include new materials, new sensors, a greater understanding of degenerating cognitive and physical processes, and methodologies for modeling and supporting these processes using technology.

There arises the need for much more sophisticated basic social science [93–97]. And there exists a link with the need to know far more about the effects of ageing on sensory, cognitive, memory, and physical skills and how to compensate for them.

It may be important to know much more about how people actually want their quality of life to improve, and a focus on people with special needs helps us learn how to ask these questions [96–99]. There are very real challenges for a whole range of social, scientific, and technological research at all levels: basic, strategic, and applied. Very real possibilities of exciting and innovative solutions may lead to technology initiatives that may include: (1) responding to the needs of ageing populations, with research into ageing and disabling diseases, coupled with technologies for maintaining independence and a reasonable quality of life and (2) neuroscience and cognitive science research into progressive degenerative disease and age-related cognitive decline [100–104]. These new technologies may include engineering, biological sciences, psychology, social sciences, mathematical techniques, artificial/computational intelligence, etc. They may offer insight and manifest our intelligence into the neural control of the locomotor system in space.

Ethical Concerns and Challenges of New Technology Initiatives

The fact is that the brain is one of the most important organs in the body. This brain is in charge of biological and neurological procedures such as memory, speech, perception, sleep, and emotion.

The technological initiatives depict a scientific approach that fuses and/or integrates electronic devices with the nervous system. These technology initiatives may give rise to some interesting but complex ethical and legal issues by virtue of the creation of an interface between the brain and computers, the brain-computer interface [1]. Whereas the advocates of these initiatives propose that the new technologies may enable humans to overcome diseases and disorders ranging from Alzheimer's, Parkinson's disease, blindness, anxiety, depression, and insomnia, the opponents argue that these initiatives may be overly invasive and could create unanticipated complications.

The technology initiative involving the transcranial magnetic stimulation (TMS) has provoked some excitement in the role of brain stimulation in psychiatric disorders. Some studies have illustrated promising but not conclusive evidence for the efficacy of TMS in treating depression. And TMS has not been shown to be effective in the treatment of obsessive-compulsive disorder, posttraumatic disorder, or schizophrenia. There may be the urgent requirement to develop further consensus on some parameters in TMS studies, including the shape of the coil, the coil-cortex distance, the motor threshold, the low-frequency versus high-frequency stimulation, and the location of the correct point of stimulation for each disorders.

Brain implants also involve the merging of microchips and artificial intelligence. This technology may be relatively new and still untested. Therefore, there may arise multiple ethical and legal issues: (1) privacy concerns may arise, (2) the identity of the individual may be stolen without permission, (3) the autonomy of the individual could be taken away, and (4) the culprits (for example, hackers) may exploit the technology and manipulate individuals for self-serving purposes.

A very significant ethical issue that these technology initiatives introduce may be exhibited through the degree of harm the users may experience through the use of the brain microchip. There is the possibility that the users would experience an immediate adverse reaction (e.g., infection and burn) as soon as the microchip is implemented into their brain. However, there could also be undetectable adverse reactions such as cognitive distortions, which may be more difficult to evaluate. For example, there could be sudden and unanticipated mood changes or emotional reactions, which may take more time to assess.

Stigmatization

Stigma may be described as the way the social status and sense of self of an individual may be tainted and devalued after being linked with a disease, disorder or impairment that has negative social connotations. There may be, at least, three types of stigma that affect persons with neurological, degenerative diseases or dementia.

The public stigma may be described as the negative or pejorative beliefs of members of the general public [19]. These beliefs may cause them to behave in discriminatory, exclusionary, or patronizing ways toward either the patients themselves, or towards persons closely associated with the patients. Public stigma may be overt, like discrimination, although other times it may be subtler, like a prejudicial belief that a patient is incompetent [20].

Individuals living with an impairment, disability or disorder may experience self-stigma. This self-stigma may be characterized by the manner in which an individual cognitively or emotionally absorbs negative beliefs, attitudes, assumptions, and stereotypes related to the disorder, such as feeling ashamed and inferior because of being associated with the disease [21]. Self-stigma may lead to depression, avoidant coping, social withdrawal, low self-esteem, hopelessness, worsened psychiatric symptoms, and decreased help-seeking behaviors.

Care-givers and family members may also experience stigma. This type of stigma, the Spillover stigma, seeks to describe the way and manner in which people who do not have any disability nor disorders, are affected by the stigma related to the disease. The spillover stigma often affects individuals who share close social proximity to those who have the disease, such as the care-givers. As a result of spillover stigma, individuals may experience many of the same social and psychological consequences as individuals with disabilities or disorders.

In general, some of the other harmful effects of stigmatization may include:

- Reluctance to seek help or treatment and less likely to stay with treatment.
- Social isolation.
- Lack of understanding by family, friends, co-workers, or others.
- Fewer opportunities for work, school or social activities or trouble finding housing.
- Bullying, physical violence or harassment.
- Health insurance that does not adequately cover your mental illness treatment.
- The belief that you will never succeed at certain challenges or that you can't improve your situation.

Stigmatization and discrimination may contribute to worsening symptoms and reduced likelihood of getting treatment. In fact, stigmatization may lead to negative effects on recovery among patients mainly due to lower self-esteem, difficulties with social relationships and reduced hope, among other effects.

References

1. Nuffield Council on Bioethics. (2013). *Novel neurotechnologies: Intervening in the brain.* ISBN: 978-1-904384-29-8.
2. The Society for Neuroscience. *Brain facts: A primer on the brain and nervous system.*
3. Sarkodie-Gyan, T. (2005). *Neurorehabiltation devices: Engineering design, measurement, and control.* McGraw-Hill Publishing Company. ISBN: 0-07-144830-6.
4. Dietz, V., Colombo, G., Jensen, L., et al. (1995). Locomotor capacity of spinal cord in paraplegic patients. *Annals of Neurology, 37*, 555–556.
5. Harkema SJ, Hurley SL, Patel UK, et al (1997) "Human lumbosacral spinal cord interprets loading during stepping." Journal of Neurophysiology 77:797–811.
6. Anastakis, D. J., et al. (2008). Cortical plasticity following nerve transfer in the upper extremity. *Hand Clinics, 24*(2008), 425–444.
7. Komitova, M., et al. (2006). On neural plasticity, new neurons and the postischemic milieu: An integrated view on experimental rehabilitation. *Experimental Neurology, 199*(1), 42–55.
8. Lledo, P. M., Alonso, M., & Grubb, M. S. (2006). Adult neurogenesis and functional plasticity in neuronal circuits. *Nature Reviews. Neuroscience, 7*, 179–193.
9. Taupin, P., & Gage, F. H. (2002). Adult neurogenesis and neural stem cells of the central nervous system in mammals. *Journal of Neuroscience Research, 69*, 745–749.
10. Komitava, M., Mattsson, B., Eriksson, P. S., & Johansson, B. B. (2005). Enriched environment increases neural stem/progenitor cell proliferation and neurogenesis in the subventricular zone of stroke-lesioned adult rats. *Stroke, 36*, 1278–1282.
11. Komitova, M., Perfilieva, E., Mattsson, B., Ericsson, P. S., & Johansson, B. B. (2002). Effects of cortical ischemia and postischemic environment enrichment on hippocampal cell genesis and differentiation in the adult rat. *Journal of Cerebral Blood Flow and Metabolism, 22*, 852–860.
12. Komitova, M., Zhao, L. R., Gido, G., Johansson, B. B., & Ericsson, P. (2005). Postischemic exercise attenuates whereas enriched environment has certain enhancing effects on lesion-induced subventricular zone activation in the adult rat. *The European Journal of Neuroscience, 21*, 2397–2405.
13. World Health Organization (WHO). *Neurological disorders: Public health challenges.*
14. Lundborg, G. (2000). Brain plasticity and hand surgery: An overview. *Journal of Hand Surgery (British and European Volume), 25B*(3), 242–252.
15. Brunelli, G. A. (2004). Sensory nerve transfers. *Journal of Hand Surgery (British and European Volume), 29B*(6), 557–562.
16. Field-Fote, E. C., Fluet, G. G., Schafer, S. D., Schneider, E. M., Smith, R., Downey, P. A., & Ruhl, C. D. (2001). The spinal cord injury functional ambulation inventory. *Journal of Rehabilitation Medicine, 2001*(33), 177–181.
17. Lam, T., Noonan, V. K., Eng, J. J., & the SCIRE Research Team. (2008). A systematic review of functional ambulation outcome measures in spinal cord injury. *Spinal Cord, 46*, 246–254. 2008 International Spinal Cord Society.
18. Capogrosso, M., Milekovic, T., Borton, D., Wagner, F., Moraud, E. M., Mignardot, J.-B., Buse, N., Gandar, J., Barraud, Q., Xing, D., Rey, E., Duis, S., Jianzhong, Y., Ko, W. K. D., Li, Q., Detemple, P., Denison, T., Micera, S., Bezard, E., Bloch, J., & Courtine, G. (2016). A brain-spinal interface alleviating gait deficits after spinal cord injury in primates. *Nature, 539*(7628), 284–288.

19. Caspermeyer, J. J., Sylvester, E. J., Drazkowski, J. F., Watson, G. L., & Sirven, J. I. (2006). *Evaluation of stigmatizing language and medical errors in neurology coverage by US newspapers*. Mayo Foundation for Medical Education and Research.
20. Beard, J. R., & Bloom, D. E. (2015). Towards a comprehensive public health response to population ageing. *HHS Public Access Lancet, 385*(9968), 658–661.
21. Executive Office of the President of the United States "Emerging technologies to support an aging population" (2019); Task force on research and development for technology to support aging adults; Committee on Technology of the National Science & Technology Council; March 2019.
22. Lundborg, G., & Rosen, B. (2007). Hand function after nerve repair. *Acta Physiologica, 2007*(189), 207–217.
23. Akay, T., & Murray, A. J. (2021). Relative contribution of proprioceptive and vestibular sensory systems to locomotion: Opportunities for discovery in the age of molecular sciences. *International Journal of Molecular Sciences, MDPI, 22*, 1467.
24. Pirela-Cruz, M. (2009). Interosseous nerve transfers for tibialis anterior muscle paralysis (foot drop): A human cadaver-based feasibility study. *Journal of Reconstructive Microsurgery, 25*(3), 203–211.
25. Briggs, D., & Hildreth, E. (2009). *Mission essential fitness (MEF) for the US army soldier by MWR*. Mission Essential Fitness Ft. Bliss, TX; For tactical Athletes.
26. Kruse, A. (2008). *Operational neuroscience: Intelligence community forum*. DARPA.
27. Greathouse, D. C., Schreck, R. C., & Benson, C. J. (1994). The United States Army physical therapy experience: Evaluation and treatment of patients with neuromusculoskeletal disorders. *Journal of Orthopaedic & Sports Physical Therapy; JOSPT, 19*(5).
28. Hauret, K. G., Taylor, B. J., Clemmons, N. S., Block, S. R., & Jones, B. H. (2006). *Frequency and causes of nonbattle injuries air evacuated from Opeartions Iraqi Freedom and Enduring Freedom, U.S. Army, 2001–2006.*
29. Department of the Army; US Army Center for Health Promotion and Preventive Medicine; Aberdeen Proving Ground MD.
30. Miller, D. P., & Lawton, C. R. (2006). *Human performance modeling for system of systems analytics: Combat performance-shaping factors* (Sandia report; Sandia2006–1111). Sandia National Laboratories.
31. McClure, S. K., Adams, J. E., & Dahm, D. L. (2005). Common musculoskeletal disorders in women. *Mayo Clinic Proceedings, 80*(6), 796–802.
32. Kaufman, K. R., Brodine, S., & Shaffer, R. (2000). Military training-related injuries surveillance, research, and prevention. *American Journal of Preventive Medicine, 18*(3S) Elsevier Science Inc.
33. Hootman, J. M., Macera, C. A., Ainsworth, B. E., Martin, M., Addy, C. L., & Blair, S. N. (2001). Association amongst physical activity level, cardiorespiratory fitness, and risk of musculoskeletal injury. *American Journal of Epidemiology, 154*(3).
34. Grossman, M. (2002). Frontotemporal dementia: A review. *Journal of the International Neuropsychological Society, 8*, 566–583.
35. Verghese, J., Lipton, R. B., Hall, C. B., Kuslansky, G., Katz, M. J., & Buschke, H. (2002). Abnormality of gait as a predictor of non-Alzheimer's dementia. *The New England Journal of Medicine, 347*(22).
36. Allali, G., Dubois, B., Assal, F., Lallart, E., de Souza, L. C., Bertoux, M., Annweiler, C., Herrmann, F. R., Levy, R., & Beauchel, O. (2010). Frontotemporal dementia: Pathology of gait? *Movement Disorders, 25*(6), 731–737.
37. Edgerton, R. V., & Roy, R. R. (1996). Neuromuscular adaptations to actual and simulated spaceflight. In American Physiological Society (Ed.), *Handbook of Physiology* (2 vols, pp. 721–763). Williams & Wilkins.
38. Fitts, R. H., Riley, D. R., & Widrick, J. J. (2000). "Physiology of a microgravity environment": Microgravity and skeletal muscle. *Journal of Applied Physiology, 89*, 823–839.

39. Jiang, B., Ohira, Y., Roy, R. R., Nguyen, Q., Illyina-Kaueva, E. I., Oganov, V., & Edgerton, V. R. (1952). Adaptation of fibers in fast-twitch muscles of rats to spaceflight and hindlimb suspension. *Journal of Applied Physiology, 73*, 58S–65S.
40. Recktenwald, M. R., Hodgson, J. A., Roy, R. R., Riazanski, S., McCall, G. E., Kozlovskaya, I., Washburn, D. A., Fanton, J. W., & Edgerton, V. R. (1999). Effect of spaceflight on rhesus quadrupedal locomotion after return to IG. *Journal of Neurophysiology*, 2451–2463.
41. Bamman, M. M., Clarke, M. S. F., Feeback, D. L., Talmadge, R. J., Stevens, B. R., Lieberman, S. A., & Greenisen, M. C. (1998). Impact of resistance exercise during bed rest on skeletal muscle sarcopenia and myosin isoform distribution. *Journal of Applied Physiology, 84*, 157–163.
42. Riley, D. A., Ellis, S., Slocum, G. R., Sedlack, F. R., Bain, J. L. W., Krippendorf, B. B., Lehman, C. T., Macias, M. Y., Thompson, J. L., Vijayan, K., & De Bruin, J. A. (1996). In-flight and post-flight changes in skeletal muscles of SLS-1 and SLS-2 spaceflown rats. *Journal of Applied Physiology, 81*, 133–144.
43. Martin, T. P., Edgerton, V. R., & Grindeland, R. E. (1988). Influence of spaceflight on rat skeletal muscle. *Journal of Applied Physiology, 65*, 2318–2325.
44. Bonin, G. (2005). *Physiological issues in human spaceflight: Review and proposed countermeasures* (MAAE 4906/MECH 5801). Biomedical Engineering and Biomechanics.
45. Oman, C. M., & Bloomberg, J. J. *Sensorimotor adaptation*. Research summary of the Sensorimotor Adaptation Team.
46. Wood, S. J. (2009). *Sensorimotor adaptation following exposure to ambiguous inertial motion cues*. Project description of the Sensorimotor Adaptation Team.
47. Young, L. R. (2012). *Sensorimotor displays and controls to enhance the safety of human/machine cooperation during lunar landing*. Project Description of the Sensorimotor Adaptation Team.
48. Bloomberg, J. J. (2011). *Development of countermeasures to enhance sensorimotor adaptation*. Project description of the Sensorimotor Adaptation Team.
49. Oman, C. M. (2008). *Advanced displays for efficient training and operation of robotic systems*. Project description of the Sensorimotor Adaptation Team.
50. Philippe, A. S., Landrock, C. K., & Blaber, A. P. (2008). Reducing incapacitating symptoms during space flight: Is the postural deficiency syndrome an applicable model? *Hippokratia, 12*, 41.
51. Harris, B. A., Bilica, R. D., Bishop, S. L., Blackwell, T., Layne, C. S., Harm, D. L., Sandoz, G. R., & Rosenow, E. C., III. (1997). Physical exam during space flight. *Mayo Clinic Proceedings, 72*, 301–308.
52. MacNeilage, P., & Angelaki, D. E. *Psychophysics and modeling of spatial orientation perception*. In Project description of the Sensorimotor Adaptation Team.
53. Small, R. L. (2011). *Modeling and mitigating spatial disorientation in low-gravity environments*. Project description of the Sensorimotor Team.
54. Songer, J. E., Merfeld, D. M., & Eaton, R. A. (2009). *Investigating high-frequency function: A potential flight diagnostic*. Project description of the Sensormotor Adaptation Team.
55. Moore, S. T. (2008). *Galvanic vestibular stimulation as an analog of post-flight sensorimotor dysfunction*. Project description of the Sensorimotor Adaptation Team.
56. Cavanagh, P., & Bloomfield, S. *Decrements in bone, connective tissue and muscle function with prolonged spaceflight: Mechanisms and countermeasures*.
57. Chi, W., Chen, S., Ishioka, A., Long, R., Chun, Y., Yan, S., Zong-Chun, Y., & Feng-Yuan, Z. (2008). Changes of muscle-related genes and proteins after spaceflight in Caenorhabditis elegans. *Progress in Biochemistry and Biophysics, 35*(10), 1195–1201.
58. Gopalakrishnam, R., Genc, K. O., Rice, A. J., Maender, C., Evans, H., & Cavanagh, P. R. Effects of long-term space flight on muscle volume. In *ISB XXth Congress-ASB 29th annual meeting*, July 31–August 5, p. 922.

59. Macho, L., Kvetnansky, R., Ficova, M., Popova, I. A., & Grigoriev, A. (2001). Effects of exposure to space flight on endocrine regulations in experimental animals. *Endocrine Regulations, 35*, 101–114.
60. Levine, B. D., Pawelczyk, J. A., Ertl, A. C., Cox, J. F., Zuckerman, J. H., et al. (2002). Human muscle sympathetic neural and haemodynamic responses to tilt following spaceflight. *Journal of Physiology, 5389*(1), 331–340.
61. Professor Gilles Clement. (2005). The maintenance of physiological function in humans during spaceflight. *International SportMed Journal, 6*(4), 185–198.
62. Iwasaki, K.-i., Levine, B. D., Rong, Z., Zuckerman, J. H., & Pawelczyk, J. A. (2007). Human cerebral autoregulation before, during and after spaceflight. *The Journal of Physiology, 579*(3), 799–810.
63. King, H. A., Moe, J. H., Bradford, D. S., & Winter, R. B. (1983). The selection of fusion levels in thoracic idiopathic scoliosis. *The Journal of Bone and Joint Surgery. American Volume, 65*, 1302–1313.
64. Lenke, L. G., Betz, R. R., Harms, J., Bridwell, K. H., Clements, D. H., Lowe, T. G., et al. (2001). Adolescent idiopathic scoliosis: A new classification to determine extent of spinal arthrodesis. *The Journal of Bone and Joint Surgery. American Volume, 83*, 1169–1181.
65. Newton, P. O., Ogilvie, J. W., Shah, S. A., & Shufflebarger, H. L. (2009). *Genetic prognostic testing for adolescent idiopathic scoliosis: Will a genetic test change scoliosis treatment?* SpineCME.Org.
66. Lincoln, T. L. (2007). Infantile idiopathic scoliosis. *The American Journal of Orthopedics, 36*, 586–590.
67. Yoshida, A., Kobayashi, K., Manya, H., Taniguchi, K., Kano, H., et al. (2001). Muscular dystrophy and neuronal migration disorder caused by mutations in a glycosyltransferase, POMGnT1. *Developmental Cell, 1*, 717–724.
68. Durbeej, M., Henry, M. D., & Campbell, K. P. (1998). Dystroglycan in developm and disease. *Current Opinion in Cell Biology, 10*, 594–601.
69. Kirschner, J., & Bonnemann, C. G. (2004). The congenital and limb-girdle muscular dystrophies: Sharpening the focus, blurring the boundaries. *Archives of Neurology, 61*, 189–199.
70. Pegoraro, E., Marks, H., Garcia, C. A., Crawford, T., Mancias, P., et al. (1998). Laminin alpha2 muscular dystrophy: Genotype/phenotype studies of 22 patients. *Neurology, 51*, 101–110.
71. Ibraghimov-Beskrovnaya, O., Ervasti, J. M., Leveille, C. J., Slaughter, C. A., Sernett, S. W., & Campbell, K. P. (1992). Primary structure of dystrophin-associated glycoproteins linking dystrophin to the extracellular matrix. *Nature, 355*, 696–702.
72. Rybakova, I. N., Patel, J. R., & Ervasti, J. M. (2000). The dystrophin complex forms a mechanically strong link between the sarcolemma and costameric actin. *The Journal of Cell Biology, 150*, 1209–1214.
73. Bigos, S. J., Bowyer, O., Braea, G., Brown, K., Deyo, R., Haldeman, S., et al. (1994). *Acute low back pain problems in adults* (Clinical practice guideline no. 14. AHCPR Publication No. 95–0642). US Department of Health and Human Services.
74. Croft, P., Rigby, A. S., Boswell, R., Schollum, J., & Silman, A. (1993). The prevalence and characteristics of chronic widespread pain in the general population. *Journal of Rheumatology, 20*, 710–713.
75. Wolfe, F., Smythe, H. A., Yunus, M. B., Bennett, R. M., Bombardier, C., Goldenberg, D. L., et al. (1990). The American College of Rheumatology 1990 criteria for the classification of fibromyalgia. *Report of the Multicenter Criteria Committee. Arthritis and Rheumatology, 33*, 160–172.
76. Bennett, R. (2007). Myofascial pain syndromes and their evaluation. *Best Practice & Research. Clinical Rheumatology, 3*, 427–445.
77. Simons, D. G., Treavell, J. G., & Simons, L. S. (1999). *Myofascial pain and dysfunction: The trigger point manual*. Williams & Wilkins.

78. Cimmino, M. A., Ferrone, C., & Cutolo, M. (2011). Epidemiology of chronic musculoskeletal pain. *Best Practice & Research. Clinical Rheumatology, 25*, 173–183.
79. Aguilera, F. J., Martín, D. P., Masanet, R. A., Botella, A. C., Soler, L. B., & Morell, F. B. (2009). "Immediate effect of ultrasound and ischemic compression techniques for the treatment of trapezius latent myofascial trigger points in healthy subjects": A randomized controlled study. *Journal of Manipulative and Physiological Therapeutics, 32*(5), 15–520.
80. Peterka, R. J. (2002). Sensorimotor integration in human control. *Journal of Neurophysiology, 88*, 1097–1118.
81. Allum, J. H. J. (1983). Organization of stabilizing reflex responses in tibias anterior muscles following ankle flexion perturbations of standing man. *Brain Research, 264*, 297–301.
82. Jeka, J. J., Schoner, G., Dijkstra, T., Ribeiro, P., & Lackner, J. R. (1977). Coupling of fingertip somatosensory information to head and body sway. *Experimental Brain Research, 113*, 475–483.
83. Johansson, R., Magnusson, M., & Akesson, M. (1988). Identification of human postural dynamics. *IEEE Transactions on Biomedical Engineering, 35*, 858–869.
84. Kavounoudias, A., Gilhodes, J. C., Roll, R., & Roll, J. P. (1999). From balance regulation to body orientation: Two goals for muscle proprioceptive information processing? *Experimental Brain Research, 124*, 80–88.
85. Day, B. L., Severac Cauquil, A., Bartolomei, L., Pastor, M. A., & Lyon, I. N. (1977). Human body-segment tilts induced by galvanic stimulation: A vestibularly driven balance protection mechanism. *Journal of Physiology (London), 500*, 661–672.
86. Hlavacka, F., & Nijiokiktjien, C. (1985). Postural responses evoked by sinusoidal galvanic stimulation of the labyrinth. *Acta Otolaryngology (Stockholm), 99*, 107–112.
87. Johansson, R., Magnusson, M., & Fransson, P. A. (1995). Galvanic vestibular stimulation for analysis of postural adaptation and stability. *IEEE Transactions on Biomedical Engineering, 42*, 282–292.
88. Nashner, L. M., & Wolfson, P. (1974). Influence of head position and proprioceptive cues on short latency postural reflexes evoked by galvanic stimulation of the human labyrinth. *Brain Research, 67*, 255–268.
89. Mergner, T., & Rosemeier, T. (1998). Interaction of vestibular, somatosensory and visual signals for postural control and motion perception under terrestrial and microgravity conditions – A conceptual model. *Brain Research Reviews, 28*(1–2), 118–135.
90. Fukuoka, Y., Nagata, T., Ishida, A., & Minamitani, H. (2001). Characteristics of somatosensory feedback in postural control during standing. *IEEE Transactions on Neural Systems and Rehabilitation Engineering, 9*(2), 145–153.
91. Ishida, A., & Imai, S. (1980). Responses of the posture control system to pseudorandom acceleration disturbances. *Medical & Biological Engineering & Computing, 18*, 433–438.
92. National Aeronautics and Space Exploration, Lyndon B. Johnson Space Center, Houston, Texas. (2008). *Impaired ability to maintain control of vehicles and other complex systems*. Human Research Program; Human Health Countermeasures Element, Evidence Report Summary.
93. Dobkin, B. H. (2003). *The clinical science of neurologic rehabilitation*. Oxford University Press.
94. Massion, J. (1994). Postural control systems. *Current Opinion in Neurobiology, 4*, 877–887.
95. Colledge, N. R., Cantley, P., Peaston, I., Brash, H., Lewis, S., & Wilson, J. A. (1994). Ageing and balance: The measurement of spontaneous sway by posturography. *Gerontology, 40*, 273–278.
96. Straube, A., Boetzel, K., Hawken, M., Paulus, W., & Brandt, T. (1988).Postural control in the early: differential effects of visual, vestibular and somatosensory input. In *Posture and gait: Development, adaptation and modulation*. Proceedingsa of the 9th international symposium on postural and gait research.
97. Diener, H. C., & Dichgans, J. (1986). Long loop reflexes and posture. In W. Bles & T. Brandt (Eds.), *Disorders of posture and gait* (pp. 41–76). Elsevier.

98. Diener, H. C., Dichgans, J., Guschlbauer, B., & Bacher, M. (1986). Role of visual and static vestibular influences on dynamic posture control. *Human Neurobiology, 5*, 105–113.
99. Allum, J. H. J., Honegger, F., & Pfaltz, C. R. (1989). The role of stretch and vestibule-spinal reflexes in the generation of human equilibrating reactions. *Progress in Brain Research, 80*, 399–409.
100. Runge, C. F., Schupert, C. L., Horak, F. B., & Zajac, F. E. (1988). Role of vestibular information in initiation of rapid postural responses. *Experimental Brain Research, 122*, 403–412.
101. Shumway-Cook, A., & Woollacott, M. H. (1995). *Motor control: Theory and practical applications* (1st ed., pp. 119–142). Williams & Wilkens.
102. Nashner, L. M. (1997). Physiology of balance, with special reference to the healthy elderly. In J. C. Masdeu, L. Sudarsky, & L. Wolson (Eds.), *Gait disorders of aging: Falls and therapeutic strategies* (pp. 37–53). Lippincott-Raven.
103. Horak, F. B. (1987). Clinical measurement of postural control in adults. *Physical Therapy, 67*, 1881–1885.
104. Johansson, R. S., & Vallbo, A. B. (1980). Spatial properties of the population of mechanoreceptive units in the glabrous skin of the human hand. *Brain Research, 184*, 353–366.
105. Magnusson, M., Enbom, H., Johansson, R., & Pyykko, I. (1990). Significance of pressor input from the human feet in anterior-posterior postural control: The effect of hypothermia on vibration-induced body-sway. *Acta Oto-Laryngologica, 110*, 182–188.
106. Toppila, E., & Pyykko, I. (2000). Chaotic model of postural stability: A position and velocity dependent system? *Automedica, 19*, 115–134.
107. Brandt, T., Paulus, W., & Straube, A. (1986). Vision and posture. In W. Bless & T. Brandt (Eds.), *Disorders of posture and gait* (pp. 157–175). Elsevier Science Publishers B.V.
108. Baloh, R. W., Jacobson, K. M., & Socotch, T. M. (1993). The effect of aging on visual-vestibulaoocular responses. *Experimental Brain Research, 95*, 509–516.
109. Naboack, C. R., & Demarest, R. J. (1981). *The human nervous system, basic principles of neurobiology*. McGraw-Hill.

Chapter 4
The Physical Determinants of Human Locomotor System

Human locomotion consists of cyclic events controlled by central pattern-generating networks (CPGs) that are located to a large extent within the spinal cord, but are under the continuous influence of peripheral and descending signals [1, 2]. CPGs output complex patterns of muscle activity and regulate phase relations among different muscle groups.

The basic unit of human locomotion is the gait cycle, which is typically recorded from the time one foot strikes the ground until that episode recurs and starts the next, repeating cycle. During one gait cycle, the body traverses a distance of one stride. Each stride is made up of one *step* by each foot, and these two steps are normally symmetrical in length.

The gait cycle is divided into segments (Fig. 4.1) that serve specific functions, each for a limited time during the gait cycle. Each lower limb supports the body during its stance phase and then leaves the floor for its swing phase, during which it advances (or steps). Although the swing phase is the "motion" portion of the gait cycle, one must remember that the horizontal push for the forward motion is the responsibility of the contralateral limb, which is in stance [2–4]. This forward push requires adequate friction between the foot and the walking surface so that hip extension torques are translated into forward propulsion of the pelvis and swing limb.

The basic function of walking involves each foot in turn either advancing forward as a step or supporting the body weight and balancing during the advancement of the contralateral lower limb [23, 25]. A period of double-limb stance (DLS) occurs between each step. During DLS, the weight is transferred from one foot to the other in a complex coordinated pattern known as weight acceptance (WA) (or loading) and weight release (pre-swing) for each of the respective limbs (see Fig.4.2).

During each gait cycle, a carefully timed pattern of acceleration and deceleration is produced by the muscles. This muscle activity must overcome gravity, as represented by the vertical component of the ground reaction force (GRF), and provide forward propulsion [2, 5, 6, 32]. The entire body is seen to be involved in the gait

© The Author(s), under exclusive license to Springer Nature Switzerland AG 2023

T. Sarkodie-Gyan, H. Yu, *The Human Locomotor System*,
https://doi.org/10.1007/978-3-031-32781-0_4

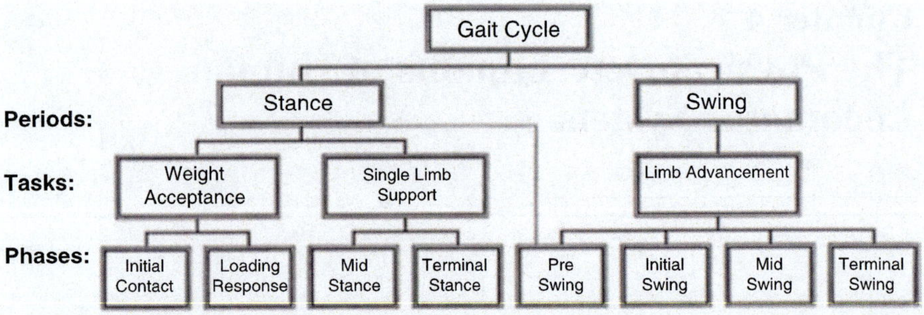

Fig. 4.1 Periods of the gait cycle

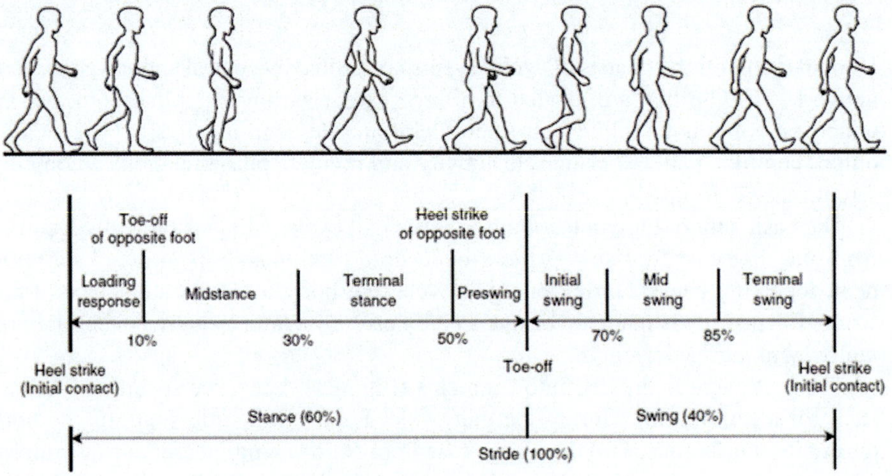

Fig. 4.2 Gait cycle and its phases. The illustration shows the representative instantaneous position of the limbs corresponding to each phase of the gait cycle

cycle, with movement occurring simultaneously in each of the three planes of movement, as well as each of the three axes of rotation. These six degrees of freedom of movement need to be considered throughout the gait cycle [2, 7, 15]. For convenience in understanding, the gait cycle is divided into natural sub-units called phases that have distinct functions in the gait cycle and have both propulsion and control aspects [16–18, 21, 23].

Description of the Gait Cycle According to Fig. 4.2

The stance phase of the gait cycle begins with the period of weight acceptance, or the loading. It constitutes a decelerating portion of the gait cycle where the foot must stop after traveling at about 4 m/s during the end of the swing phase. This sudden stop requires controlled braking involving simultaneous action of the ankle, knee, and hip (At the same time, the left leg is involved in the equally complex task of lift-off to initiate its swing phase). The ankle is in a few degrees of dorsiflexion at the heel strike and then rapidly plantar flexes under the control of an eccentric (lengthening) contraction of the anterior tibialis muscle, as well as the lesser dorsiflexors, until the foot is flat on the ground. Simultaneously, the knee begins to flex under the eccentric control of the quadriceps (the loading response), and the trunk reaches its lowest point during the cycle [5, 7, 19, 35]. The hip, which was flexed approximately 40 degrees at the heel strike, immediately begins to extend on the pelvis as the trunk smoothly continues forward. The forward momentum of the trunk is similarly controlled by contraction of the hip extensors, both the gluteus maximus and the long hamstrings, resulting in controlled hip extension as the trunk moves forward. The hip and pelvis rotate opposite to one another at the same time whereby, the hip rotation of the stance limb is internal.

As the left foot leaves the ground, the right hip continues to rotate internally as that limb enters the single-limb support portion of the stance phase. The right ankle begins to passively dorsiflex as the tibia tilts forward and the pelvis lowers on the left in the familiar Trendelenburg tilt. The pelvis is also moving laterally to the right so that the center of mass (COM) of the body is aligned over the right foot to allow balance. During this period, the pelvis and COM of the body are rising. To achieve this, some of the kinetic energy of the forward motion is converted into potential energy as the trunk rises against the force of gravity [31].

In the middle of the single-limb stance (SLS) phase, the ankle reaches maximum dorsiflexion and the heel begins to rise from the floor. This heel rise signals a major shift in the function of the right leg in the gait cycle as it ends its decelerating role and begins to serve as the acceleration for its step phase. The foot now rocks over to the forefoot as the knee extends, and the pelvis (not the hip) maximally rotates externally and the right hip maximally extends in preparation for opposite heel strike [2, 5, 7]. The left lower limb is fully stretched out for its heel strike so that the pelvis is supported and the drop of the COM of the body is minimized.

The weight-release phase on the right must now shift weight support to the left lower limb while accelerating the right leg for its forward propulsion or swing phase. During weight release, the ankle is actively plantar flexed by concentric action of the gastrocnemius, soleus, posterior tibialis, and lesser plantar flexors.

These forces contribute primarily to the vertical components of the ground reaction force (GRF). At the same time, right hip flexion is being produced by the iliacus, psoas, and tensor fascia lata muscles, and the left hip internal rotators are causing forward rotation of the right pelvis. These left hip muscles contribute much to the horizontal forces or forward propulsion of the right lower limb [2, 5, 7]. As the toe finally leaves the ground as a result of this combined push and pull on the right lower limb, the swing phase is initiated, causing a step to occur.

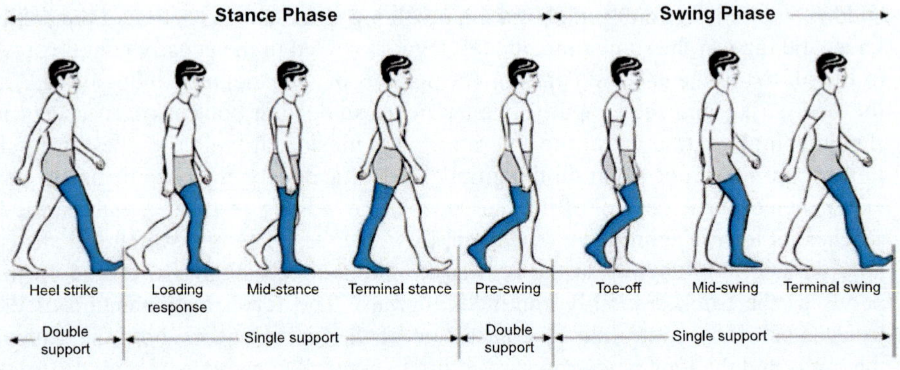

Fig. 4.3 Functional tasks of the gait cycle

Weight Acceptance (WA) (Fig. 4.3): This is the period when weight is rapidly loaded on the outstretched limb. The impact of the floor-reaction force is absorbed and the body continues in a forward path while stability is maintained (Initial Contact and Loading Response).

Single-Limb Support (SLS) (Fig. 4.3): This is the period when the body progresses over a single, stable limb. Weight is transferred onto the metatarsal heads and the heel comes off the ground (Mid Stance and Terminal Stance).

Swing Limb Advancement (SLA) (Fig. 4.3): This is the time when the limb is unloaded and the foot comes off the ground. The limb is moved from behind to in front of the body, reaching to take the next step (pre-swing, initial swing, mid-swing, and terminal swing).

During the beginning of the swing phase, the right leg continues to accelerate as the hip flexion, the knee flexion, and the ankle dorsiflexion combine to cause the toe to pass over the ground cleanly. Typically, the toe reaches a minimum height of less than 2.5 cm. in the middle of the swing phase. This minimal elevation is energy-conserving by reducing the work done to elevate the body parts against gravity. This closeness to the ground can be a safety problem, causing stumbles on uneven ground, but the gains of energy efficiency caused by reducing the step height are important enough for this safety risk.

The second half of the swing phase returns the right leg into a role of decelerating as the forward motion is slowed in preparation for the heel strike. The ankle is held in dorsiflexion while the hip flexes and the knee extends. This combination, along with the forward rotation of the right pelvis and the external rotation of the right hip, results in maximal length of the step. Any injury or dysfunction of a joint or muscle-tendon unit that reduces the step length will have a major impact on the efficiency of walking. At the end of swing, the hip extensors serve to brake the forward flexion of the hip; and, at faster gait speeds, the hamstrings slow and control

Fig. 4.4 Heel strike

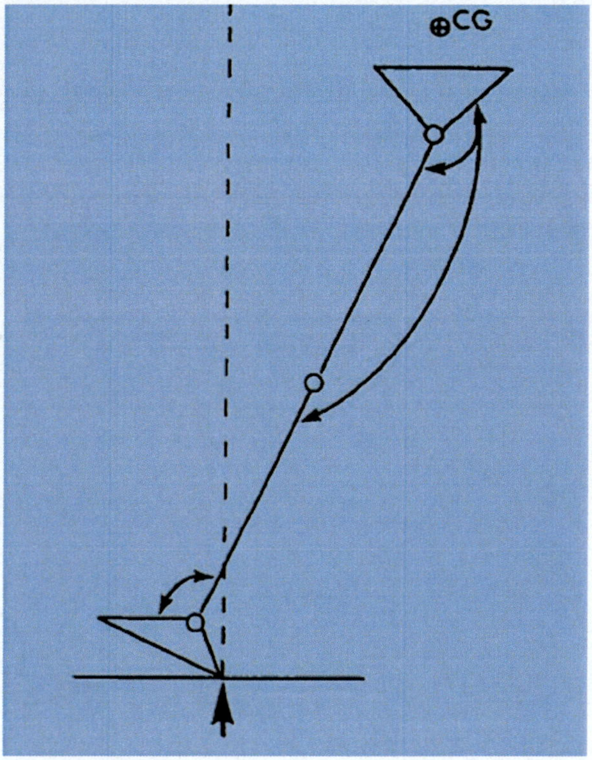

the knee extension [2, 5, 7, 8]. The speed of the leg and foot must be controlled to prevent slipping at heel strike.

The summary of the determinants of the human locomotion is given below:

Initial Contact (IC) Constitutes the moment when the foot contacts the ground, heel strike (heel strike) (Fig. 4.4).

The characteristic reactions with respect to the heel strike include:

- The Anterior to the hip causes flexion moment.
- The Anterior to the knee causes extension moment.
- The Posterior to the ankle causes plantarflexion moment.
- The Hip is flexed to 25-degrees. The gluteus maximus and hamstrings are active in preventing further flexion.
- The Knee is in full extension at the heel-strike. The extension moment is overcome by the action of the hamstrings which controls the knee extension and initiates flexion.
- The Ankle is in the neutral position, and then it begins to plantarflex. This plantarflexion is controlled by action ***of*** the pretibial muscles (Fig. 4.5).

Fig. 4.5 Shortly after heel strike

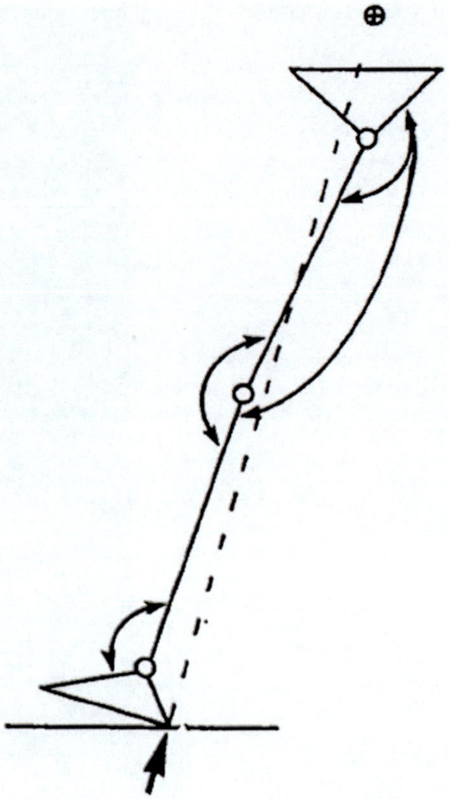

The characteristic reactions with respect to the after heel strike (Fig. 4.5) include:

- The Anterior to the hip causes flexion moment.
- The Posterior to the knee causes flexion moment.
- The Posterior to the ankle causes plantarflexion moment.
- The Hip is held in 25-degrees of flexion by the action of the gluteus maximus and the hamstrings.
- The Knee is in 5-degrees of flexion and continues to flex.
- The rate of flexion is controlled by action of the quadriceps.
- The Ankle is in 5-dgrees of plantarflexion and continues to plantarflex under the control of the pretibial muscles.

Loading Response (LR) The weight is rapidly transferred onto the outstretched limb, the first period of double-limb support (foot flat) (Fig. 4.6).

The characteristic reactions with respect to the foot flat include:

- The Anterior to the hip causes flexion moment.
- The Posterior to the knee causes flexion moment.
- The Posterior to the ankle causes plantarflexion moment.

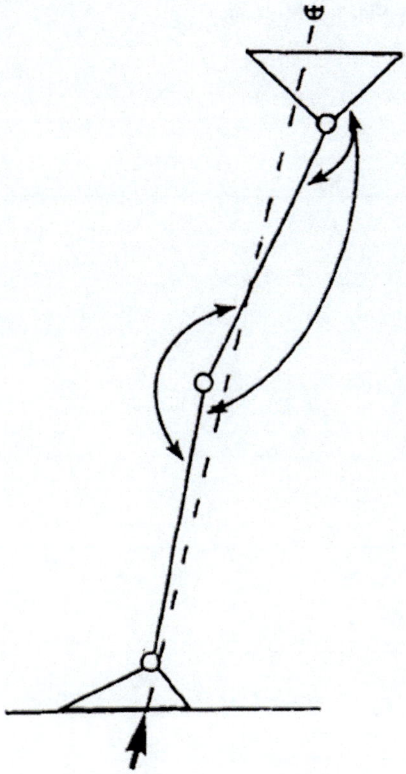

Fig. 4.6 Foot flat

- The Hip is in 25-degrees of flexion then begins to extend by the action of the gluteus maximus and the hamstrings.
- The Knee reaches 15-degrees of flexion and continues to flex until it reaches 20-degrees shortly after foot flat.
- It then begins to extend. The quadriceps are active in controlling the angle of flexion.
- The Ankle is in 10-degrees of plantarflexion. The plantarflexion moments reduce as the reaction moves along the foot and the pretibial muscle activity falls off. As the ground reaction passes the anterior to the ankle joint the segments of the supporting limb begin to rotate over the fixed foot.

Midstance (MSt) The body progresses over a single, stable limb (Fig. 4.7).

The characteristic reactions with respect to the midstance include:

- The Posterior to the knee causes a flexion moment.
- The Anterior to the ankle causes dorsiflexion moment.
- The Hip is in 10-degrees of flexion and continues to extend as the ground reaction moves posterior to the hip joint shortly after midstance.

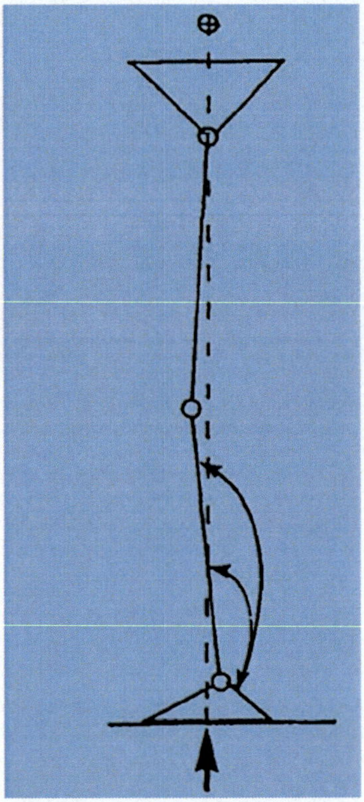

Fig. 4.7 Midstance

- The Knee reaches 10-degrees of flexion and continues to extend. Quadriceps action has fallen off and it is suspected that the soleus is active in controlling knee flexion.
- The Ankle exhibits 5-degrees of dorsiflexion and continues to dorsiflex due to ground reaction. The dorsiflexion is controlled by the calf group of muscles which begins to display activity.

Terminal Stance (TSt) Progression over the stance limb continues. The body moves ahead of the limb and weight is transferred onto the forefoot (Heel off) (Fig. 4.8).

The characteristic reactions with respect to the Heel off include:

- Posterior to hip causing extension moment.
- The Anterior to the knee causes extension moment.
- The Anterior to the ankle causes dorsiflexion moment.
- The Hip reaches about 13-degrees of extension then begins to flex. The iliacus and psoas major are active in controlling extension and initiating flexion.
- The Knee is flexed to about 2-degrees, which is the maximum extension reached at this point in the gait cycle. The gastrocnemius may be active in preventing further extension.

Fig. 4.8 Heel off

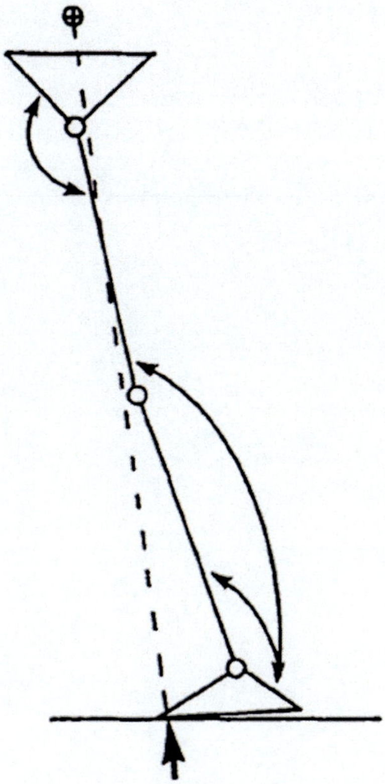

- The Ankle reaches 15-degrees of dorsiflexion after which it plantarflexes due to a powerful contraction of the calf muscles which counteracts the dorsiflexion moment and assists in propelling the body forward.

Pre-Swing (PSw) A rapid unloading of the limb occurs as weight is transferred onto the forefoot (toe-off) (Fig. 4.9).

The characteristic reactions with respect to the Toe-off include:

- By toe-off, the reaction has lost most of its significance as the majority of weight is borne by the other foot.
- The Hip is in 10-degrees of extension and continues to flex due to the plantarflexion of the foot and activity of the rectus femoris.
- The Knee is flexed to about 40-degrees and continues **to** flex under the small ground reaction moment and plantarflexion of the foot.
- The Ankle has reached 20-degrees of plantarflexion due **to** contraction of the calf muscles. These muscles become inactive directly after toe-off.

Initial Swing (ISw) The thigh begins to advance as the foot comes up from the floor (acceleration) (Fig. 4.10).

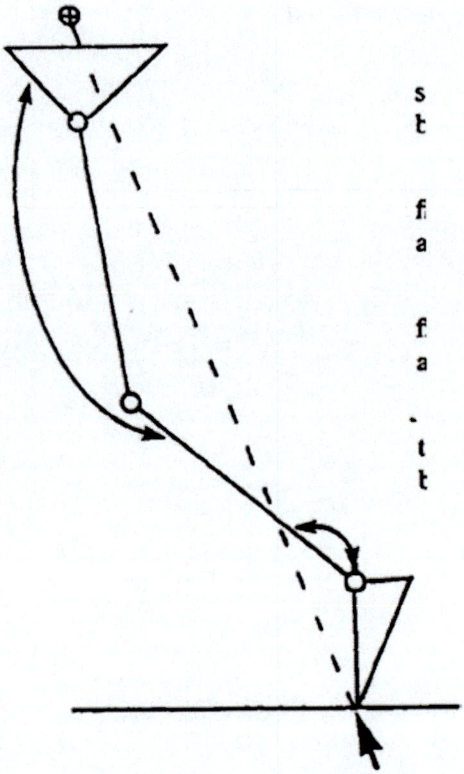

Fig. 4.9 Toe-off

The characteristic reactions with respect to the Initial Swing (acceleration) include:

- The Hip is in 10-degrees of extension and flexes as the hip flexors accelerate the limb forward.
- The Knee **is** in 40-degrees of flexion and continues to flex under pendulum action as the limb accelerates.
- The Ankle is in 20-degrees of plantarflexion directly after toe-off**.** It then begins to dorsiflex under action of the pretibial group of muscles.

Mid-Swing (MSw) The thigh continues to advance as the knee begins to extend; the foot clears the ground (Fig. 4.11).

The characteristic reactions with respect to the Mid-swing include:

- The Hip is flexed to about 20-degrees and continues to flex.
- The Knee reaches about 65-degrees of flexion then begins.
- The Ankle has reached its neutral position and is held there by slight activity of the pretibial muscles.

Terminal Swing (TSw) The knee extends; the limb prepares to contact the ground for initial contact (deceleration) (Fig. 4.12).

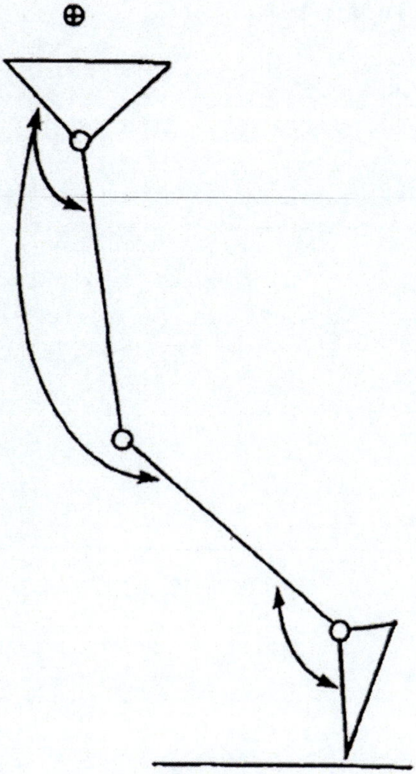

Fig. 4.10 Initial swing

The characteristic reactions with respect to the terminal swing (deceleration) include:

- The Hip reaches 25-degrees of flexion and is restrained by gluteus maximus and the hamstrings.
- The Knee is in full extension and restrained by the hamstrings.
- Ankle is still held in the neutral position by action of the pretibial muscles.
- The hip, knee, and ankle are now positioned for the following heel strike. All the muscle groups that are necessary to counteract the ground reaction force are now active.

Significance of Measuring Physiological/ Biomechanical Quantities

The reliable measurement of biomechanical work performed by humans during locomotion is critical for understanding the muscle-tendon function, joint-specific contributions, and energy-saving mechanisms [7, 10, 11, 22]. Biomechanical work

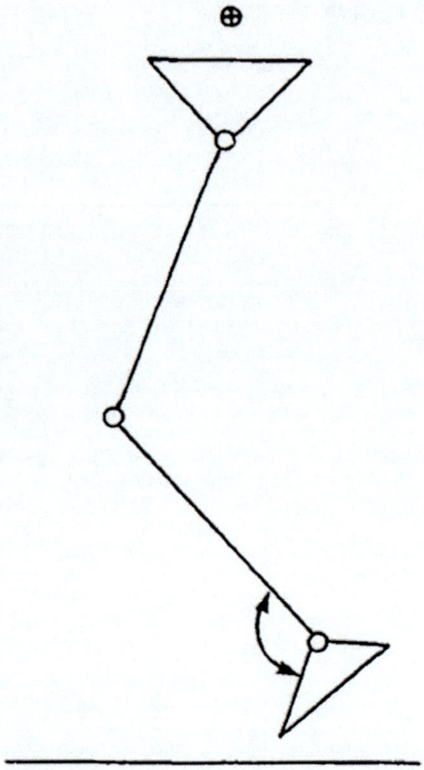

Fig. 4.11 Mid-swing

is often measured at the level of specific joints and body segments, representing the net contributions from underlying muscles, tendons, and other tissues [30, 39, 42].

Movements of segments are generally described with references to anatomical planes of motion (e.g., the sagittal, coronal, and transverse planes) and associated axes of rotation (e.g., the transverse, medial, and longitudinal axis). This provides a reference frame for the description of the displacement of body segments in a three-dimensional space [11, 12]. It is possible to describe these movements with reference to the foot (Fig. 4.13).

The Major Determinants of Human Gait

The mechanisms that constitute the major determinants of human gait include the pelvic rotation, the pelvic list, the stance knee flexion, the foot and knee mechanisms, and the hip adduction.

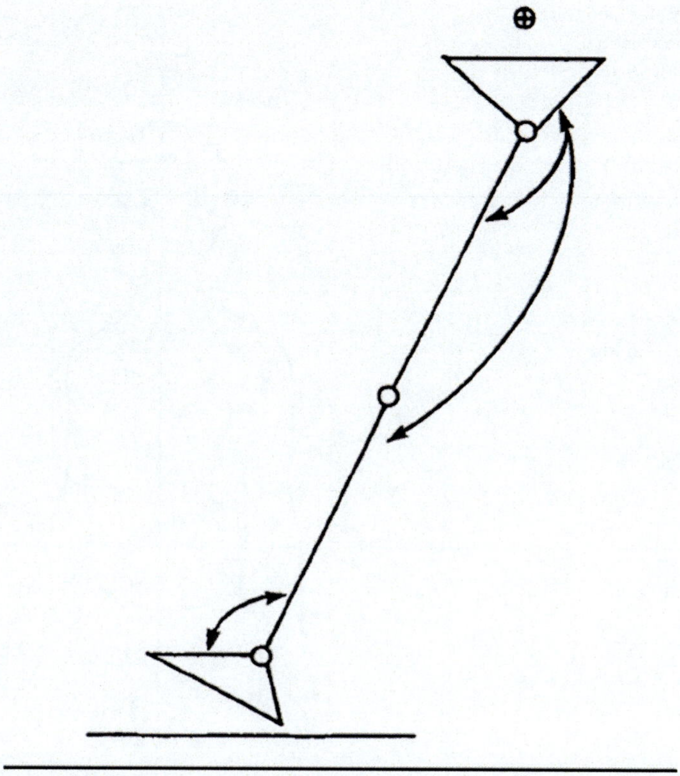

Fig. 4.12 Terminal swing

The Human Pelvis

There is perhaps no other postcranial bone that yields as much information about the biology of a living organism as the pelvis [6, 10, 13]. The pelvis plays important functions in (1) locomotion, as body weight is transmitted to the lower limbs through the pelvic girdle, (2) childbirth, as the human neonate must pass through the birth canal, which lies within the pelvic girdle as the baby exits the body, and (3) support of abdominal organs which are held up by both the pelvic floor musculature and the pelvis itself. Figure 4.14 illustrates the muscular pelvic anatomy.

The human pelvis includes the sacrum, the coccyx, and two os coxae. Each os coxae consists of three parts; the ischium, the ilium, and the pubis. The articulations within the pelvis are; inferiorly between the sacrum and the coccyx (sacrococcygeal symphysis), posteriorly between the sacrum and each ilium (sacroiliac (SI) joint), and anteriorly between the pubic bodies (pubic symphysis). The SI joint offers some movement during childhood, but transitions to a modified synarthrodial joint, which allows little to no movement, by adulthood. The pubic symphysis is a cartilaginous synarthordial joint with a fibrocartilaginous interpubic disc [6, 10, 14]. This

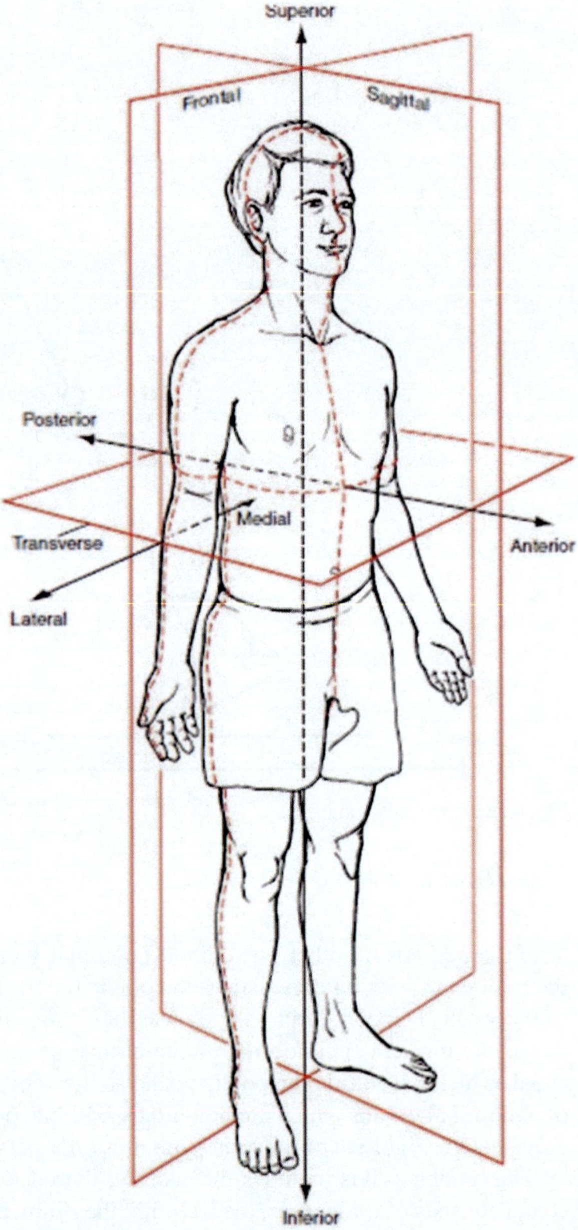

Fig. 4.13 Reference frame for three-dimensional representation of the human body

articulation allows a small amount of translation and rotation. As the pelvic ring creates a closed chain, movement at the pubic symphysis requires simultaneous movement at the SI joint, and vice versa.

The articulation between the pelvis and the femur is the acetabulum, which is formed by portions of the ilium, ischium, and pubis. This synovial ball-and-socket

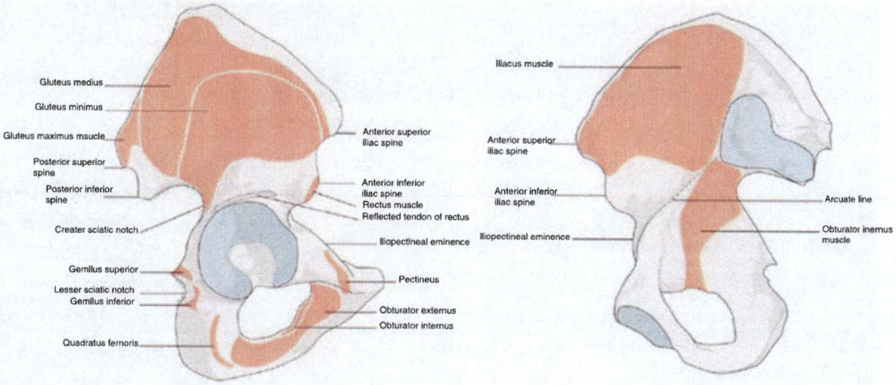

Fig. 4.14 Muscular pelvic anatomy

joint allows substantial motion within the sagittal plane, with additional motion in the frontal and transverse planes. The acetabulum has articular cartilage in the shape of a horseshoe where the femoral head articulates with the os coxae. The normal orientation of the acetabulum is described as being rotated approximately 20–40 degrees off vertical in the frontal plane, and 20–30 degrees anteriorly [6, 10, 14]. This orientation positions the bone to provide the most stability medially, superiorly, and posteriorly. Fig. 4.15 illustrates the ligamentous pelvic anatomy.

Ligamentous Pelvic Anatomy

Ligaments	Functions
● Anterior SI ligaments	Resist external rotation
● Posterior SI ligament	Resist cranial-caudal Translation
● Dorsal sacroiliac ligament	Resist A-P translation
● Iliolumbar ligament	Resist rotation
● Sacrospinous ligament	Resist external rotation
● Sacrotuberous ligament	Resist cranial-caudal translation and flexion

Constituent Mechanisms of Human Gait

Figure 4.16 illustrates a hypothetical compass of gait. The pelvis is represented by a single bar with a small cuboid representing the center of mass (COM) of the body [8–10]. The legs are rigid bars articulating only at the hip. No foot, ankle, or knee joints are present. The pathway of the COM is a series of interconnecting arcs.

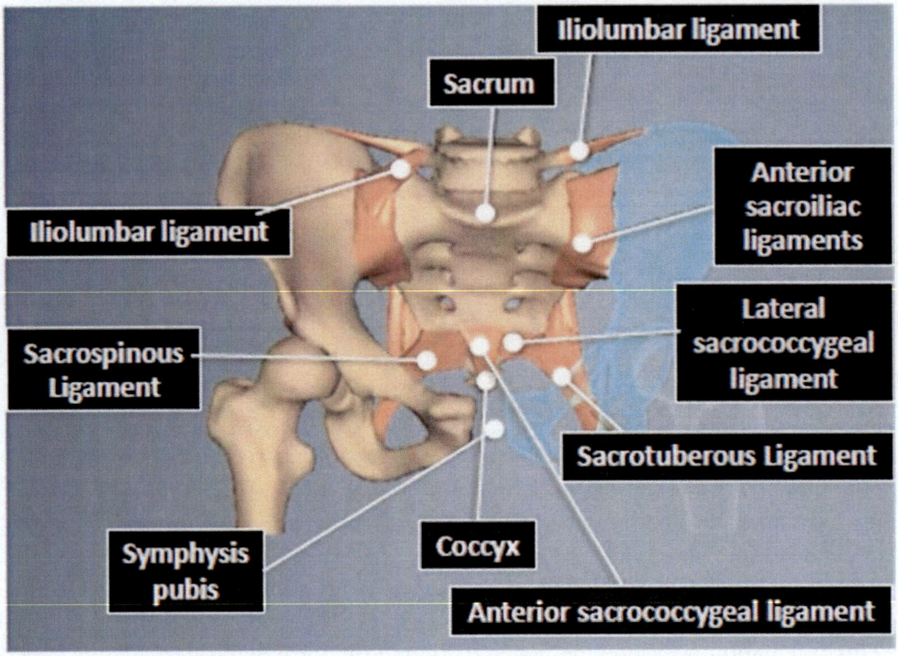

Fig. 4.15 Ligamentous pelvic anatomy

Pelvic Rotation (Fig. 4.17): The pelvis rotates in the transverse plane 4 degrees forward on the swing side and 4 degrees backward on the stance limb (total of 8 degrees); this decreases the vertical displacement of the center of gravity by 9.5 mm (3/8in).

Pelvis rotates in transverse plane:

- Forward on swing side.
- Backward (relatively) on stance side.
- The pelvis moves forward with swing limb.
- Trails behind with the following limb.
- Flattens the Arc of CM motion by increasing the effective leg length at these times.

Figure 4.18 illustrates the dynamic behavior of the pelvic rotation [5].

Pelvic List (Lateral Tilt (Pelvic Drop) (Fig. 4.19): The pelvis tilts or drops down an average of 5 degrees in relation to the horizontal plane on the side opposite to that of the weight-bearing limb during midstance; this decreases the vertical displacement of the center of gravity by an average of just over 3 mm (1/8 in).

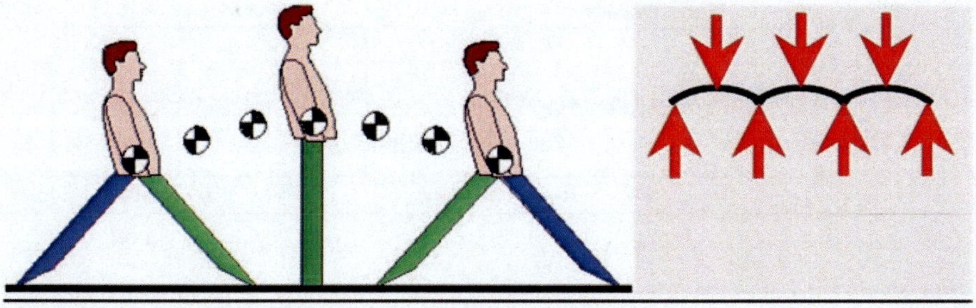

Fig. 4.16 Smoothing out CM excursion

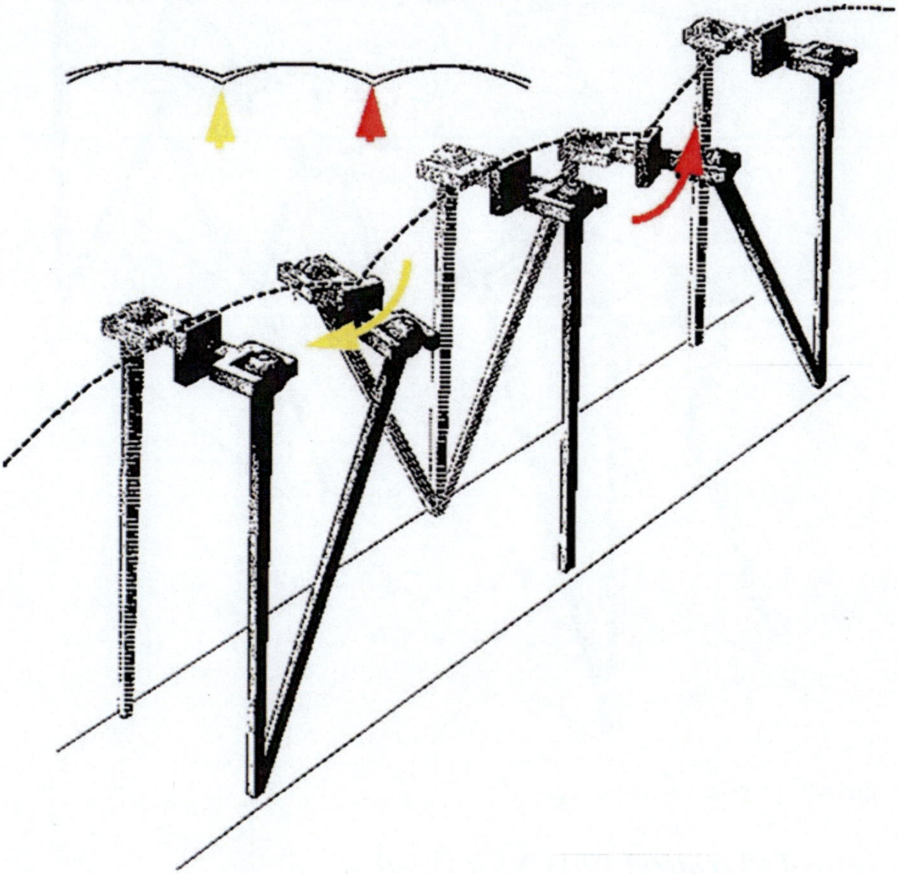

Fig. 4.17 Pelvic rotation

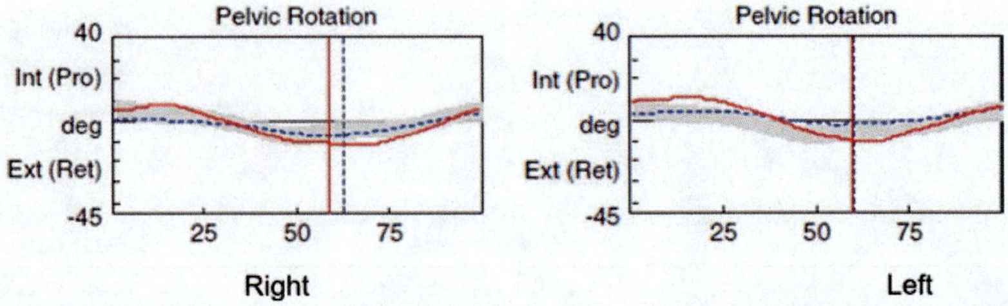

Fig. 4.18 (Left) and 1–18 (Right) illustrate the dynamic behavior of the pelvic rotation

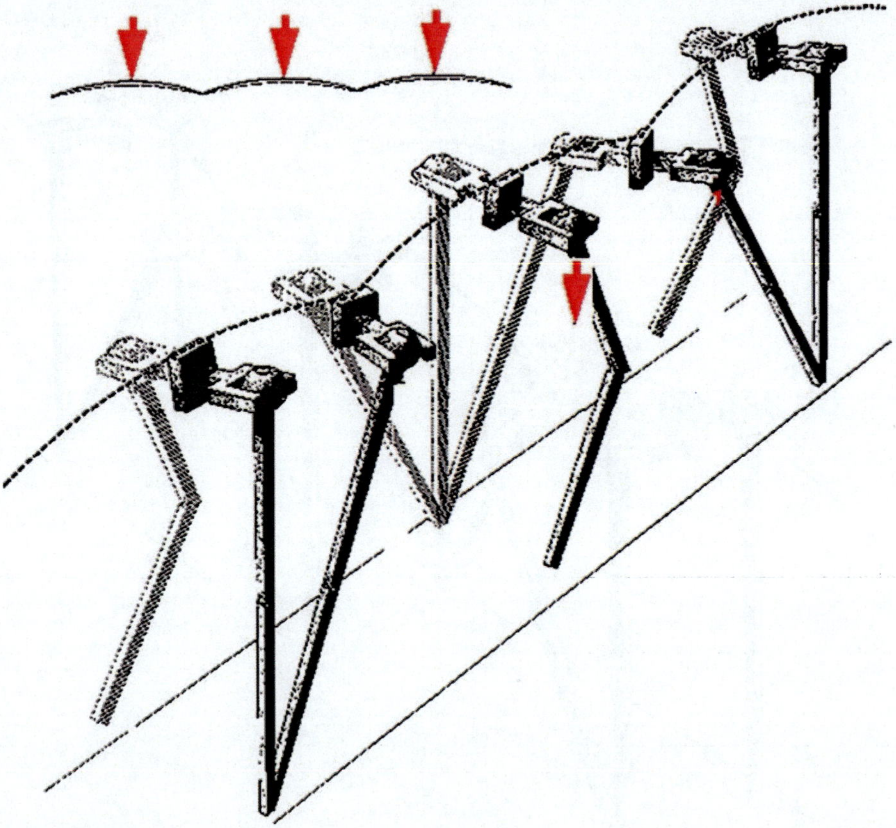

Fig. 4.19 Pelvic list

Pelvic List (Lateral Tilt) (Pelvic Drop)

- Pelvis dips down on swing side during swing.
- Lowers CM and flattens arc.
- Lateral pelvic tilting (dropping on the unsupported side) during midstance prevents an excessive rise in the center of gravity of the body [1, 5].

Fig. 4.20 Dynamic characteristics of the pelvic tilt [5]

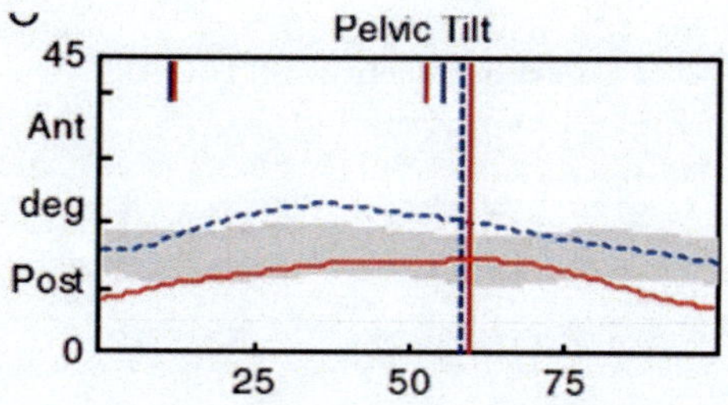

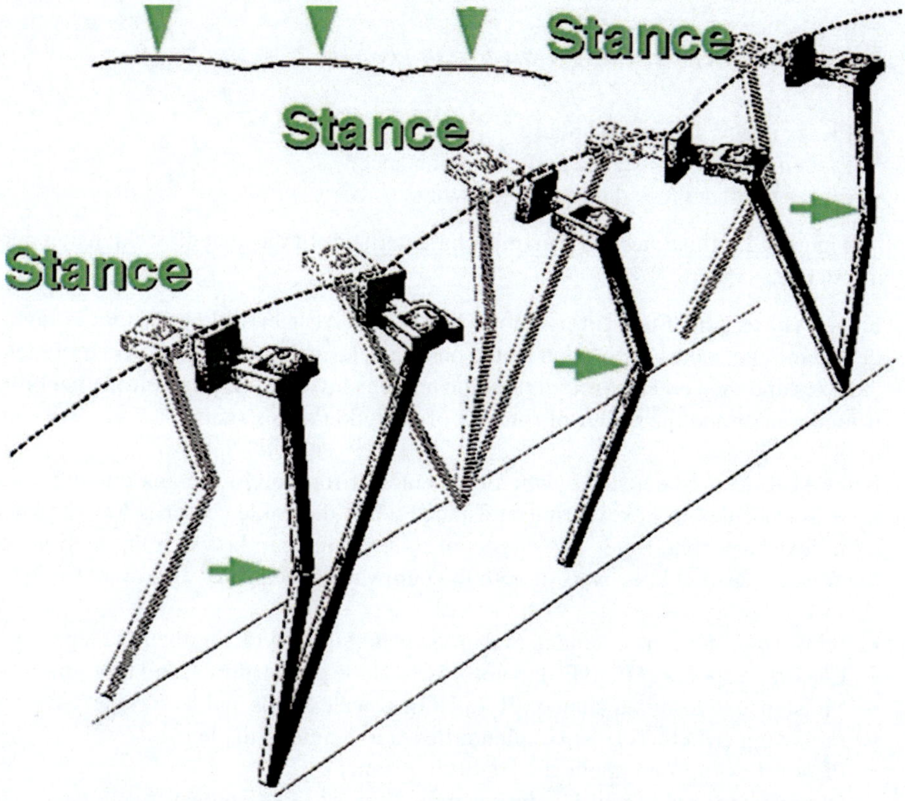

Fig. 4.21 Stance knee flexion

Figure 4.20 illustrates the dynamic characteristics of the Pelvic Tilt.

Knee Flexion (Stance Knee Flexion) (Fig. 4.21): After heel strike in the stance phase; at heel strike, the knee is fully extended and at foot flat, the knee is flexed 15–20 degrees; this absorbs shock and helps minimizes the displacement of the center of gravity.

- Shortens the leg during stance.
- Flexion at the beginning and end of stance smoothens the abrupt changes in CM.
- Flattens the arc.

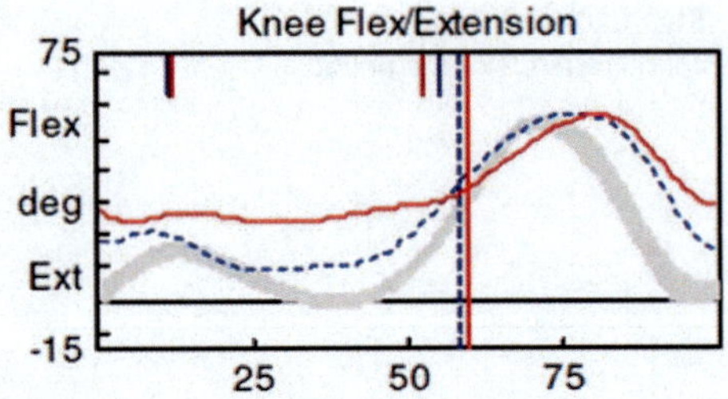

Fig. 4.22 illustration of knee flex/extension during stance [5]

Knee Movement During Gait Cycle [1, 5]

- Flexes during loading response.
- Extends during midstance and terminal stance.
- Flexes to 60 degrees during initial swing.

Figure 4.22 illustrates the dynamic characteristics of the knee flex/extension during stance.

Knee, Ankle, and Foot Interaction (Fig. 4.23) At heel strike, the foot is dorsiflexed and the center of rotation of the ankle is elevated. At foot flat, the foot plantarflexes and the center of rotation on the ankle is lowered. At push off the heel lifts from the floor and the center of rotation of the ankle raises again.

Knee Motion At heel strike, when the center of rotation of the ankle is high, the knee joint begins to flex. During midstance when the ankle center is low the knee joint flexes a second time. The effects of this relationship between the motions of the foot, ankle, and knee is to smooth the pathway of the center of gravity [1, 5].

- Heel-strike: Knee is extended and ankle is dorsiflexed to lengthen the leg.
- Loading response (HS to FF): Knee flexes, ankle plantarflexes, and foot pronates.
- Midstance to terminal stance (FF to HO): Knee extends, ankle dorsiflexes.
- Pre-swing (HO to TO): ankle plantarflexes to lengthen the leg.
- Plantar flexion: Lengthens the leg during stance.
- Dorsiflexion: Lengthens leg during swing, prior to foot contact.

The output signals illustrated in Fig. 4.24 illustrate the dynamic behavior of the knee, ankle, and foot within the gait cycle.

Lateral Displacement from Hip Adduction and Genu Valgum (Genu Valgum and Hip Adduction) (Fig. 4.25) The center of gravity must shift over the stance foot to provide balance, otherwise the person will fall over the unsupported limb [1, 5].

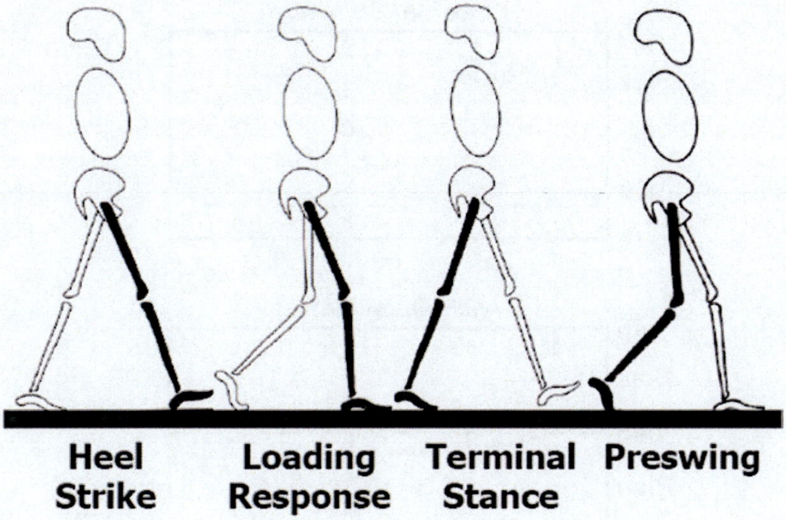

Fig. 4.23 Illustration of knee, ankle, and foot interactions

- Valgus at the knee permits a narrower walking base and, thus, a smaller lateral shift.
- Tibia about vertical.
- Femur articulates adducts to shift CM in the frontal plane, toward the line of progression.

The output signals of Fig. 4.26 depict the dynamic functions of the hip flexion/extension within the gait cycle.

Muscle Activation in the Gait Cycle

The pattern of movements in the lower limbs during walking results from the interaction between external forces (joint reaction and ground reaction forces) and internal forces (these are produced by muscles and other soft tissues). Knowledge about the ground reaction force is essential toward the understanding of muscle activity and timing contributions for stability and propulsion of the human body [6, 7, 10, 20, 24].

There are basically three main muscle types in the human body. These include the skeletal, the cardiac, and the smooth muscles, respectively. Each of these muscles exhibit specific cellular components, physiology, specific functions, and pathology. Whereas the skeletal muscles primarily control movement and posture, cardiac muscles encompass the heart. The smooth muscles are always present throughout

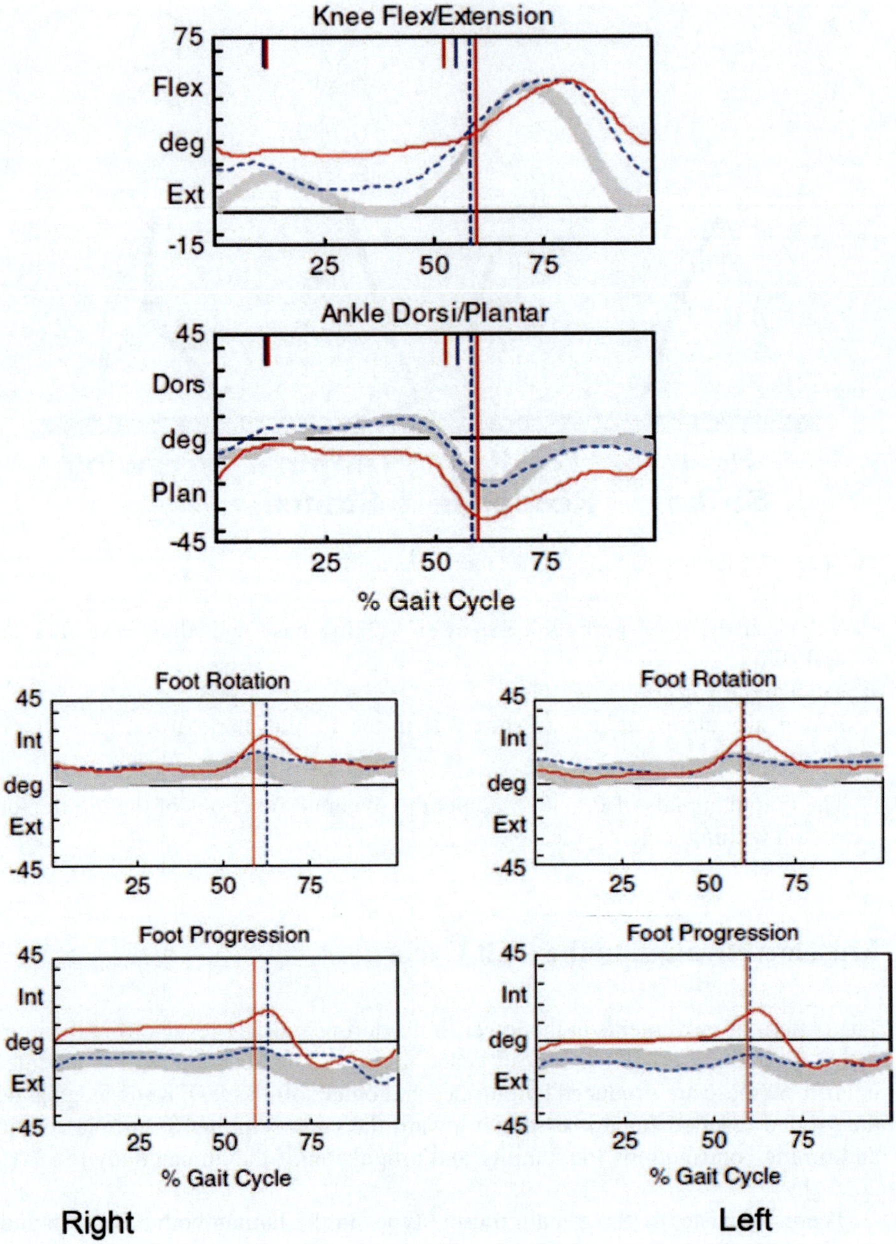

Fig. 4.24 Dynamic behavior of the knee, ankle, and foot [5]

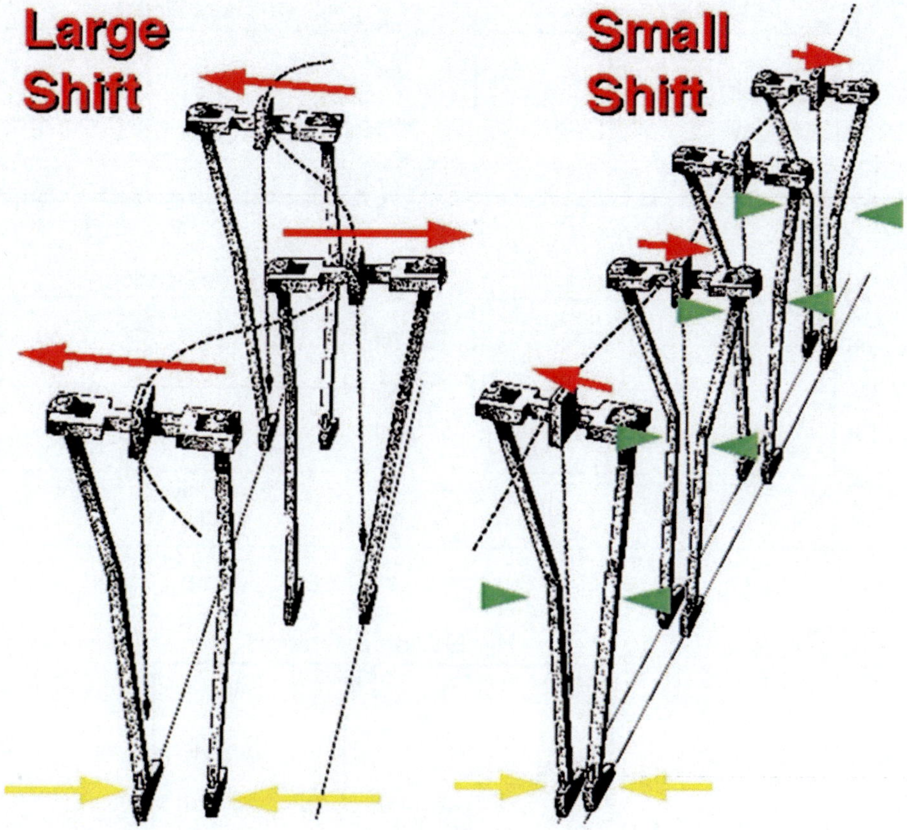

Fig. 4.25 Lateral displacement

the gastrointestinal, reproductive, urinary, vascular, and respiratory systems [36–38]. Figure 4.24 illustrates the muscle activation pattern in the lower extremity during locomotion of the Able-bodied subject [7, 8, 27–29], (see Fig. 4.27).

Force application may affect displacement of body segments and affect biological structures (positively or negatively).

In human locomotion biomechanics, the assessment of forces between the foot and the ground are of particular interest. This interaction is typically assessed using the ground reaction forces (GRFs). In accordance with Newton's third law of motion, the GRF is the force exerted by the ground on the foot, as the foot collides

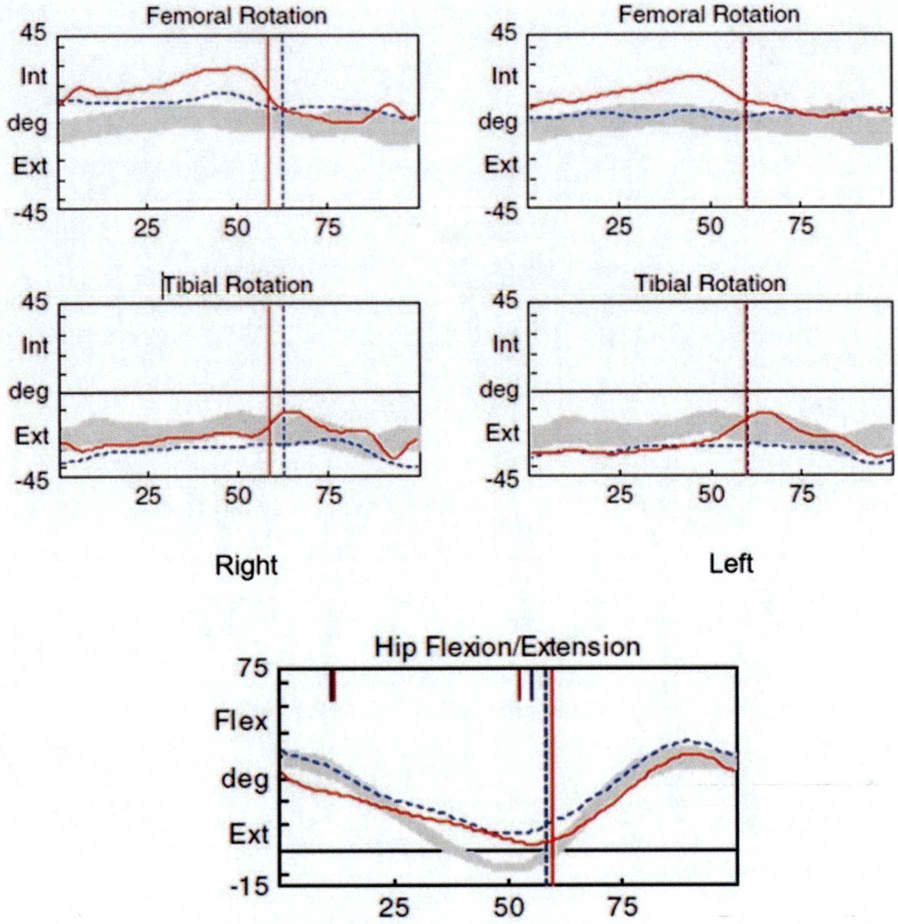

Fig. 4.26 Dynamic functions of the hip flexion/extension [5]

with the ground. Moreover, the GRF is a result of the accelerations of all segments of a body contacting the ground [14, 26, 33, 34].

The GRF can be described in three dimensions (Fig. 4.28). The terms vertical, anterior/posterior, and medial/lateral are used to describe force components in relation to the orientation of the foot on the ground [3, 4, 8, 40, 41]. Furthermore, the terms normal force and shear force are used to describe a force that is applied

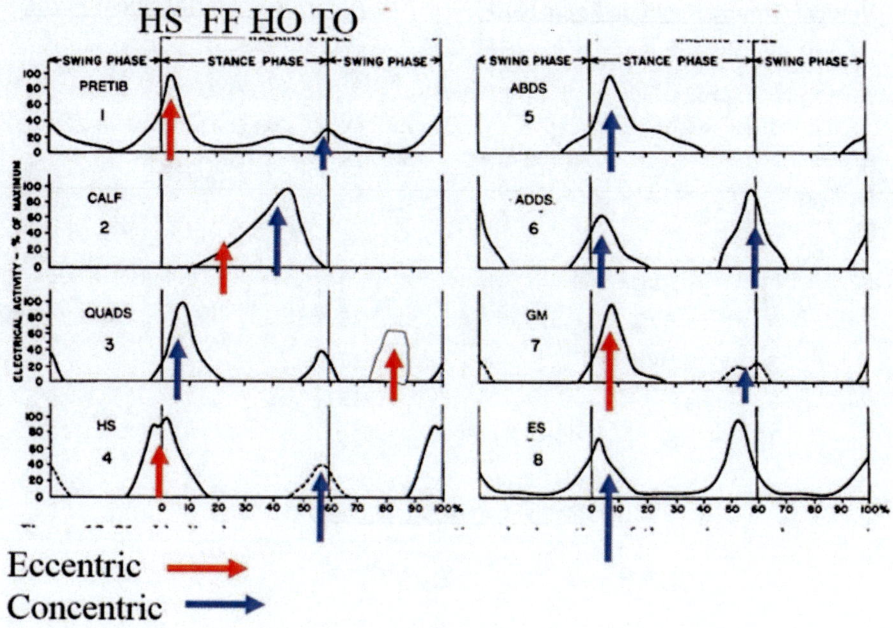

Fig. 4.27 Muscle activation in the gait cycle

perpendicular to the surface of interest (normal force) and a force that is applied parallel to the surface of interest (shear force).

Figure 4.29 illustrates prototypical ground reaction loads (forces and vertical torques) during stance of one foot for an adult woman walking at 1.18 m/s. The key events are marked as follows:

B, *E*—vertical forces exceed body weight at weight acceptance and release (force means acceleration is occurring).

G, *H*—posterior shear forces decelerate during loading and reverse to accelerate for push-off.

J, *K*—the ground reaction torque is in response to rotations of the limb.

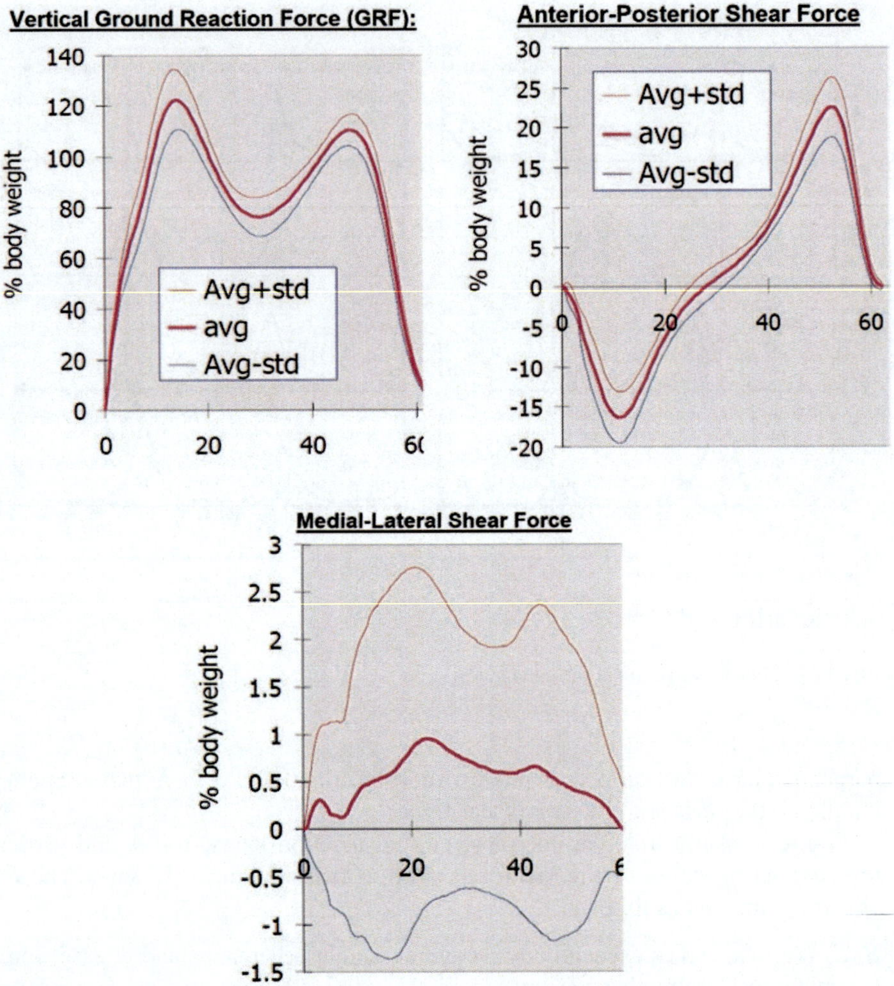

Fig. 4.28 Illustration of the GRFs

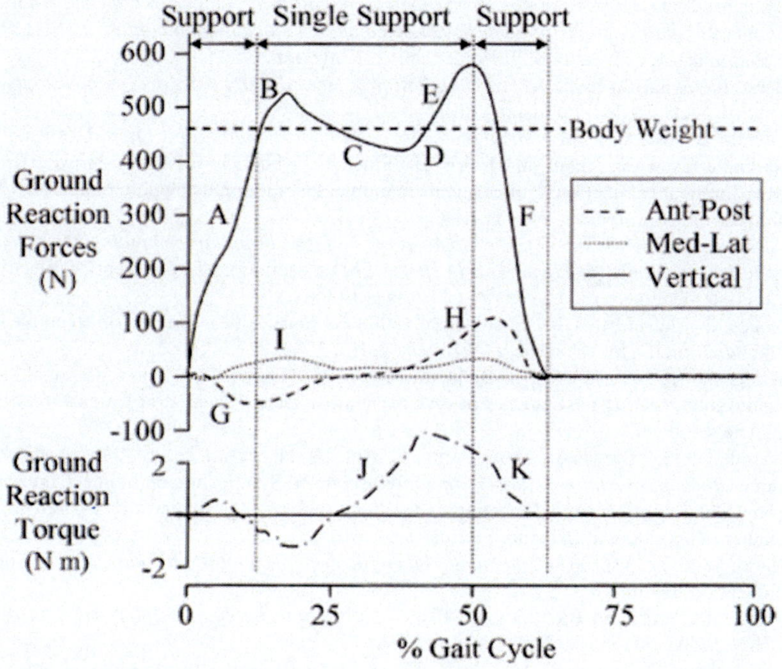

Fig. 4.29 Ground reaction loads (forces and vertical torque)

References

1. Aoi, S., & Tsuchiya, K. (2011). Generation of bipedal walking through interactions among the robot dynamics, the oscillator dynamics, and the environment: Stability characteristics of a five-link planar biped robot. *Autonomous Robots, 30*, 123–141.
2. Jensen, K. J. (2005). *Diagnostics of human locomotion.* M.S. Thesis, New Mexico Institute of Mining and Technology, Department of Mechanical Engineering, Mechatronics Program.
3. Holmes, P., Full, R. J., Koditschek, D., & Guckenheimer, J. (2006). The dynamics of legged locomotion: Models, analyses, and challenges. *SIAM Review, 48*, 207–304.
4. Nilson, J., Thorstensson, A., & Halbertsma, J. (1985). Changes in leg movements and muscle activity with speed of locomotion and mode of progression in humans. *Acta Physiologica Scandinavica, 123*, 457–475.
5. Aoi, S., Ohasshi, T., Bamba, R., Fujiki, S., Tamura, D., Funato, T., Senda, K., Ivanenko, Y., & Tsuchiya, K. (2019). *Neuromusculoskeletal model that walks and runs across a speed range with a few motor control parameter changes based on the muscle synergy hypothesis.* Scientific Reports: www.nature.com/scientificreports
6. Jo, S., & Massaquoi, S. G. (2007). A model of cerebrocerebello-spinomuscular interaction in the sagittal control of human walking. *Biological Cybernetics, 96*, 279–307.
7. Yu, H. (2010). *Categorization of functional impairments in human locomotion using the methods of the fusion of multiple sensors and computational intelligence.* PhD dissertation, Department of Electrical and Computer Engineering, University of Texas at El Paso.

8. Uhlenbrock, D. (1999). *An advanced biomedical gait training machine for the rehabilitation of non-ambulatory stroke patients based on computer-aided human motion analysis.* PhD dissertation at the University of Teesside, United Kingdom.
9. Sarkodie-Gyan, T. (2005). *Neurorehabiltation devices: Engineering design, measurement, and control.* McGraw-Hill Publishing. ISBN:0-07-144830-6.
10. Dobkin, B. H. (2003). *The clinical science of neurologic rehabilitation.* Oxford University Press.
11. Sarkodie-Gyan, T., Alaqtash, M., Yu, H., Spier, E., Abdelgawad, A., & Brower, R. (2011). Measurement of functional impairments in human locomotion using pattern analysis. *Elsevier Journal of Measurement, 44*, 181–191.
12. Alaqtash, M. M. S. (2012) *The application of fuzzy granular computing for the analysis of human dynamics behavior in 3D space.* PhD dissertation, Department of Electrical and Computer Engineering, University of Texas at El Paso.
13. Yogev, G., Hausdorff, J. M., & Giladi, N. (2008 February 2008). The role of executive function and attention in gait. *Movement Disorders, 23*(3), 329–472.
14. Kriegeskorte, N., Mur, M., & Bandetini, P. (2008, November). Representational similarity analysis-connecting the branches of systems neuroscience. *Frontiers in Systems Neuroscience, 2*, Arte 4.
15. Nishikawa, K., Biewener, A. A., Aerts, P., Ahn, A. N., et al., (2006). *Neuromechanics: An integrative approach for understanding motor control.* Symposium on biomechanics and neuromuscular control; annual meeting of the Society for Integrative and Comparative Biology, January 4–8, 2006, at Orlando, Florida.
16. Sarkodie-Gyan, T. (2004). *2nd annual bioengineering technology seminar.* ASME Continuing Education Institute.
17. Nuffield Council on Bioethics. (2013). *Novel neurotechnologies: Intervening in the brain.* ISBN: 978-1-904384-29-8.
18. Uhlenbrock, D., Hesse, S., & Sarkodie-Gyan, T. (1999). Development of an advanced mechanised gait-trainer, controlling movement of the centre of mass, for the restoration of gait in non-ambulatory subjects. *Journal of Biomedical Technik, 44*(7–8/1999), 194–201.
19. The Society for Neuroscience. *Brain facts: A primer on the brain and nervous system.*
20. Heylighten, F., & Joslyn, C. (2001). Cybernetics and second-order cybernetics. In R. A. Meyers (Ed.), *Encyclopedia of physical science & technology* (3rd ed.). Academic.
21. Mergner, T. (2002). The Matryoshka Dolls principle in human dynamic behavior in space: A theory of linked references for multisensory perception and control of action. *Cahiers de Psychologie Cognitive/Current Psychology of Cognition, 2002, 21*(2–3), 129–212.
22. Sarkodie-Gyan, T., Yu, H., Bogale, M., Hernandez, N. V., & Pirela-Cruz, M. (2017). Application of multiple sensor data fusion for the analysis of human dynamic behavior in space: Assessment and evaluation of mobility-related functional impairments. *Journal of Biomedical Science and Engineering, 2017*(10), 182–203.
23. Vaughan, C. L., Davis, B. L., & O'Connor, J. C. (1992). *Dynamics of human gait* (2nd ed.).
24. Hesse, S., Uhlenbrock, D., & Sarkodie-Gyan, T. (1999). Gait pattern of severely disabled hemiparetic subjects on a new controlled gait trainer as compared to assisted treadmill walking with partial body weight support. *Journal of Clinical Rehabilitation, 13*, 401–410. 0269–2155(99) CR2710A.
25. Hesse, S., Reiter, F., Jahnke, M., Dawson, M., Sarkodie-Gyan, T., & Mauritz, K.-H. (1997). Asymmetry of gait initiation in hemiparetic stroke subjects. *Official Journal of the American Congress of Reha Medicine & the American Academy of Physical Med. and Reha, 78*, 719–724.
26. Uhlenbrock, D., Sarkodie-Gyan, T., Hesse, S., & Reiter, F. (1997). Development of a gait trainer for the rehabilitation of non-ambulatory patients. *Journal of Biomedical Technik, 42*(7-8/1997), 196–202.
27. U.S. Department of Health and Human Services, Public Health Service, Centers for Disease Control and Prevention, National Center for Chronic Disease Prevention and Health Promotion, Division of Nutrition and Physical Activity. Promoting physical activity: A guide to community action. Human Kinetics (1999).

28. Fleischer, C. (2007). *Controlling exoskeletons with EMG signals and a biomechanical body model.* Doktor der Ingenieurwissenschaften Dissertaion, Fakultaet IV_Electrotechnik und Informatik, Technische Univ Berlin.
29. Garate, V. R., et al. (2016). Walking assistance using artificial primitives: A novel bioinspired framework using motor primitives for locomotion assistance through a wearable cooperative exoskeleton. *IEEE Robotics & Automation Magazine, 23*, 83–95.
30. Sarkodie-Gyan, T., Alaqtash, M., Yu, H., MacDonald, C., Spier, E., & Brower, R. (2010). Determination of functional impairments using the methods of computational intelligence. In *Proceedings of the 6th world congress for neurorehabilitation* (pp. 111–117).
31. Kahn, E. B., Ramsey, L. T., Brownson, R. C., et al. (2002). The effectiveness of interventions to increase physical activity: A systematic review. *American Journal of Preventive Medicine, 22*(4S), 73. Elsevier Science.
32. Dawson, M., Reiter, F., Sarkodie-Gyan, T., Provinciali, L., & Hesse, S. (1996). Gait initiation, development of a measurement system for use in a clinical environment. *Journal of Biomedizinische Technik, 41*(7–8), 213–217.
33. Jensen, K., & Sarkodie-Gyan ,T. (2004). The paradigm of a smart gait emulator. In *Proceedings of the 2004 Japan-USA Symposium on Flexible Automation*, Denver.
34. Jensen, K., & Sarkodie-Gyan, T. (2004). Experimental Investigations on a Gait Emulator for Neurological Rehabilitation. In *World Automatic Control Congress, Proceedings, WAC Seville, Spain*, 2004.
35. Sarkodie-Gyan, T. (2005). *Biomedicine miniaturization, Irvine, California*. ASME Continuing Education Institute.
36. Alexander, N. B., Mollo, J. M., Giordani, B., Ashton-Miller, J. A., Schultz, A. B., Grunawalt, J. A., & Foster, N. L. (1995). Maintenance of balance, gait patterns, and obstacle clearance in Alzheimer's disease. *Neurology, 45*, 908–914.
37. DiCarlo, A., Baldereschi, M., Amaducci, L., Maggi, S., Grogoletto, F., Scarlato, G., & Inzitari, D. (2000). Cognitive impairment without dementia in older people: prevalence, vascular risk factors, impact on disability. *The Italian Longitudinal Study on Aging. Journal of The American Geriatrics Society, 48*, 775–782.
38. Della Sala, S., Spinnler, H., & Venneri, A. (2004). Walking difficulties in patients with Alzheimer's disease might originate from gait apraxia. *Journal of Neurology Neurosurgery and Psychiatry, 75*, 196–220.
39. Erickson, C. A., & Barnes, C. A. (2003). The neurobiology of memory changes in normal aging. *Experimental Gerontology, 38*, 61–69.
40. Bowler, J. V. (2002). The concept of vascular cognitive impairment. *Journal of the Neurological Sciences, 203 and 204*, 11–15.
41. Scherder, E., Eggermont, L., Swaab, D., van Heuvelen, M., Kamsma, Y., de Greel, M., van Wijck, R., & Mulder, T. (2007). Gait in ageing and associated dementias: Its relationship with cognition. *Elsevier J Neuroscience and Biobehavioral Reviews, 31*(2007), 485–497.
42. Grillner, S., McClelland, A., & Perret, C. (1981 August 3). Entrainment of the spinal pattern generators for swimming by mechano-sensitive elements in the lamprey spinal cord in vitro. *Brain Research, 217*(2), 380–386.

Chapter 5
Measurement in Human Locomotor System

The intelligent sensory-motor functions account for the human ability to exhibit adaptive behaviors both in complex terrain and also when confronted with unexpected perturbations. It is significant to understand the interaction of the muscles, sense organs, motor pattern generators, and the central nervous system towards the production of this adaptively coordinated movement. Insight into this complexity inherent in the human dynamic behavior in space may provide meaning into the neural control of locomotion, the neuromusculoskeletal system, and would offer some significant aid to ameliorate human health problems that may include the design of prosthetics, the restoration of movement following brain or spinal cord injury, as well as the design, actuation, and control of assistive neurological devices [1–3].

There is an emerging significance towards measurement and instrumentation for the proper monitoring and assessment of human locomotion and quality of life (QoL). The fields of application encompass both healthy and pathological locomotion as encountered in rehabilitation medicine, orthopedics, kinesiology, sports medicine, and the like [4, 5]. Measurement of human locomotion provides an up-to-date description of the instrument systems used for measurement of: (1) kinematics of human movement; (2) kinetic quantities experienced by the human body in contact with the ground; and, (3) myoelectric changes associated with locomotion activity, respectively.

It is significant to understand the physical principles underlying the operation of various measurement systems, as well as signal processing issues that must be addressed in order to obtain and use quantitative measurement variables in biomechanics and physiology.

The human locomotor system is very complex in that it involves many physical, perceptual, and motor aspects. Therefore, the description and understanding of this complex system may be accomplished by applying systems analysis approach, psychophysics, as well as the measurement of the complex motor behavior using those multiple sensors/transducers that monitor and control posture [3, 5].

© The Author(s), under exclusive license to Springer Nature Switzerland AG 2023

T. Sarkodie-Gyan, H. Yu, *The Human Locomotor System*,
https://doi.org/10.1007/978-3-031-32781-0_5

Measurement and Instrumentation

The application of measurement and instrumentation in medicine and surgery is rapidly growing, driven by the capabilities of instrument technology, by the development of other technologies, and by social and economic demands. Measurements of physiological, biochemical, physical, and other patient-related variables seek to provide essential information for the diagnosis and prognosis of patients that lead towards critical decision-making in the clinical environment. Should the measurement of biomechanics/physiological variables exhibit inaccuracies, therapeutic interventions targeted at effecting changes in those variables may be either marginal or even impossible [4–6]. It is crucially important that clinicians, therapists, physicians, and other medical practitioners share with other scientists the same understanding of objectives, terms, units, and criteria attributed to the measurements of critical medical variables.

Most areas of technology rely to a substantial extent on instrumentation and could not function effectively without it. Instrumentation is the key enabler of modern medical technology, through automation, which enhances reliability, efficiency repeatability, and ensures consistent quality. The demand for improved and assured quality means ever-better instrumentation [9, 10]. An important growing point is the integration of sensors, information processing, actuators, and control systems into medical products such as neurodiagnostic devices, neurodevices, assistive devices (orthotics, prosthesis), and in neuropharmaceutical products.

Understanding the concepts of measurement and instrumentation, and recognizing the corresponding limitations and constraints, may help to interpret clinical results and in technological assessment of new medical devices.

Measurement constitutes the process of obtaining the magnitude of a quantity relative to an agreed standard [9, 10, 12]. And the measurement of any quantity involves comparison with some precisely defined unit value of the quantity. Standard units of measure need to be identified and defined as accurately as possible. Advances in technology would require increasingly precise and accurate measurements.

The interest in the measurement of health-related quality of life has increased considerably in recent years. Disease-specific and generic (non-disease-specific) measures are increasingly applied in different medical research contexts. Existing measures are being developed further and new ones created in a search for 'better' instruments [9, 11]. However, at least as far as generic measures are concerned, none of the measures and approaches developed so far can claim to have established a position as the measure, either as a standardized system of describing health states or as a method of valuing them.

The human body may be viewed as a stack of superimposed segments with a kinematic hierarchy in which each lower segment tends to transfer its motion to the other superimposed ones. It may stand to reason that the internal reconstruction of the physical links between segments could be established through the fusion of the acquired information from the array of multiple sensor data [3, 14, 20].

The skeletal system is the internal framework of the human body made up of bones. This consists of two divisions; the axial skeleton and the appendicular skeleton. The axial skeleton is the main framework of the body and consists of the spine, skull, and chest. The appendicular skeleton consists of the bones of the upper and lower extremities.

The skeleton is responsible for the support, movement, and protection of the human body, as well as production of blood cells, storage of minerals, and endocrine regulation. Its movement is dependent upon joints (where two bones meet). Movement is powered by skeletal muscle, which connects to various sites on the bones. The development of blood cells also occurs in the skeletal system which primarily occurs in the bone marrow of long bones such as the femur and the tibia.

As depicted in Figs. 5.1 and 5.2, there are three perpendicular anatomical planes, in the human body, which are referred to as the cardinal planes [13, 14]. The sagittal plane is formed by the SI (Superior/Inferior) and AP (Anterior/Posterior) axes, the

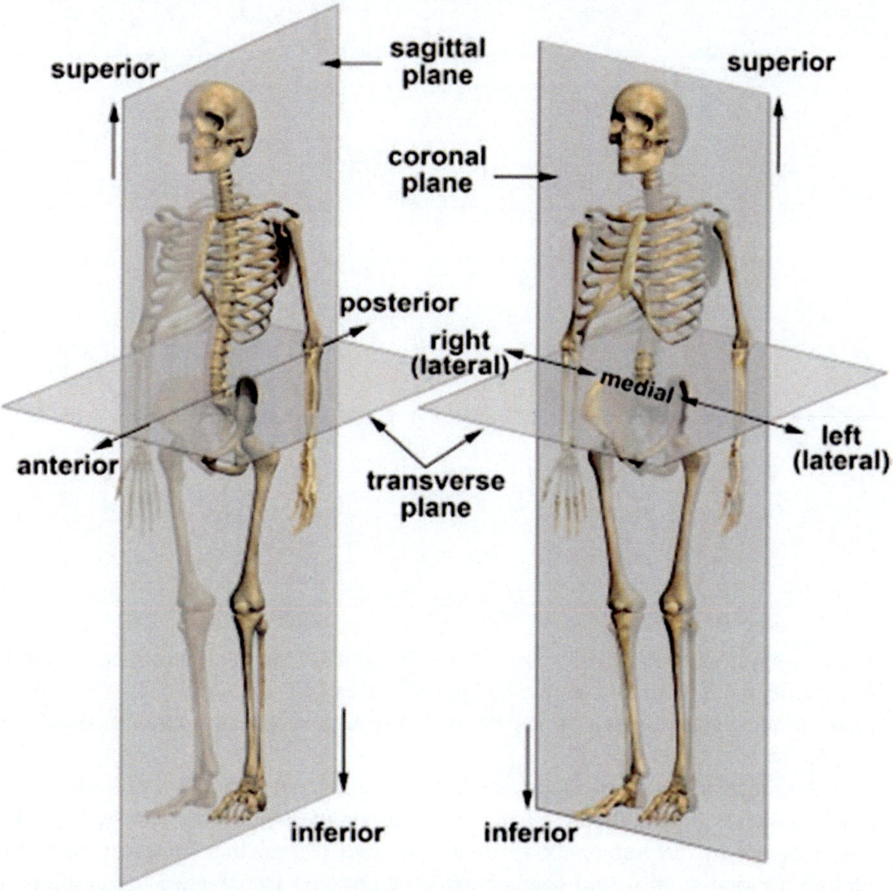

Fig. 5.1 Planes of reference and directional terms for the human skeleton

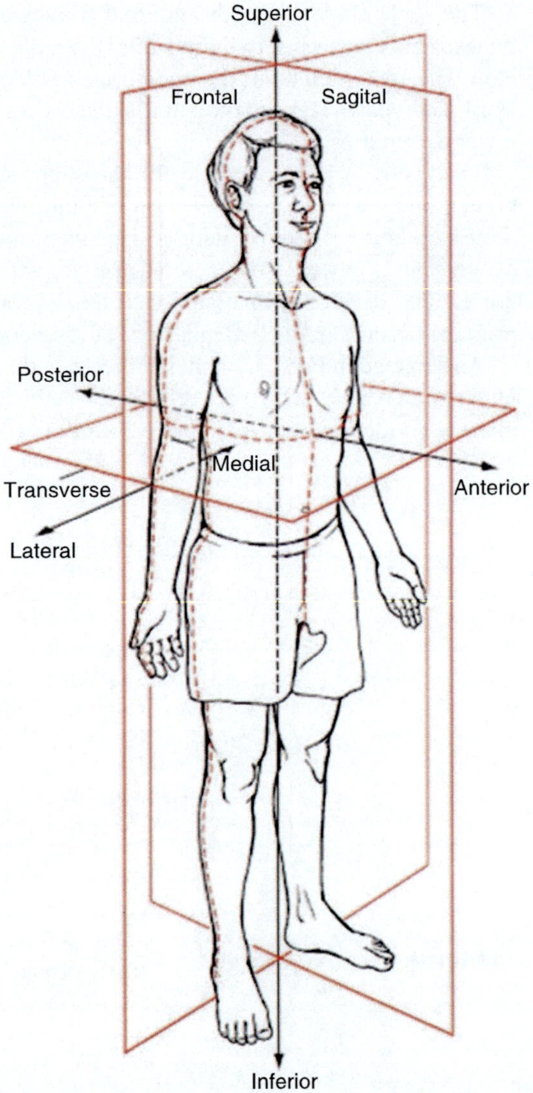

Fig. 5.2 The cardinal planes as reference frame for three-dimensional representation of the human body

frontal (or coronal) plane is formed by the SI and ML (Medial Lateral) axes, and the transverse plane is formed by the AP and ML axes (Figs. 5.1 and 5.2). Therefore, a three-dimensional analysis is necessary for a complete representation of human locomotion.

The human being is a vertebrate whose skeletal system consists of bones, cartilage, ligaments, and tendons and accounts for about 20% of the body weight. The living bones in the human body contain active tissues that consume nutrients, require a blood supply, and change shape or remodel in response to variations in mechanical stress. Figure 5.3 illustrates the human skeletal system.

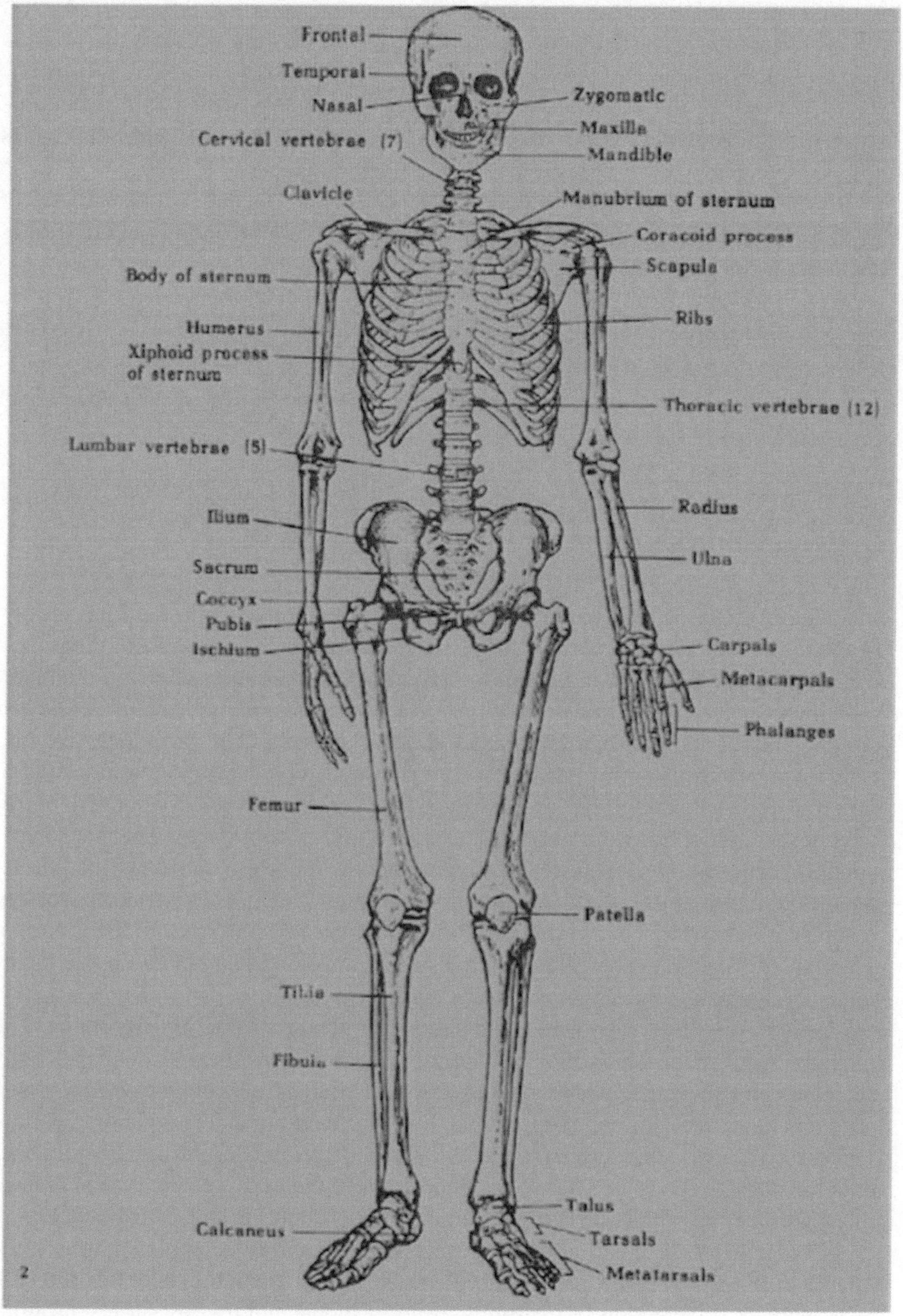

Fig. 5.3 The human skeletal system

Bones provide a rigid framework, the skeleton, that supports and protects the soft organs of the body. The skeleton also protects the soft body parts. The fused bones of the cranium surround the brain to make it less vulnerable to injury. Vertebrae surround and protect the spinal cord and bones of the ribcage, which help protect the heart and lungs of the thorax [16, 17, 19]. Bones work together with muscles as simple mechanical lever systems to produce body movement.

Bones contain more calcium than any other organ. The intercellular matrix of bone contains large amounts of calcium salts where the most important constitutes calcium phosphate.

When blood calcium levels decrease below normal, calcium is released from the bones so that there will be an adequate supply for metabolic needs [17, 19]. When blood calcium levels are increased, the excess calcium is stored in the bone matrix. The dynamic process of releasing and storing calcium goes on almost continuously.

Systems Analysis Approach

Innovations towards the quantitative analysis of the biomechanics and physiology of the human dynamic system in space are essential to truly understand the principles of adaptive behavior in human locomotion. Human motor control fundamentally involves a series of transformations of information among different levels and components of the neuromuscular and skeletal systems. The proprioceptive and exteroceptive sensory information data are transduced by sensory structures that in turn transfer a subset of their information to the central nervous system which, following yet another transformation, issues a set of motor commands. The mechanical coupling between musculoskeletal elements and the muscles controlling them is yet another transformation of information in the system [1, 2, 7]. The control problem revolves around the specific transfer functions that describe each transformation. The transfer functions depend on the rules of organization and operation that determine the dynamic behavior of each subsystem.

The active control of posture and balance is crucial for tasks of daily living. The neural control of postural orientation and equilibrium involves virtually all body segments and several sensory systems. Postural equilibrium refers to balance, that is, resisting external forces acting on the body to maintain a desired body position. On earth, gravity is the dominant external force affecting equilibrium. Postural orientation refers to body geometry, or the position of the body segments with respect to each other and with respect to various reference frames in the environment.

In order to study postural control, psychophysics may be applied towards the quantitative response that relates the physical stimuli (which constitutes the subject's subjective selection of treadmill speed as input) and the response sensations and perceptions they affect (balance, posture). Thereby, the human dynamic behavior may be measured in terms of parameters involving the ground reaction forces under the feet, the positions of the body segments, and the electromyographic data (EMG) activity [3, 18, 19]. The combination of these measured attributive features may allow us to infer the active neural processes underlying balance reactions.

Psychophysics and Its Significance

Psychophysics constitutes the study of quantitative relations between psychological events and physical events or, more specifically, between sensations and the stimuli that produce them.

By determining the stimulus magnitude that is just sufficient to produce a sensation (or a response), it is possible to specify the minimum perceptible stimulus, or the absolute stimulus threshold (stimulus limen), for the various senses [6–8]. The central inquiry of psychophysics pertains to the search for a lawful, quantitative relation between stimulus and sensation for the range of stimuli between these limits.

Analysis of Different Types of Joints and Segments

Figure 5.4 illustrates the complexity of the musculoskeletal system that consists of a stack of superimposed kinematic hierarchical segments. The hierarchical system is structured and consists of linked references which are anchored in gravito-inertial space [3, 13, 14]. The perceptual notion of space is derived mainly through the vestibular sensors, the gyroscopes, and accelerometers, as illustrated in Fig. 5.5.

The Acquisition of Biomechanics/Physiological Information

The behavior of the human locomotor system in space constitutes a complex phenomenon that involves many physical, perceptual, and motor aspects. The understanding of this complexity may be assisted through the application of a systems analysis approach, psychophysics as well as the measurement of the complex motor behavior using an array of multiple sensors/transducers [3, 22, 26, 28]. Psychophysics accounts for the study of quantitative relations between sensations and the stimuli that produce them.

Experimental systems identification approach is used for the understanding of the sensorimotor control of the human dynamic behavior in space. The selected speed of the instrumented treadmill constitutes the stimulus magnitude that is just sufficient to produce the dynamic responses or sensations of the musculoskeletal system. Figure 5.5 illustrates the concept of the acquisition of the dynamic response using an array of multiple sensors/transducers [26].

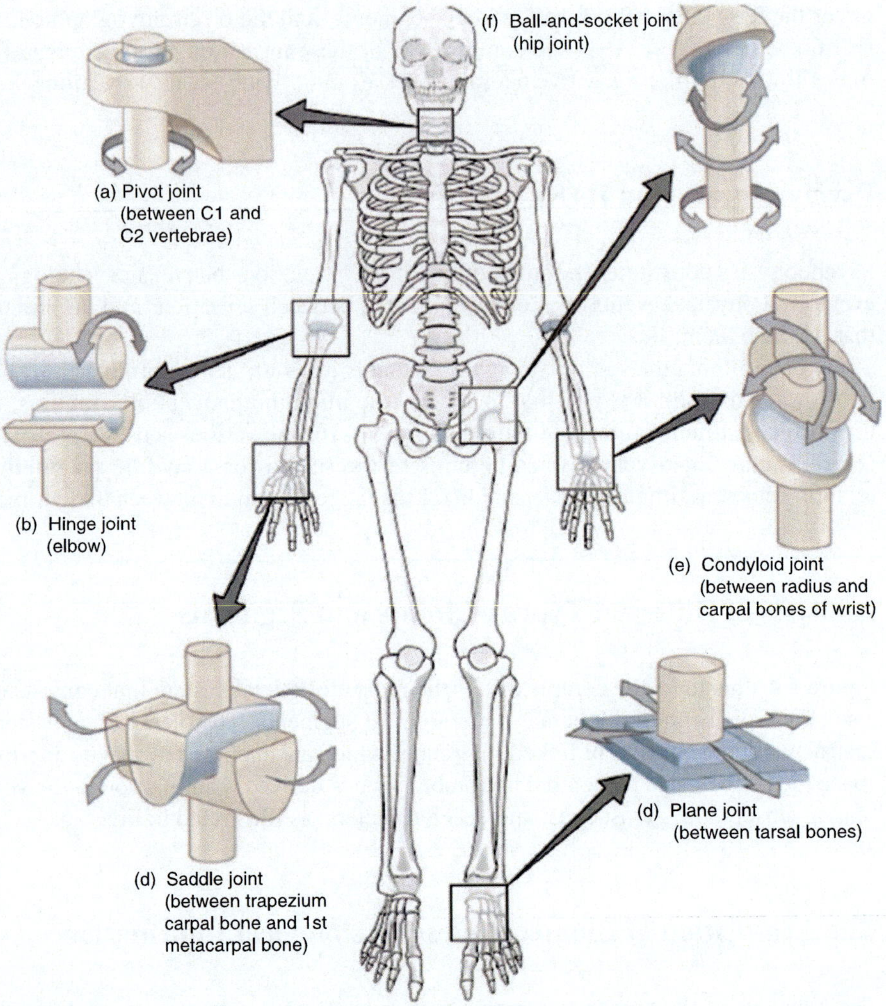

Fig. 5.4 Types of Joints

3D Determination of the Head and Neck

The human neck is a very complex mechanism because it contains vital neurological, vascular, and respiratory structures as well as the cervical vertebrae and spinal cord.

A change in the neck movement pattern of an individual may result from disturbances in proprioception which involves disturbances of the vestibular system or from changes in the movement strategy due to pain or also from physical changes in the joints and musculature [14, 21, 29]. Therefore, a detailed description of neck movement characteristics is very important for the understanding of neck disorders.

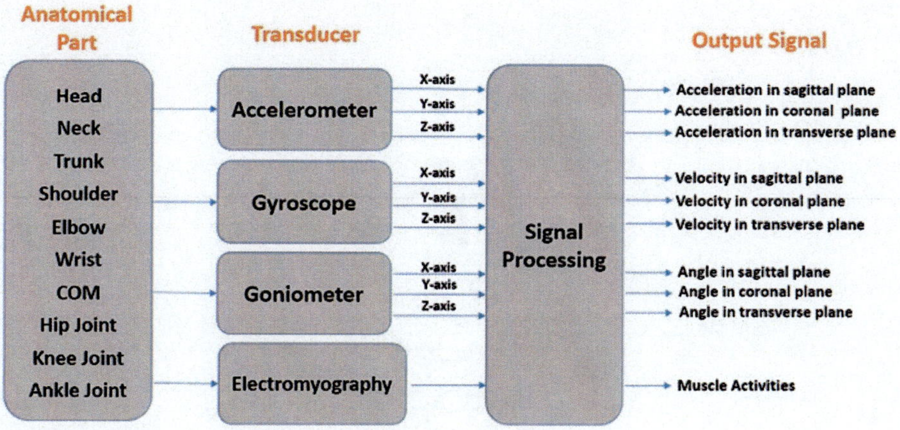

Fig. 5.5 Knowledge elucidation of kinematics data

The head and neck system consists of seven vertebrae and is a complex system from a kinematical point of view.

The neck and shoulder region consists of several complex muscle arrangements and multi-segmental cervical joints that move and stabilize the head and neck. Neuronal pathways run through the neck and are involved in daily functions such as speech, vision, swallowing, and breathing.

The joint of the first and second vertebrae of the neck that allows the head to move back and forth constitutes the pivot joint. Pivot joints consist of the rounded end of one bone fitting into a ring formed by the other bone [13, 14]. This structure allows rotational movement, as the rounded bone moves around its own axis (Fig. 5.6).

The movement of the head from side to side exhibits rotational movement which may be described as the movement of a bone as it rotates around its longitudinal axis. Rotation can be towards the midline of the body, which is referred to as medial rotation, or away from the midline of the body, lateral rotation.

Determination of the Shoulder Joint

The shoulder joint is a ball-and-socket type of joint (Fig. 5.4). Ball-and-socket joints possess a rounded, ball-like end of one bone fitting into a cuplike socket of another bone [13, 14]. This organization allows the greatest range of motion, as all movement types are possible in all directions.

The shoulder joint is the junction between the chest and the arm. The range of movements and the strength of the muscles around the shoulder depend on two shoulder joints, many muscles, tendons, and bones of the upper arm and the upper back.

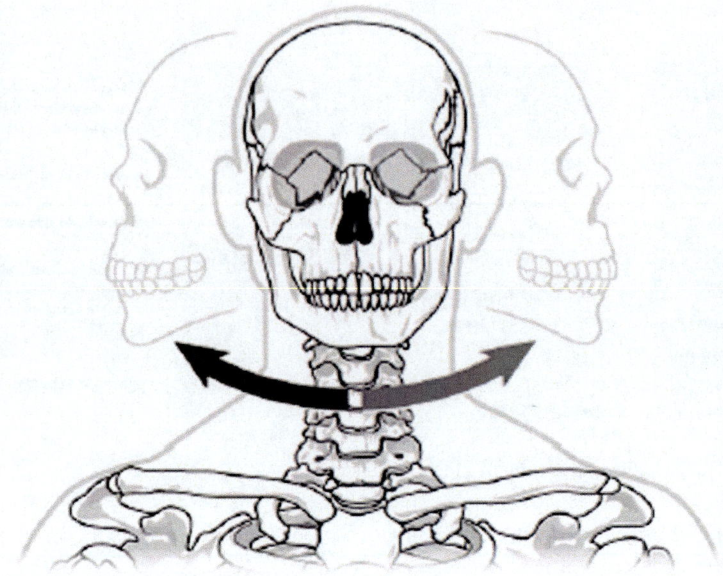

Fig. 5.6 Joint in the neck and pivot joint

There are two joints at the shoulder.

- *The glenohumeral joint* constitutes the ball-and-socket junction of the top of the arm bone and the socket of the shoulder blade.
- *The acromioclavicular joint* constitutes the junction of the collarbone with the shoulder blade.

Most shoulder motion occurs at the ball-and-socket glenohumeral joint, but for full motion of the shoulder, the acromioclavicular joint needs to function normally.

3D Determination of the Elbow Joint

The elbow joint is a synovial joint found in the upper limb between the arm and the forearm. It is the point of articulation of three bones: the humerus of the arm and the radius and the ulna of the forearm.

The elbow joint is classified structurally as a synovial joint (Fig. 5.4). It is also classified structurally as a compound joint, as there are two articulations in the joint [13, 14]. Synovial joints, also called diarthroses, are free movable joints. The articular surfaces of the bones at these joints are separated from each other by a layer of hyaline cartilage. Smooth movement at these joints is provided by a highly viscous synovial fluid, which acts as a lubricant.

A fibrous capsule encloses the joint, and is lined internally by a synovial membrane. Synovial joints can be further categorized based on function. The elbow joint is functionally a hinge joint, because it allows movement in only one plane (uniaxial).

3D Determination of the Wrist Joint

The joint of the wrist that allows the palm of the hand to be turned up and down is a pivot joint (Fig. 5.4). Pivot joints consist of the rounded end of one bone fitting into a ring formed by the other bone. This structure allows rotational movement, as the rounded bone moves around its own axis [13, 14].

The joints of the wrist and fingers exhibit angular movement along two axes, which can move side to side and up and down. This type of joint, the condyloid joints, consists of an oval-shaped end of one bone fitting into a similar oval-shaped hollow of another bone. This is also sometimes called an ellipsoidal joint.

3D Determination of the Hip Joint

The hip joint is a ball-and-socket type of joint (Fig. 5.4). Ball-and-socket joints possess a rounded, ball-like end of one bone fitting into a cuplike socket of another bone. This organization allows the greatest range of motion, as all movement types are possible in all directions.

The hip joint is the junction where the hip joins the leg to the trunk of the body. It consists of two bones, namely, the thighbone or femur, and the pelvis, which is made up of three bones called ilium, ischium, and pubis [13, 14]. The ball of the hip joint is made by the femoral head while the socket is formed by the acetabulum. The acetabulum is a deep, circular socket formed on the outer edge of the pelvis by the union of three bones; ilium, ischium, and pubis. The lower part of the ilium is attached to the pubis while the ischium is considerably behind the pubis. The stability of the hip is provided by the joint capsule or acetabulum and the muscles and ligaments that surround and support the hip joint.

The head of the femur rotates and glides within the acetabulum. A fibrocartilaginous lining called the labrum is attached to the acetabulum which further increases the depth of the socket.

The femur is one of the longest bones in the human body. The upper part of the thighbone consists of the femoral head, femoral neck, and greater and lesser trochanters [14, 15]. The head of the femur joins the pelvis (acetabulum) to form the hip joint. Next to the femoral neck, there are two protrusions known as greater and lesser trochanters which serve as sites of muscle attachment.

Articular cartilage is the thin, tough, flexible, and slippery surface lubricated by synovial fluid that covers the weight-bearing bones of the body. This enables smooth movements of the bones and reduces friction.

Lumbo-Pelvic Hip Complex

Figure 5.7 illustrates the lumbo-pelvic hip complex. The human pelvis includes the sacrum, the coccyx, and two os coxae. Each os coxae consists of three parts which include the ischium, the ilium, and the pubis, respectively. The articulations within the pelvis constitute inferiorly between the sacrum and the coccyx (sacrococcygeal symphysis), posteriorly between the sacrum and each ilium (sacroiliac (SI) joint), and anteriorly between the pubic bodies (pubic symphysis). The pubic symphysis includes a cartilaginous synarthordial joint with a fibrocartilaginous interpubic disc [30–32]. This articulation allows a small amount of translation and rotation. As the pelvic ring creates a closed chain, movement at the pubic symphysis requires simultaneous movement at the SI joint, and vice versa.

The pelvis constitutes the center of the body that connects the upper body with the lower body. It transmits and absorbs forces from both the upper and lower body, and it connects to the lumbar spine through the sacrum and it establishes the hip joint where the femur meets the pelvis at the acetabulum.

All forces that act on the spine, ribcage, head, shoulder girdle, and arms (upper body) are transmitted through the pelvis to the lower body. It, therefore, stands to reason that injuries, misalignment, and deviations in the upper body, thus, affect the pelvis and the lower body.

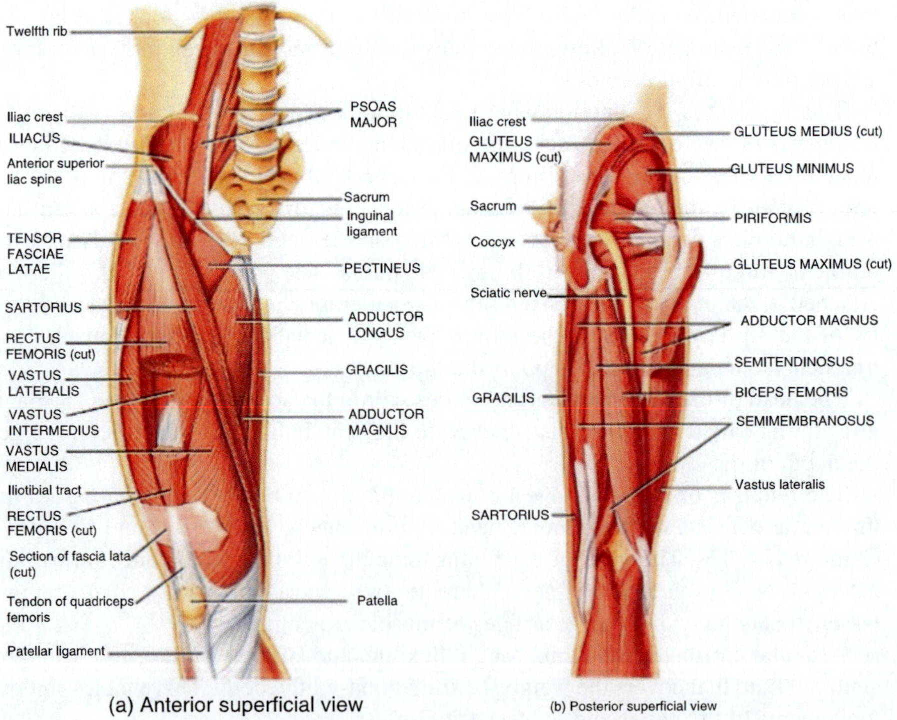

Fig. 5.7 The lumbo-pelvic hip complex

Just as a complex interplay exists among the head, neck, and shoulders, so too does a dynamic interrelationship exist among the lumbar spine, the pelvis, and the hips. The lumbar vertebrae transfer weight to the pelvis. The sacroiliac joint which constitutes the joint between the sacrum and the two iliac bones, is an important site of force transfer. The pelvis then transfers weight/force to the legs via the femur. The head of the femur sits high up in the deep hip socket.

The articulation between the pelvis and the femur is the acetabulum, which is formed by portions of the ilium, ischium, and pubis [31, 32]. This synovial ball-and-socket joint allows substantial motion within the sagittal plane, with additional motion in the frontal and transverse planes.

During human locomotion, the pelvis has motions about all three axes. The magnitude of these motions is partially dependent on walking speed, with larger motions occurring at faster walking speeds.

In the sagittal plane, the pelvis is typically maintained in anterior pelvic tilt throughout gait, and completes two full cycles of a sinusoidal wave for each gait cycle. Following initial contact, the pelvis tilts posteriorly for less than 20% of the gait cycle. It then begins to tilt anteriorly again until the contralateral foot contacts the ground at approximately 50% of the gait cycle. The cycle then repeats itself, tilting posteriorly, and anteriorly again.

In the frontal plane, the pelvis completes one cycle of motion throughout each gait cycle. At initial contact, the pelvis is approximately neutral with left and right ASIS being level (or zero degrees of obliquity). Following contact, the pelvis drops for less than 20% of the gait cycle, after which it begins to raise again. When the other foot contacts the ground, the pelvis is approximately neutral again. If viewed relative to the initial foot, the pelvis would be described as continuing to hike.

In the transverse plane, the pelvis completes one cycle of motion as well. At initial contact of the right foot, the pelvis is in backward rotation with the left ASIS posterior to the right ASIS. This backward rotation continues briefly during the double support phase, and then changes to forward rotation of the pelvis. The forward rotation continues until just after contralateral heel strike (around 50%). From the viewpoint of the swing leg, the pelvis is then rotating backward again until just after ipsilateral heel strike.

The muscles in the lumbo-pelvic hip region exhibit the ability to slow and regulate the force with which the bones of the pelvis and spine "lock" together during movement [30, 31].

Muscles perform this function by lengthening under tension to ensure that the skeletal structures of the pelvis and lower back do not "crash" into each other as they move. However, musculoskeletal imbalances and muscle dysfunction or other soft-tissue dysfunction can prevent muscles from functioning as they should. When that happens, the bony structures of the body are not controlled as they move, causing stress to the joints. Persistent muscle dysfunction can affect the integrity of the joints, leading to inflammation, degeneration, pain, and injury.

The muscles of the Lumbo-Pelvic Girdle include the Gluteus Maximus, the Hip Flexors, and the Hamstrings (Fig. 5.7).

Contractions of the gluteus maximus muscle extend the hips, pushing them forward and rotating the legs outward. The glutes also slow the hips as they flex and the pelvis as it rotates forward during movements. The glutes are an important muscle for transitional movements like sit-to-stand transfers and ascending stairs during ambulation. If there is weakness in the glutes the individual may have difficulty with these activities, limiting independent living.

Contractions of the hip flexors (psoas and iliacus) bend the hips so the torso comes towards the leg, or vice versa, and rotate the leg outward. However, weight-bearing movements cause the hip flexors to function very differently. As the hips travel behind the body during walking, running, and lunging, the hip flexors lengthen under tension to slow extension of the hips and legs. However, if the hip flexors cannot lengthen effectively during this movement, then the pelvis rotates forward excessively while the lower back overarches. These movements can cause tremendous stress to the sacroiliac joints (where the pelvis meets the lower back) and to the intervertebral disks of the lumbar spine.

The major function of hamstring contractions is to assist with bending the knees and extending the hips. However, another important function of the hamstrings is to lengthen under tension in order to slow down the hips as they flex and the pelvis as it rotates forward. Lengthening of these muscles also helps slow down the knees as they straighten [30, 31]. These additional functions of the hamstrings decelerate stress to the hip/leg, spine, and pelvis joints during movements such as lifting or rowing. Biomechanically, tight hamstrings can lead to increased knee and hip flexion, leading to increased anterior pelvic tilt and increased force on the lumbar spine. Simply put, if the hamstrings cannot lengthen effectively during these movements, pain and injury can result from these restrictions in the lumbo-pelvic posture.

3D Determination of the Knee Joint

The knee exhibits the function of a hinge joint (Fig. 5.4) whereby hinge joints have a slightly rounded end of one bone that fits into the slightly hollow end of the other bone. In this way, one bone moves while the other remains stationary, similar to the hinge of a door. The knee is one of the largest and most complex joints in the body. The knee joins the thigh bone (femur) to the shin bone (tibia). The smaller bone that runs alongside the tibia (fibula) and the kneecap (patella) are the other bones that make the knee joint [33, 34]. Tendons connect the knee bones to the leg muscles that move the knee joint whereas ligaments join the knee bones and provide stability to the knee.

3D Determination of the Foot Joint

The foot is a complex biomechanical structure of the human body that consists of 33 joints, 26 bones, and more than a hundred muscles, tendons, and ligaments that all work together to bear weight, allow for locomotion, and transmit forces [36, 37]. The joints of the foot are made wherever two or more of the foot bones meet.

The joints between the tarsal bones of the foot are known as the intertarsal joints, and the specific intertarsal joints of the foot include the subtalar joint, the talocalcaneonavicular joint, the calcaneocuboid joint, the cuneonavicular joint, the cuboideonavicular joint, and the intercuneiform joint, respectively.

All the toes of the foot except the big toe have three joints which include the metatarsophalangeal (MCP) joint that is found at the base of the toe, the proximal interphalangeal joint (PIP) that is found in the middle of the toe, and the distal phalangeal joint (DP) that is found near the tip of the toe. The big toe has the metatarsophalangeal joint and one interphalangeal joint. A layer of cartilage covers the surfaces of the joints where the bones meet and this allows them to glide smoothly with each other as they move. A thin membrane called the synovium encloses the joint and secretes a fluid to lubricate them.

3D Determination of the Ankle Joint

The ankle joint (Fig. 5.4) is a hinged synovial joint with primarily up-and-down movement (plantarflexion and dorsiflexion). However, when the range of motion of the ankle and subtalar joints (talocalcaneal and talocalcaneonavicular) is taken together, the complex functions as a universal joint [35, 36].

The ankle joint connects the bones of the leg, the fibula, and tibia, with the talus of the foot. It is a complex hinge joint that consists of two articulations.

3D Determination of the Ground Reaction Forces (GRFs)

The ground reaction force is equal in magnitude and opposite in direction to the force that the body exerts on the supporting surface through the foot [38, 39]. Figure 5.8 illustrates the principle underlying the ground reaction forces (GRFs)..

The ground reaction force vector (GRFV) passes upward from the foot and produces movement at each lower extremity joint. The GRFV is a reflection of the total mass-times-acceleration product of all body segments and therefore represents the total of all net muscle and gravitational forces acting at each instant of time over the stance period [39, 40].

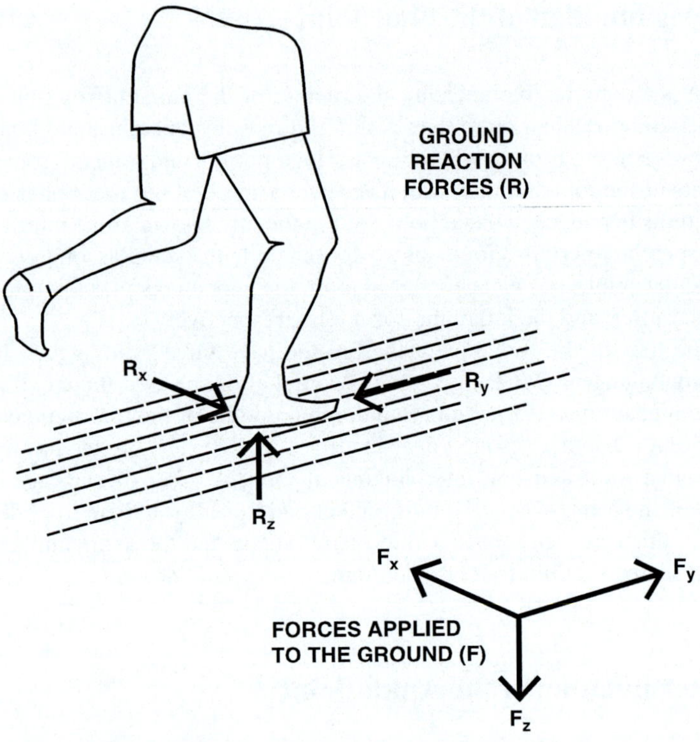

Fig. 5.8 The ground reaction force

By common convention, the directions of these forces are defined along the X-axis (Anterior-Posterior), Y-axis (Mediolateral), and Z-axis (Vertical), as illustrated in Figs. 5.9, 5.10, and 5.11, respectively [38–40].

In order to be able to measure quantitatively the three-dimensional ground reaction forces (3D GRFs), the experimental set-up as illustrated in Fig. 5.12 may be preferred.

Regenerative Medicine Procedures

Annually, very many individuals worldwide suffer injury to the peripheral neuromuscular circuitry that can cause different degrees of limb function loss, and even ulceration and articular deformity. The peripheral nerve system is responsible for relaying electrochemical messages between the brain-spinal cord unit and the muscles, joints, and skin. Because a number of these patients are subjected to life-long disabilities, this type of injury has become a serious health problem. Regenerative medicine procedures, such as nerve transfers, seek to prolong and improve the

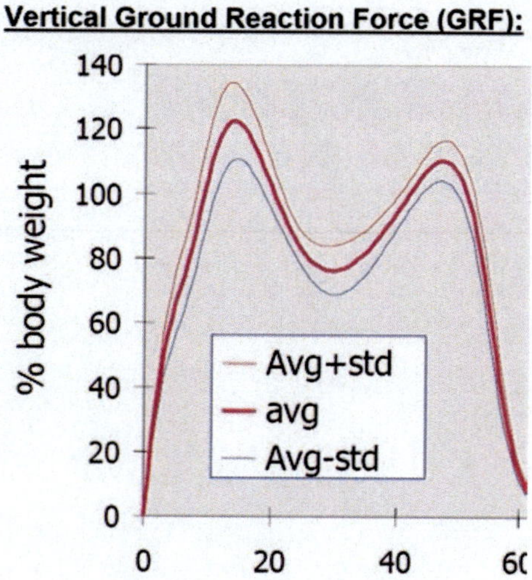

Fig. 5.9 Vertical GRF

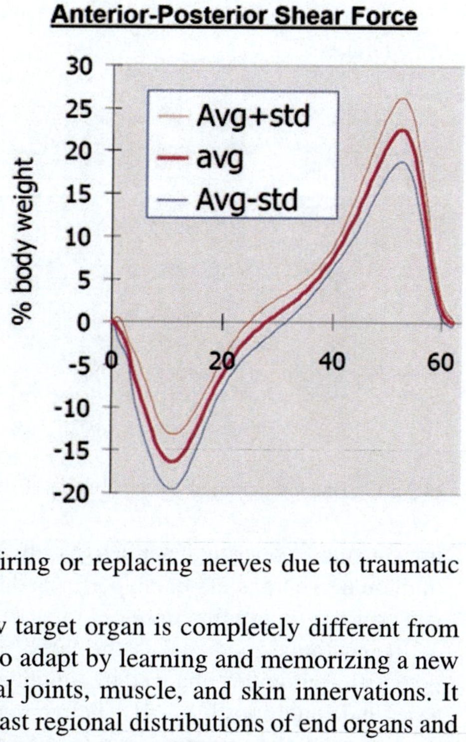

Fig. 5.10 Anterior-posterior shear force

quality of life of these patients by repairing or replacing nerves due to traumatic injuries [23–25].

In peripheral nerve transfers, the new target organ is completely different from the previous one, challenging the brain to adapt by learning and memorizing a new method for operating these new regional joints, muscle, and skin innervations. It also requires unlearning and forgetting past regional distributions of end organs and structures by creating a new synaptic circuitry among participating sensory motor neurons and deactivating the old neural circuitry.

In peripheral nerve surgery, there are clinical situations when a simple repair of the severed nerve is not possible. In injuries that are not reconstructable by traditional methods, the peripheral nerve surgeon may have the surgical option of

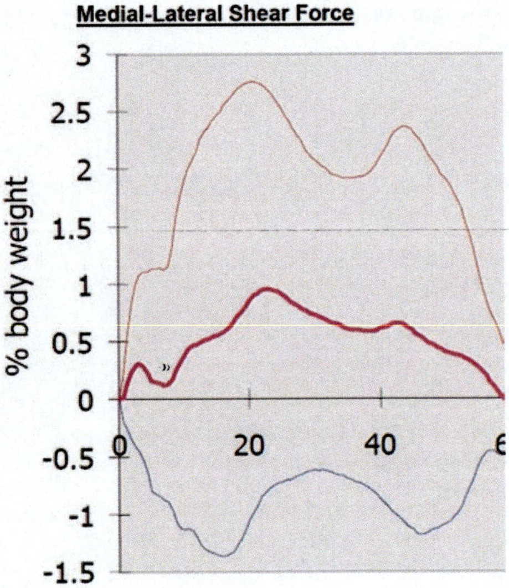

Fig. 5.11 Medial-lateral shear force

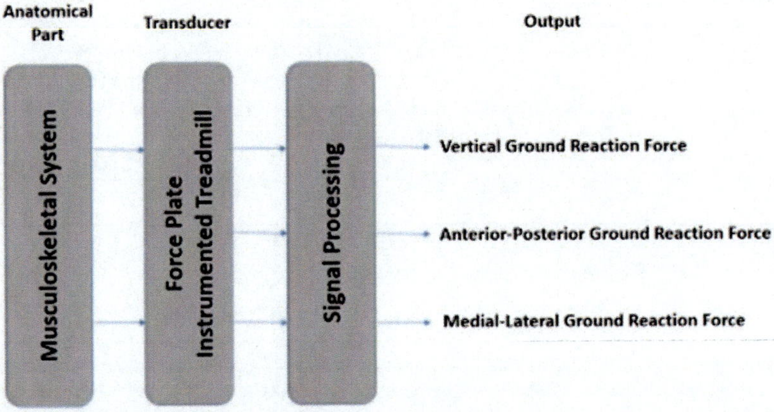

Fig. 5.12 Block diagram of ground reaction forces (GRFs)

transferring a noncritical nerve to a critical role. There are several clinical examples of these nerve transfers being performed to provide reinnervation of motor function and critical sensory function.

Regenerative medicine is the latest advancement in medical care, having the potential to provide enhanced treatment and therapy for previously untreatable diseases and conditions [24, 25]. Though regenerative medicine is a new field of medicine, peripheral nerve regeneration dates back centuries. With current technologies, regenerative medicine will have a prominent role in revitalizing health outcomes for chronic diseases and traumatic injuries. However, because regenerative medicine is a relatively new field, there is limited understanding concerning the underlying mechanisms involved in successful outcomes. In order to better understand the science behind the medicine, it is necessary to explore the adaptations of the body that occur with regenerative medicine. It is crucial to understand the interactions that occur between the body and regenerative medicine technologies with the purpose of helping clinicians provide optimal treatment and care for each patient.

Over the last few years, the technique of nerve transfers has emerged as an alternative to repair traction nerve injuries. Moreover, injured long nerve segments that previously had a poor prognosis with traditional nerve repairs are now being addressed through new nerve transfer and regenerative procedures. These new concepts are also being used to improve function in patients with specific central or spinal focal (CNS) injuries. The success of a nerve transfer is very much dependent on the degree of the cerebral reparative plasticity of the individual.

The motivation for the study of the peripheral nerve transfer is that for sensory reeducation to take place, the process of cerebral plasticity must occur. Therefore, it may be significant to illuminate our understanding regarding the fact that the cerebral cortex undergoes adaptations to changes in inputs for the establishment of specific connections with other regions of the brain in order to develop unique networks as capabilities. The adaptation process involves a reeducation of itself to learn the new task of the substituted peripheral nerve and to be able to continue its processing of sensation, movement, and cognition. By understanding factors that influence cerebral plasticity, these factors can be optimized to provide maximal recovery for peripheral nerve transfers and ultimately improve the outcome of this type of regenerative surgery.

The classic indication for peripheral nerve transfer surgery is a high median nerve injury of the hand whereas traditional repair of the lesion would result in a poor outcome with loss of function. With respect to the sensory loss incurred in a high median nerve injury, the restoration of sensation to the ulnar or dominant side of the thumb and the radial or dominant side of the index finger is of utmost importance in order to preserve the function of the hand [23–25]. Moreover, patients tend to bypass parts of their hand that do not have sensation. In this situation, sensation can be provided to these critical regions by using the superficial branch of the radial nerve (SBR) from the radial side of the wrist or proper digital nerves from noncritical areas supplied by the ulnar nerve.

Should the injured nerve be located in the lower extremity, there are several repair and reconstruction options available. These injuries commonly involve branches of the sciatic nerve, the largest nerve in the body that has been used as a model for peripheral nerve regeneration. In clinical situations, we often experience mixed nerves that contain motor, sensory, and also autonomic fibers that are affected. The sciatic nerve is an excellent model to study the complexity of innervations by evaluating motor and sensory recovery [23]. When an injury is far from its target organ, nerve transfers are considered as a reconstruction option. In these cases, the restoration of motor function is the usual goal. Nevertheless, the brain has to undergo the process of reeducation or plasticity to learn the new skill that it has acquired to fully integrate this function with the other available motors.

Injuries to the peripheral nerve system occur due to various causes including, but not limited to, myelin or axonal degeneration from systemic diseases, acute compression traumatic wounds, and chemical and thermal injuries. Common sites of traumatic nerve injury are the wrist (median, ulnar, and radial nerves), the hand (common and proper digital nerves to the fingers), and the lower leg (sciatic, peroneal, sural, and tibial nerves). Some neurological systemic disease states can also cause interruption in the neuromuscular circuitry at various levels.

The main techniques for the repair or reconstruction of sensory, motor, and mixed nerves after traumatic injury include: (1) direct microsurgical nerve repair, (2) use of a neuroconduit (hollow guides), (3) nerve grafting using either autogenous or allograft tissue, and (4) nerve transfer neurorrhaphy using less critical nearby nerves and rarely (5) direct implantation of the distal nerve stump into a muscle to restore innervation [23–25].

For the measurement of the kinematics, kinetics, and the EMG of the lower extremity, we suggest the use of the instrumented gait analysis set-up as depicted in Fig. 5.13. It consists of an array of inertial motion units (IMUs) sensor system, a dual-belt instrumented treadmill, and an electromyography [10, 11, 26, 27].

Figure 5.14 depicts the stimulation of the motor nerve, the peroneal nerve, which allows the recording of electrical signals of the anterior ankle, the fibular nerve, and the popliteal fossa. This recorded electrical signal following motor nerve stimulation is the compound muscle action potential, CMAP. The (orthodromic or antidromic) stimulation of the sensory nerves enables the recording of the radial, median, ulnar, sural, and the superficial peroneal nerves. This recording of the evoked response, the sensory nerve action potential (SNAP), is made directly from the sensory nerve.

Figure 5.15 illustrates the stimulation of the tibial nerve at the ankle resulting in recordings of the somatosensory-evoked potentials made with respect to the sciatic nerve action potential, the lumbar potential (N20), and also with respect to the somatosensory-evoked potentials generated in the cerebral cortex, cortical potential (P37), and recorded by scalp electrodes.

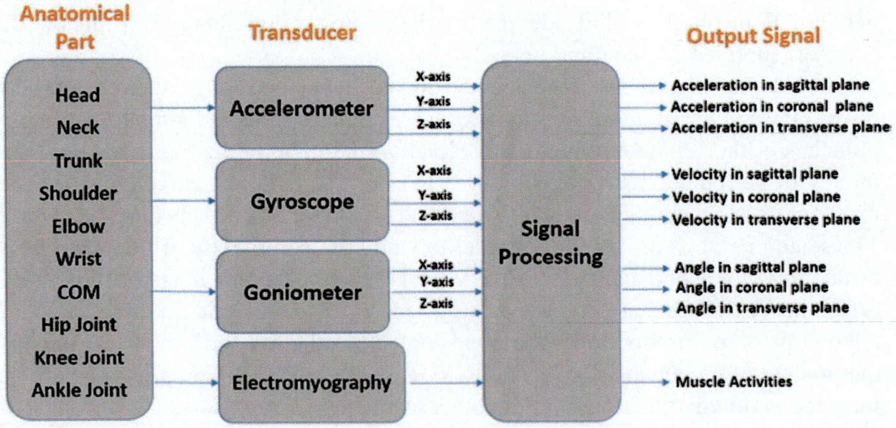

Fig. 5.13 Instrumented gait analysis setup

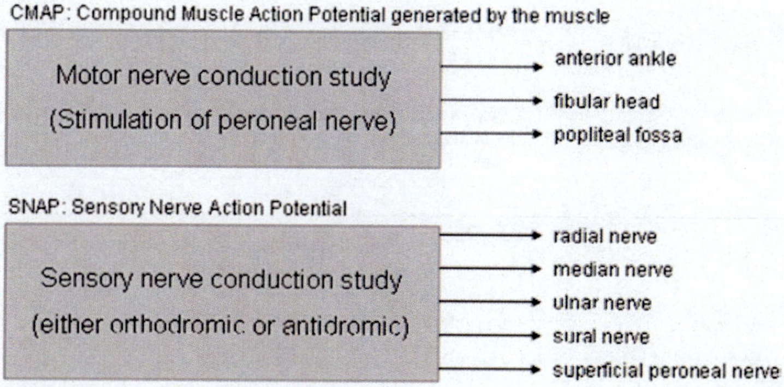

Fig. 5.14 Sensory/motor nerve conduction test

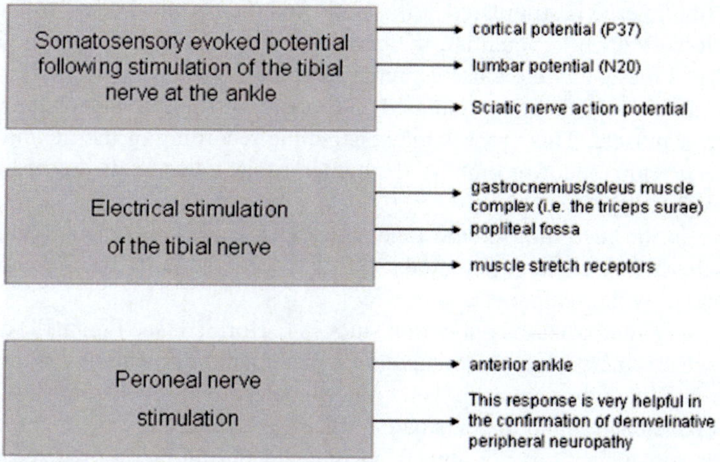

Fig. 5.15 Somatosensory-evoked potentials test

The electrical stimulation of the tibial nerve in Fig. 5.15, gives recordings from the gastrocnemius/soleus muscle complex (i.e. the triceps surae). The stimulation of the peroneal nerve results in an orthodromic electrical signal being conducted in the normal (orthodromic) direction from the spinal cord to the muscles innervated by the nerve. This response is also helpful in the confirmation of demyelinative peripheral neuropathies.

In Fig. 5.16, the needle EMG evaluates the integrity of the motor unit, which means, the motor neuron, motor axon, and the muscle to which it is attached. The thermo-needle and pressure-specified sensory units determine the temperature and pressure thresholds of the nerves.

For the lower extremity, gait analysis is performed, as described in Fig. 5.13 and the designs illustrated in Figs. 5.14 and 5.15, respectively.

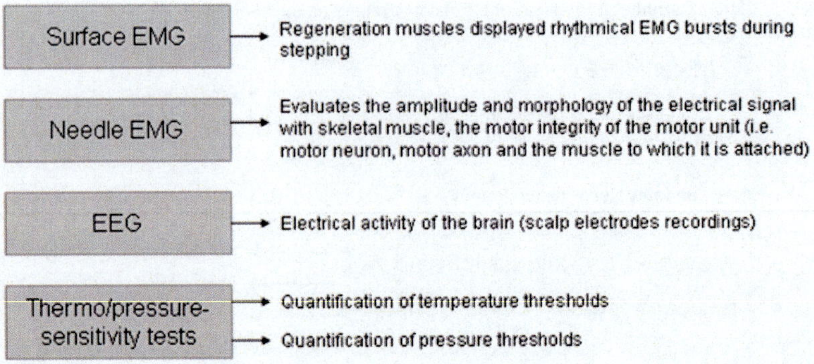

Fig. 5.16 Experimental design for diagnosis

- The tibial nerve is stimulated at the ankle and the peroneal nerve to monitor the sciatic nerve action potential, the cortical and lumbar potentials, the sural, the popliteal fossa, and the muscle stretch receptors.
- The sural/tibial nerve is stimulated and recorded from the biceps femoris/tibial anterior muscle. This process may enable the recording of the flexion or flexor reflex (FR) in the lower limbs in the human subject, that is, the lower limb flexor reflex (LLFR).
- There is the recording of the kinematics and kinetic parameters and also the somatosensory-evoked potentials (SSEPs) generated in the cerebral cortex recorded by the scalp electrodes.
- A sensory/motor nerve conduction study is performed (see Fig. 5.14) and record the sensory nerve and the compound muscle action potentials. In fact, several afferents, including nociceptive and non-nociceptive fibers from skin and muscles, have been found to contribute to LLFR activation.
- Further investigations are always encouraged into the motor-evoked potentials from repetitive electrical stimulation of the common peroneal nerve and determine its association with changes in the motor response of the tibialis anterior muscle before, during, and after repetitive electrical stimulation.

Figures 5.13, 5.14, and 5.15 introduce diagnostic processes for the continuous conditioned monitoring of the state of the subject following nerve transfer. The monitoring and assessment of the regaining of axonal regeneration involving remyelination and reinnervation or the formation of synapses of the appropriate muscle and sensory targets is a prerequisite for functional recovery after nerve transfer.

In Fig. 5.14, the stimulation of the motor nerve, the peroneal nerve, allows the recording of electrical signals of the anterior ankle, the fibular nerve, and the popliteal fossa. This electrical signal recorded following motor nerve stimulation is the compound muscle action potential, CMAP. The (orthodromic or antidromic) stimulation of the sensory nerves enables the recording of the radial, median, ulnar, sural, and the superficial peroneal nerves. The recording of the evoked response, the sensory nerve action potential (SNAP) is made directly from the sensory nerve.

There should occur recordings of the kinematic, kinetic, and EMG data from the gait analysis in conjunction with the recordings of the somatosensory evoked potentials. These results will deliver insight into the relevance of exercise and enforced physical training (EPT), with respect to its association with neurogenesis after nerve transfer.

Evaluation may take place to determine the effects of enforced physical training (EPT) on neurogenesis in the subventricular zone (SVZ) in the wall of the lateral ventricles and the subgranular zone (SGZ) of the hippocampus regions.

References

1. Mergner, T. (2007). Modeling sensorimotor control of human upright stance. *Elsevier, Progress in Brain Research, 165*. ISBN: 0079-6123: P. Cisek, T. Drew, & J. F. Kalaska (Eds).
2. Nishikawa, K., Biewener, A. A., Aerts, P., Ahn, A. N., Chiel, H. J., et al. (2006). *Neuromechanics: An integrative approach for understanding motor control*. From the symposium 'biomechanics and neuromuscular control' presented at the annual meeting of the Society for Integrative and Comparative Biology, January 4–8, 2006.
3. Mergner, T. (2002). The Matryoshka Dolls principle in human dynamic behavior in space: A theory of linked references for multisensory perception and control of action. *Cahiers de Psychologie Cognitive; Current Psychology of Cognition, 21*(2–3), 129–212.
4. Houle, D., Pelabon, C., Wagner, G. P., & Hansen, T. F. (2011). Measurement and meaning in biology. *The Quarterly Review of Biology, 86*(1), 3–34.
5. Zhan, M., & Xie, X. (2020). Precision measurement physics: Physics that precision matters. *National Science Review, 7*, 1795.
6. Keller, A., & Vosshall, L. B. (2004). Human olfactory psychophysics. *Current Biology, 14*(20), R876.
7. Narens, L., & Mausfeld, R. (1992). On the relationship of the psychological and the physical in psychophysics. *Psychological Review, 99*(3), 467–479.
8. Gottschaldt, K., Lersch, F. S., & Thomae, H. (1966). *Handbuch der Psychologie*. Verlag fuer Osychologie, Dr. C. J. Hogrefe.
9. de Groot, S., & Nieuwenhuizen, M. G. (2013). Validity and reliability of measuring activities, movement intensity and energy expenditure with the DynaPort MoveMonitor. *Elsevier Journal on Medical Engineering & Physics, 35*(2013), 1499–1505.
10. Sarkodie-Gyan, T., Huiying, Y., Murad Alaqtash, A., Abdelgawad, E., & Spier, R. B. (2011). Measurement of functional impairments in human locomotion using pattern analysis. *Elsevier Journal of Measurement, 44*, 181.
11. Alaqtash, M., Huiying, Y., Brower, R., Abdelgawad, A., & Sarkodie-Gyan, T. (2011). Application of wearable sensors for human gait analysis using fuzzy computational algorithm. *Elsevier Journal of Engineering Applications of Artificial Intelligence, 24*, 1018.
12. Maurer, C., Mergner, T., Bolha, B., & Hlavacka, F. (2000). Vestibular, visual, and somatosensory contributions to human control of upright stance. *Elsevier Journal of Neuroscience Letters, 281*(2000), 99–102.
13. Karduna, A. R. *Introduction to biomechanical analysis*. Chapter 1.
14. Uhlmann, K. (1996). *Lehrbuch der Anatomie des Bewegungsapparates*. UTB für Wissenschaft.
15. Yoon, J., Novandy, B., Yoon, C.-H., & Park, K.-J. (2010). A 6-DOF gait rehabilitation robot with upper and lower limb connections that allows walking velocity updates on various terrains. *IEEE Transactions on Mechatronics, 15*(2), 201–215.
16. Kruse, A. A. (2007). Operational neuroscience: Neurophysiological measures in applied environments. *Aviation, Space, and Environmental Medicine, 78*(5, section II), 191–119.

17. Silver, R., Boahen, K., Grillner, S., Kopell, N., & Olsen, K. L. (2007). Neurotech for neuroscience: Unifying concepts, organizing principles, and emerging tools. *The Journal of Neuroscience, 27*(44), 11807–11819.
18. Berkowitz, A. (2005). Physiology and morphology indicate that individual spinal interneurons contribute to diverse limb movements. *Journal of Neurophysiology, 94*, 4455–4470.
19. Daley, M. A. (2005). The role of limb posture in stabilizing bipedal locomotion over uneven terrain. *Comparative Biochemistry and Physiology Part A, 141*, S127.
20. Hood, S., McBain, T., Portas, M., & Spears, I. (2012). Measurement in sports biomechanics. *Measurement and Control, 45*, 182.
21. Winter, D. A. (1995). Human balance and posture control during standing and walking. *Gait & Posture, 3*, 193–214.
22. Aoi, S., Ohashi, T., Bamba, R., Fujiki, S., Tamura, D., Funato, T., Senda, K., Ivanenka, Y., & Tsuchiya, K. (2019). Neuromusculoskeletal model that walks and runs across a speed range with a few motor control parameter changes based on the muscle synergy hypothesis. *Scientific Report*; Published Online: 23 January 2019: www.nature.com/scientificreports
23. Pirela-Cruz, M. (2009). Interosseous nerve transfers for tibialis anterior muscle paralysis (foot drop): A human cadaver-based feasibility study. *Journal of Reconstructive Microsurgery, 25*(3), 203–211.
24. Lundborg, G. (2000). Brain plasticity and hand surgery: An overview. *Journal of Hand Surgery (British and European Volume), 25B*(3), 242–252.
25. Brunelli, G. A. (2004). Sensory nerve transfers. *Journal of Hand Surgery (British and European Volume), 29B*(6), 557–562.
26. Sarkodie-Gyan, T., Huiying, Y., Bogale, M., Hernandez, N. V., & Pirela-Cruz, M. (2017). Application of multiple sensor data fusion for the analysis of human dynamic behavior in space: Assessment and evaluation of mobility-related functional impairments. *Journal of Biomedical Science and Engineering, 10*, 182–203.
27. Yu, H. (2010). *Categorization of functional impairments in human locomotion using the methods of the fusion of multiple sensors and computational intelligence*. PhD dissertation, Department of Electrical and Computer Engineering, University of Texas at El Paso.
28. Vaughan, C. L., Davis, B. L., & O'Connor, J. C. (1992). *Dynamics of human gait* (2nd ed.). Kiboho Publishers.
29. Kriegeskorte, N., Mur, M., & Bandetini, P. (2008). Representational similarity analysis-connecting the branches of systems neuroscience. *Frontiers in Systems Neuroscience, 2*, Article 4.
30. Beales, D. J., O'Sullivan, P. B., & Briffa, N. K. (2010). The effects of manual pelvic compression on trunk motor control during an active straight leg raise in chronic pelvic girdle pain subjects. *Manual Therapy, 15*, 190–199.
31. Sado, N., Yoshioka, S., & Fukashiro, S. (2020). Three-dimensional kinetic function of the lumbo-pelvic-hip complex during block. Research Article. *PLOS ONE*. https://doi.org/10.1371/journal.pone.0230145
32. Sado, N., Yoshioka, S., & Fukashiro, S. (2017). A non-orthogonal joint coordinate system for the calculation of anatomically practical joint torque power in three-dimensional hip joint motion. *International Journal of Sport and Health Science, 15*, 111–119. https://doi.org/10.5432/ijshs.201712
33. Prathap Kumar, J., Arun Kumar, M., & Venkatesh, D. (2018). Healthy gait: Review of anatomy and physiology of knee joint. *International Journal of Current Research and Review, 12*, 1.
34. Blackburn, T. A., & Craig, E. (1980). Knee anatomy: A brief review. *Journal of American Physical Therapy Association, 60*, 1556–1560.
35. Lu, T.-W., & Chang, C.-F. (2012). Biomechanics of human movement and its clinical applications. *Kaohsiung Journal of Medical Sciences, 28*, 13–25.
36. Kuo, M. Y., Hsu, H. C., Chang, L. Y., Li, G. J., & Lu, T. W. (2002). A method for the measurement of the three-dimensional kinematics of the ankle-foot complex with footwear during activities. *Journal of Chinese Medical Science, 3*, 43–52.

37. Lu, T. W., & O'Connor, J. J. (1996). Lines of action and moment arms of the major force-bearing structures crossing the human knee joint: Comparison between theory and experiment. *Journal of Anatomy, 189*, 575–585.
38. Winiarski, S., & Rutkowska-Kucharska, A. (2009). Estimated ground reaction force in normal and pathological gait. *Acta of Bioengineering & Biomechanics, 11*(1), 53.
39. Vaverka, F., Elfmark, M., Svoboda, Z., & Janura, M. (2015). System of gait analysis on ground reaction force assessment. *Acta Gymnica, 45*, 193.
40. Wells, R. P. (1981). The projection of the ground reaction force as a predictor of internal joint moments. *Bulletin of Prosthesis Research*, 15–19.

Suggested Readings

Addae-Mensah, K. A., & Wikswo, J. P. (2008). Measurement techniques for cellular biomechanics in vitro. *Experimental Biology and Medicine (Maywood), 233*(7), 792–809.

Alsaadi, S., Mcauley, J., Hush, J., & Maher, C. (2011). Prevalence of sleep disturbance in patients with low back pain. *European Spine Journal, 20*, 737–743.

Anastakis, D. J., et al. (2008). Cortical plasticity following nerve transfer in the upper extremity. *Hand Clinics, 24*(2008), 425–444.

2007 Annual Report of the U.S. Organ Procurement and Transplantation Network and the Scientific Registry of Transplant Recipients (OPTN/SRTR): Transplant Data 1997–2006. Health Resources and Services Administration, Healthcare Systems Bureau, Division of Transplantation.

Arendt-Nielsen, L., Graven-Nielsen, T., Svarrer, H., & Svensson, P. (1996). The influence of low back pain on muscle activity and coordination during gait: A clinical and experimental study. *Pain, 64*, 231–240.

Beitz, W., & Kuettner, K.-H. (1983). *Dubbel Taschenbuch fuer den Maschinenbau* (15th ed.). Springer.

Fong, D. T., Chan, Y. Y., Hong, Y., Yung, P. S., Fung, K. Y., & Chan, K. M. (2008). Estimating the complete ground reaction forces with pressure insoles in walking. *Journal of Biomechanics, 41*(11), 2597–2601. https://doi.org/10.1016/j.jbiomech.2008.05.007. PMID:18571656.

Fox, B. D., Judge, L. W., Dickin, D. C., & Wang, H. (2020). Biomechanics of military load carriage and resulting musculoskeletal injury: A review. *Journal of Orthopedics & Orthopedic Surgery, 1*(1), 6–11.

Gray's anatomy for students, Second Edition.

Guan, X., Kuai, S., Song, L., Liu, W., Liu, Y., Ji, L., & Wang, R. (2019). Effects of ankle joint motion on pelvis-hip biomechanics and muscle activity patterns of healthy individuals in knee immobilization gait. *Journal of Healthcare Engineering, 2019*, Article ID 3812407.

Hesse, S., Reiter, F., Jahnke, M., Dawson, M., Sarkodie-Gyan, T., & Mauritz, K.-H. (1997). Asymmetry of gait initiation in hemiparetic stroke subjects. *Official Journal of the American Congress of Rehabilitation Medicine & the American Academy of Physical Medicine and Rehabilitation, 78*, 719–724.

Hesse, S., Uhlenbrock, D., & Sarkodie-Gyan, T. (1999). Gait pattern of severely disabled hemiparetic subjects on a new controlled gait trainer as compared to assisted treadmill walking with partial body weight support. *Journal of Clinical Rehabilitation, 13*, 401–410. 0269-2155(99) CR2710A.

Kim, M. K., & Lee, Y. S. (2013). Kinematic analysis of the lower extremities of subjects with flat feet at different gait speeds. *Journal of Physical Therapy Science, 25*, 531–533. https://doi.org/10.1589/jpts.25.531

Komitava, M., Mattsson, B., Eriksson, P. S., & Johansson, B. B. (2005). Enriched environment increases neural stem/progenitor cell proliferation and neurogenesis in the subventricular zone of stroke-lesioned adult rats. *Stroke, 36*, 1278–1282.

Komitova, M., et al. (2006). On neural plasticity, new neurons and the postischemic milieu: An integrated view on experimental rehabilitation. *Experimental Neurology, 199*(1), 42–55.

Lafortune, M. A., Cavanagh, P. R., Sommer, H. J., 3rd, & Kalenak, A. (1992). Three-dimensional kinematics of the human knee during walking. *Journal of Biomechanics, 25*, 347–357.

Lebel, K., Boissy, P., Hamel, M., & Duvar, C. (2015). Inertial measurements of motion for clinical biomechanics: Comparative assessment of accuracy under controlled conditions-changes in accuracy over time. *PLoS One*. https://doi.org/10.1371/journal.pone.0118361

Lledo, P. M., Alonso, M., & Grubb, M. S. (2006). Adult neurogenesis and functional plasticity in neuronal circuits. *Nature Reviews. Neuroscience, 7*, 179–193.

Lundborg, G., & Rosen, B. (2007). Hand function after nerve repair. *Acta Physiologica, 2007*(189), 207–217.

Mergner, T. (2010). A neurological view on reactive human stance control. *Elsevier Journal, Annual Reviews in Control, 34*(2010), 177–198.

Nuffield Council on Bioethics. (2013). *Novel neurotechnologies: Intervening in the brain*. ISBN: 978-1-904384-29-8.

Roriz, P., Carvalho, L., Frazao, O., Santos, J. L., & Simoes, J. A. (2014). From conventional sensors to fibre optic sensors for strain and force measurements in biomechanics applications: A review. *Journal of Biomechanics, 47*(2014), 1251–1261.

Sarkodie-Gyan, T. (2005). *Neurorehabilitation devices: Engineering design, measurement, and control*. McGraw-Hill Company. ISBN: 0071448306.

Taupin, P., & Gage, F. H. (2002). Adult neurogenesis and neural stem cells of the central nervous system in mammals. *Journal of Neuroscience Research, 69*, 745–749.

Wolpaw, J. R., & Tennissen, A. M. (2001). Activity-dependent spinal cord plasticity in health and disease. *Annual Review of Neuroscience, 2001*(24), 807–843.

Woo, S. L.-Y., Fox, R. J., Sakane, M., Livesay, G. A., Rudy, T. W., & Fu, F. H. (1998). Biomechanics of the ACL: Measurements of in situ force in the ACL and knee kinematics. *Elsevier Journal The Knee, 5*(1998), 267–288.

Chapter 7
Technology Initiatives in the Human Locomotor System

Review of the Nervous System

The nervous system controls the functioning of the human body both internally and externally. It consists of the brain, spinal cord, cranial nerves, and spinal nerves. Along with the endocrine system, it is also responsible for the regulation and maintenance of the body's homeostasis [1–4, 6]. It uses millions of sensory receptors to monitor changes occurring both inside and outside the body. These changes or stimuli are considered sensory input. The nervous system processes and interprets the sensory input and makes decisions about what should be done at each moment (integration). It then causes a response by activating effector organs such as muscles or glands in a response called motor output.

The nervous system consists of the central nervous system (including the brain and spinal cord) and the peripheral nervous system (including the nerve fibers extending from the brain and spinal cord that carry information between the central nervous system and the rest of the body). The central nervous system (CNS) occupies the dorsal body cavity, and constitutes the integrating and control center of the nervous system. It interprets sensory input and dictates motor output based on reflexes, current conditions, and past experience.

The peripheral nervous system (PNS) is the part of the nervous system outside the CNS. The PNS consists mainly of nerves (bundles of axons) that extend from the brain and spinal cord, and ganglia (collections of neuron cell bodies). Spinal nerves carry impulses to and from the spinal cord, and cranial nerves carry impulses to and from the brain. These peripheral nerves serve as communication lines that link all parts of the body to the CNS.

The peripheral nervous system has two functional subdivisions as illustrated in Fig. 7.1. The afferent (sensory) system consists of nerve fibers (axons) that convey impulses to the central nervous system from the sensory receptors located throughout the body [1, 2, 4–6]. The somatic sensory fibers convey impulses from the skin,

© The Author(s), under exclusive license to Springer Nature Switzerland AG 2023

T. Sarkodie-Gyan, H. Yu, *The Human Locomotor System*,
https://doi.org/10.1007/978-3-031-32781-0_7

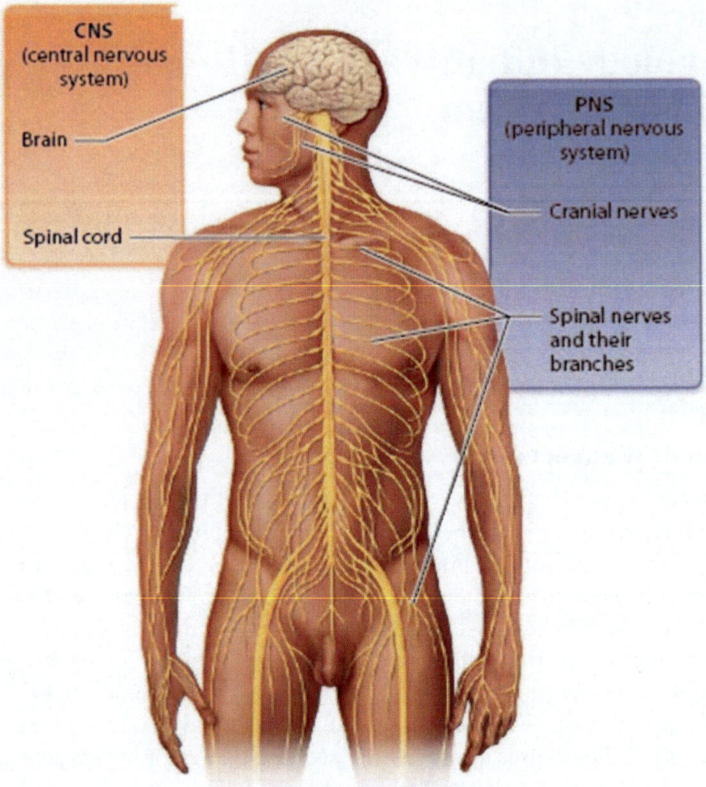

Fig. 7.1 Structure of the nervous system

skeletal muscles, and joints (soma means body), and the visceral sensory fibers transmit impulses from the visceral organs (organs within the ventral body cavity).

The sensory division keeps the central nervous system (CNS) constantly informed of events going on both inside and outside of the body.

The efferent (or motor) division of the peripheral nervous system (PNS) transmits impulses from the CNS to the effector organs, which consist of the muscles and glands. These impulses activate the muscles to contract and the glands to secrete. This means that they effect a motor response).

The motor (efferent) division exhibits two main functions; (1) The somatic nervous system consists of somatic motor nerve fibers that conduct impulses from the CNS to the skeletal muscles. It is often referred to as the voluntary nervous system because it allows us to consciously control our skeletal muscles, and (2) The autonomic nervous system (ANS) that involves the visceral motor nerve fibers that regulate the activity of the smooth muscles, cardiac muscles, and glands, respectively. Autonomic means "a law unto itself," and because we generally cannot control such activities as the pumping of our heart or the movement of food through our digestive tract, the ANS is also called the involuntary nervous system [1, 2, 6].

There are basically three overlapping functions of the nervous system. These include; (1) Sensory input: The nervous system uses its millions of sensory receptors to monitor changes occurring both inside and outside the body. The gathered information is called sensory input; (2) Integration: The nervous system processes and interprets sensory input and decides what should be done at each moment, and this process is called integration; and (3) Motor output: The nervous system activates the effector organs (the muscles and the glands) in order to cause a response known as the motor output [1, 2, 6].

The structure of the nervous system is depicted in Fig. 7.1.

The brain is the control center of the human body. The CNS has various centers located within it that carry out functions including the sensory, motor and the integration of data. These centers may be subdivided into the Lower Centers (including the spinal cord and brain stem) and the Higher Centers communicating with the brain via the effectors. The peripheral nervous system (PNS) is a vast network of spinal and cranial nerves that are linked to the brain and the spinal cord. It contains sensory receptors which help in the processing of changes in the internal and external environment. This information is transmitted to the CNS through the afferent sensory nerves. The PNS is then subdivided into the autonomic nervous and the somatic nervous systems, respectively [1, 2, 6]. The autonomic has involuntary control of the internal organs (blood vessels, smooth and cardiac muscles). The somatic nervous system has voluntary control of the skin, the bones, the joints, and the skeletal muscles [1, 5, 6]. The two systems function together by way of the nerves from the PNS entering and becoming part of the CNS, and vice versa (Fig. 7.2).

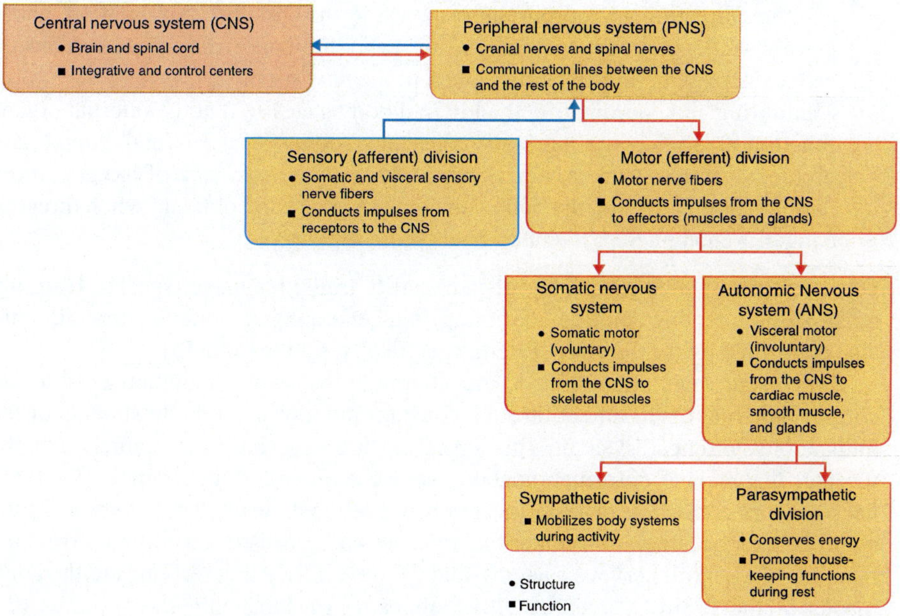

Fig. 7.2 Structural and functional divisions of the nervous system

The brain and spinal cord are protected by bony structures, membranes, and fluid. The brain is held in the cranial cavity of the skull and it consists of the cerebrum, cerebellum, and the brain stem. The nerves involved are cranial and spinal nerves.

The three main functions of the nervous system include sensory input, integration of data, and motor output. The sensory input involves the gathering of information or data by the body by way of neurons, glia, and synapses. It consists of excitable nerve cells (neurons) and synapses that form between the neurons and connects them to centers throughout the body or to other neurons. These neurons operate on excitation or inhibition, and although the nerve cells may vary in size and location, their communication with each other determines their function. These nerves conduct impulses from the sensory receptors to the brain and the spinal cord. The data is then processed through the integration of data, which occurs only in the brain. After the brain has processed the information, impulses are then transmitted from the brain and spinal cord to the muscles and to the glands, which is known as the motor output. Glia cells are found within tissues and are not excitable but help with myelination, ionic regulation, and extracellular fluid [1, 5, 6].

The central nervous system exhibits significant functions. These functions may include:

(a) Organizing and directing the motor responses of the voluntary muscle system, enabling the body to move more effectively as a whole and to achieve purposeful movement. This coordination of muscles makes possible complex activities, such as walking, running, playing a piano, and using a computer, as well as simple activities, such as maintaining muscle tone and posture while at rest.
(b) Monitoring and recognizing stimuli (and information) within the environment, and then directing an appropriate response to the stimuli. This function makes possible reflex actions, such as pulling away the hand from a hot surface, as well as perceiving music being played in the next room.
(c) Monitoring and coordinating the internal body states so that the internal organs function as a unit, internal body constancy (homeostasis) is maintained, and protective action is taken. For example, in response to a lack of oxygen, more rapid breathing occurs, the body shivers in response to cold; and when threat or danger is encountered, the heart beats more rapidly.

Other functions, such as display of personality traits, language, speech, learning, remembering, feeling emotion, reasoning, and generating and relaying thoughts, are also controlled by the nervous system, specifically, by the brain [3].

Homeostatic regulation refers to the attempt of the body to maintain a stable and constant internal environment through constant monitoring and adjustments of the human physiological system. This involves three mechanisms, namely (1) the receptor that receives information about some change in the environment (the feedback sensors consisting of proprioception, visual, and vestibular systems), (2) the control center or integration center that receives and processes information from the receptor (the central nervous system), and (3) the effector that responds to the commands of the control center by either opposing or enhancing the stimulus (the muscles as actuators). The healthy functioning of the central nervous system plays the central role in the regulations of our bodies, our capacities for perception and action, for language and emotions, our conceptions of ourselves and our relationships with others and, in fact, in our abilities to lead fulfilling lives [5, 6]. The skeletal and the muscular systems work hand in hand in the process of homeostatic regulation in a combined system otherwise known as the muscular-skeletal system (or the musculoskeletal system). In the musculoskeletal system, coordinated movement emerges from the interplay among descending output from the central nervous system, sensory input from the whole body and environment, and the dynamics of the muscles. Hence, human dynamic behavior in space becomes very complex due to its composition of many physical, perceptual, and motor aspects. Fundamentally, motor control involves a series of transformations of information that communicates from the brain and spinal cord to the muscles and to the body, and then back to the brain. Hence, the brain plays an irreducible role in the maintenance and performance of the body, including controlled movements and autonomic functions, as well as thoughts, emotions, memories, and behavior [3, 6, 7].

Unfortunately, several intractable medical conditions have emanated. Some of these intractable medical conditions include injuries, infections, or diseases as a result of damage to the brain and to the spinal cord. Many of these intractable conditions and/or types of damages lead to serious disorders that affect memory, cognition, movement, or consciousness. They cause conditions such as chronic pain, and many of the most intractable medical conditions [3].

Over the years, our understanding of the brain has improved considerably partly due to the development of the functional magnetic resonance imaging (fMRI) scanners which produce familiar striking images of the brain activity and provides suggestive hints concerning the basis of neurological disorders. It, therefore, appears that insight into neuroscience may lead to new, innovative, and technological initiatives towards the treatment of neurological disorders through interventions in the brain [3]. Some of the new technological initiatives involve invasive physical intrusion into the brain. Other initiatives depend on methods of interacting with the brain from outside, typically by exposing the brain to electromagnetic fields.

Functional Magnetic Resonance Imaging (fMRI)

Functional magnetic resonance imaging (fMRI) constitutes an imaging method for the demonstration of regional, time-varying changes in brain metabolism [8, 17, 19, 21, 22]. These metabolic changes can be consequent to task-induced cognitive state changes or the result of unregulated processes in the resting brain.

Functional magnetic resonance imaging is based on the principles of the magnetic resonance imaging (MRI), which in turn is based on the physics principle of Nuclear Magnetic Resonance (NMR) [13, 20, 23–25].

It may be borne in mind that a significant part of these technological initiatives is currently at their infant stages (such as the introduction of human neural stem cells into the brain), while the other initiatives try to take advantage of existing techniques (such as deep brain stimulation) to provide treatments of further neurological and psychiatric disorders.

Technology initiatives may seek to discover how the brain generates behavior [18], and even focus on the challenge that will help towards the development of tools that may be applied to elucidate fundamental operations of the brain [14–16].

Principle of the Nuclear Magnetic Resonance (NMR)

If a sample of certain atomic nuclei is placed in a magnetic field and subjected to an appropriate radiofrequency (RF) radiation (energy), they will re-emit some of the absorbed energy in the form of radio signals. This phenomenon is known as nuclear magnetic resonance (NMR). When the emitted signals are spatially encoded, information relating to the structure of the chemicals as well as the nature of the static and dynamic interactions between the nuclei inherent in a wide variety of materials and biological systems may be elucidated that could be used towards the creation of an image of the distribution of the nuclei in the sample [9, 10, 61].

NMR images are in some ways similar to X-ray computer tomographic (CT) scans, but because the physical processes underlying NMR phenomenon are entirely different from those related to X-rays, different kinds of information are contained in NMR images. It is necessary to understand NMR phenomenon in order to appreciate the implications of the images.

The nuclear magnetic resonance (NMR) phenomenon is based on the fact that nuclei of atoms have magnetic properties that can be utilized to yield chemical information. Based on quantum mechanics, subatomic particles (protons, neutrons, and electrons) possess spin. In some atoms, these spins are paired and cancel each other out so that the nucleus of the atom has no overall spin. However, in many atoms, the nucleus does possess an overall spin [9–11, 62, 66]. Figure 7.3 illustrates the charged nucleus rotating with angular frequency.

Nuclei exhibit positive charges and many of them behave as though they were spinning. Any material object that may be charged and moves possesses a magnetic moment and, therefore, produces a magnetic field. By virtue of this characteristic, a spinning nucleus acts as a tiny bar magnet oriented along the spin rotation axis as illustrated in Fig. 7.3.

This tiny magnet is often known as a nuclear spin. Should this small magnet be placed in the field of a much larger magnet, its orientation will no longer be random. There will be one most probable orientation. However, should the tiny magnet be oriented precisely 180° in the opposite direction, that position could also be maintained. Scientifically, the most favorable orientation would be the low-energy state and the less favorable orientation the high-energy state. The description of these two states may be appropriate for most nuclei of biological interest that includes ^{1}H, ^{13}C, ^{15}N, ^{19}F, and ^{31}P; this implies that all those which have nuclear spin quantum number

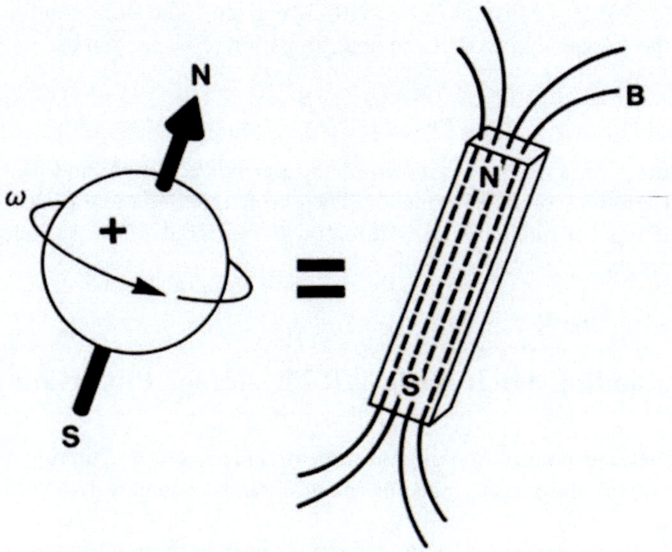

Fig. 7.3 The charged nucleus rotating with angular frequency

I = l/2. This is a quantum mechanical requirement that any individual nuclear spins of a nucleus with I = l/2 be in one of the two states (and nothing in between) whenever the nuclei are in a magnetic field [69–72]. It is important to note that the most common isotopes of carbon, nitrogen, and oxygen (^{12}C, ^{14}N, and ^{16}O) do not have a nuclear spin.

The small nuclear magnet may spontaneously "flip' 'from one orientation (energy state) to the other as the nucleus sits in the large magnetic field [10, 11, 73–75]. However, should the energy that is equal to the difference in energies of the two nuclear spin orientations be applied to the nucleus (or more realistically, group of nuclei), much more flipping between the energy levels is induced. The irradiation energy lies within the RF range and is typically applied as a short (e.g., many microseconds) pulse. The absorption of energy by the nuclear spins causes transitions from higher to lower energy as well as from lower to higher energy. This two-way flipping is a hallmark of the resonance process. The energy absorbed by the nuclear spins induces a voltage that may be detected by a suitably tuned coil of wire, amplified, and the signal displayed as a free induction decay (FID). Relaxation processes eventually return the spin system to the thermal equilibrium, which occurs in the absence of any further perturbing RF pulses [9, 11, 76, 77]. The energy required towards the induction of the flipping and to obtain an NMR signal is just the energy difference between the two nuclear orientations. It depends on the strength of the magnetic field B_0 in which the nucleus is placed, expressed as:

$$\Delta E = \gamma h B_0 / 2\pi \tag{7.1}$$

where h is Planck's constant (6.63 × 10–27 erg sec). The Bohr condition ($\Delta E = h\nu$) enables the frequency νo of the nuclear transition to be written as

$$\nu o = \gamma B_0 / 2\pi \tag{7.2}$$

Equation 7.2 is often referred to as the Larmor equation, and $\omega_0 = 2\pi\nu o$ is the angular Larmor resonance frequency. The gyromagnetic ratio g is a constant for any particular type of nucleus and is directly proportional to the strength of the tiny nuclear magnet.

Understanding MRI: Basic MR Physics for Physicians [12]

The magnetic resonance imaging machine includes a set of main magnet coils, three gradient coils, shim coils, and an integral radiofrequency (RF) transmitter coil (Fig. 7.4).

Figure 7.4 illustrates the relative positions of the different magnetic coils that are included in the magnetic resonance imaging (MRI) machine [12].

The patient is positioned within the bore of the machine and is surrounded by coils that lie concentric to each other and in the following order (from furthest to closest to the patient): main magnet coils, gradient coils, and radiofrequency (RF) coils. For neuroimaging, a further RF coil is placed around the patient's head to improve signal-to-noise (SNR) ratio.

Due to the necessary use of RF electromagnetic waves or radio waves, it is required that the physical arrangement depicted in Fig. 7.4 maintains potential sources of electromagnetic noise out and its own radiofrequency (RF) in. This is achieved by enclosing the magnet and its associated coils within a Faraday shield, which constitutes a special, copper-lined examination room [12].

The principles of the Maxwell equations state that when an electric current flows through a wire, a magnetic field is induced around the wire. Resistance to the flow

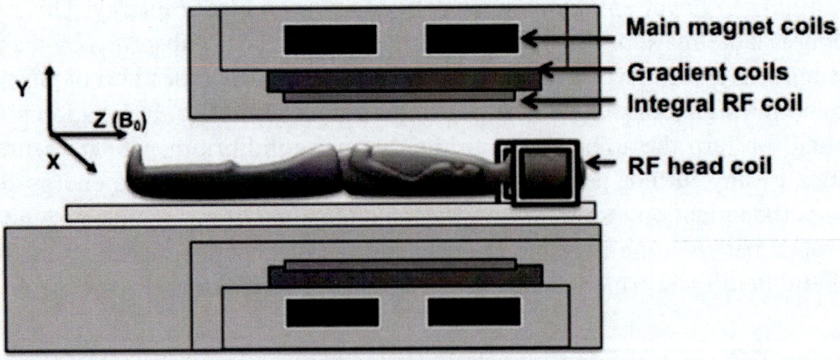

Fig. 7.4 Schematic representation of the MRI machine [12]

of the electric current can be reduced to negligible levels if a special metal conductor is cooled substantially. In this situation, lower resistance allows the use of high electric currents to produce high-strength magnetic fields, with little heat disposition. This principle is employed in the generation of superconducting magnets which is the type of magnet typically used in clinical magnetic resonance (MR) systems. The main magnet coils, made of a superconducting metal alloy, are cooled (close to absolute zero, ~4°K or −269 °C) using expensive cryogenic liquid helium.

The main magnet coils generate a strong, constant magnetic field (B_0) to which the patient is exposed. The strength of the magnetic field is measured in units of Tesla (T). One T is equivalent to approximately 20,000 times the magnetic field of the earth. Currently, most clinical MR systems are superconducting and operate at 1.5 T or 3 T. Field strengths reaching 9.4 T have been used in human imaging, albeit typically in research.

The MR system uses a set of coordinates to define the direction of the magnetic field. Gradient coils representing the three orthogonal directions (x, y, and z) lie concentric to each other within the main magnet (Fig. 7.4). They are not super-cooled and operate relatively close to room temperature. Each gradient coil is capable of generating a magnetic field in the same direction as B_0, but with a strength that changes with position along the x, y, or z directions, depending on which gradient coil is used. The magnetic field generated by the gradient coils is superimposed on top of B_0 so that the main magnetic field strength varies along the direction of the applied gradient field (Fig. 7.5).

The Image shown in Fig. 7.5 depicts the generation of a gradient in B_0 in the Z-direction. For a standard clinical MR system, this is accomplished using two coils in which the current flowing through them runs in opposite directions to each other (the so-called Maxwell pair type). The magnetic field at the center of one coil adds to the B_0 field while the magnetic field at the center of the other subtracts from B_0, thus creating a gradient in the B_0 field. The X and Y gradient coils (not shown in the

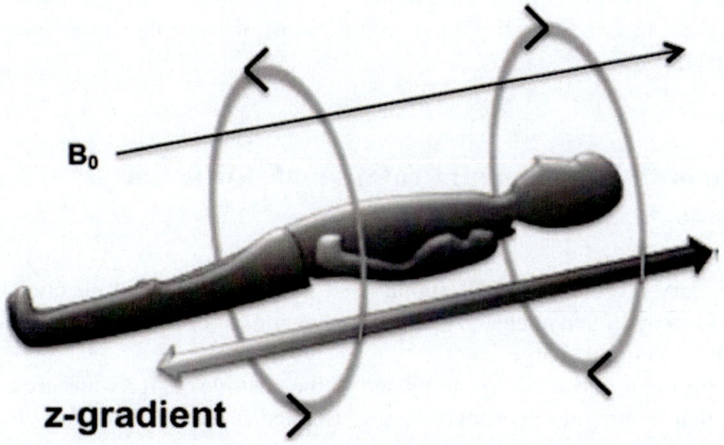

Fig. 7.5 Image to illustrate the generation of a gradient magnetic field, B_0, in the z-direction

Figure) generally produce the desired gradients over the patient, but minimize any interfering field interacting with the main magnet's coils and containers [12].

The radiofrequency coils (RF coils), so named because the frequency of the electromagnetic energy generated by them lies within the megahertz range, are mounted inside the gradient coils and lie concentric to them and to each other. They may be thought of as the 'antenna' of the MRI system and accordingly, they have two main purposes: (1) to transmit RF energy to the tissue of interest and (2) to receive the induced RF signal back from the tissue of interest [12].

Some RF coils perform the dual role of transmission and reception of RF energy whereas others transmit or receive only. For neuroimaging, a separate RF receiver coil that is tailored to maximize the signal from the brain is usually applied around the head of the patient in order to detect the emitted MR signals. The RF field is also referred to as the B_1 field. When switched on, the B_1 field combines with B_0 to generate MR signals that are spatially localized and encoded by the gradient magnetic fields towards the creation of an MRI. The output signal picked up by the receiver coil is digitized and then sent to a reconstruction computer processor to yield the image after some complex mathematical manipulation.

Localization of the MR signal requires good homogeneity within the local magnetic field. In other words, the more uniform the magnetic field the better. However, the placement of an object (including a patient) within the main B_0 field creates local susceptibility effects and reduces homogeneity. Shimming refers to adjustments made to the magnet to improve its homogeneity. Shimming can be passive or active [12]. Passive shimming is achieved during magnet installation by placing sheets or little coins of metal at certain locations at the edge of the magnet bore (close to where the RF and gradient coils lie). Active shimming provides additional field correction around an object of interest through the use of shim coils, which are activated by the currents controlled by the host computer, under the guidance of the scanner application software and the operator. Homogeneity of and, hence variation in B_0, is quoted in parts per million (ppm), that is, a fraction, of the static magnetic field over a specified spherical volume. For a 1 T system with a homogeneity of 1 ppm over 40 cm, no two points within 20 cm of isocenter differ by more than 0.000001 T.

Origin of the MR Signal: Protons and 'Little Bar Magnets' [12]

The primary origin of the MR signal used to generate almost all clinical images comes from hydrogen nuclei. Hydrogen nuclei consist of a single proton that carries a positive electrical charge.

The proton is constantly spinning and so the positive charge spins around with it. Recall that a moving electrical charge is called a current and that an electrical

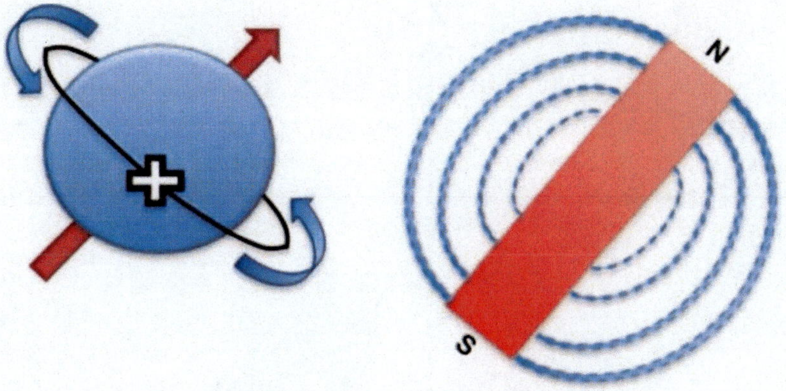

Fig. 7.6 Protons possess a positive charge and are constantly spinning around their own axes. This generates a magnetic field making protons similar to bar magnets

current generates a magnetic field [12]. Thus, protons have their own magnetic fields and behave like little bar magnets (Fig. 7.6).

The magnetic field for each proton is known as a magnetic moment. Magnetic moments are normally randomly orientated.

However, when an external magnetic field (B_0) is applied they align either parallel or against (anti-parallel) with the external field. The preferred state of alignment is the one that requires the least energy, this means, parallel to B_0. Accordingly, more protons align with B_0 than against it. The difference in the number of protons aligning parallel and antiparallel to B_0 is typically very small but ultimately depends on the strength of B_0 as well as the temperature of the sample. As a rough estimate, for about ten million protons aligning parallel to B_0, there are approximately 10,000,007 aligning antiparallel to the external magnetic field [12].

Precession [12]

When put in an external static magnetic field, the overall effect on a group of protons (that individually are aligned either parallel or antiparallel to B_0) means that the group of spins classically move in a particular way called precession. Precession can be likened to the movement of a spinning top. When spun, the top wobbles but does not fall over and the axes of the top circles form a cone shape (Fig. 7.7).

The movement of precession can be likened to the wobbling motion seen when a spinning top is spun. The handle of the spinning top follows a circular path.

The speed of precession, that is, how many times the protons precess per second, is measured as the precession frequency (also named the Larmor frequency, ω_0, in MHz) and determined by the Larmor equation:

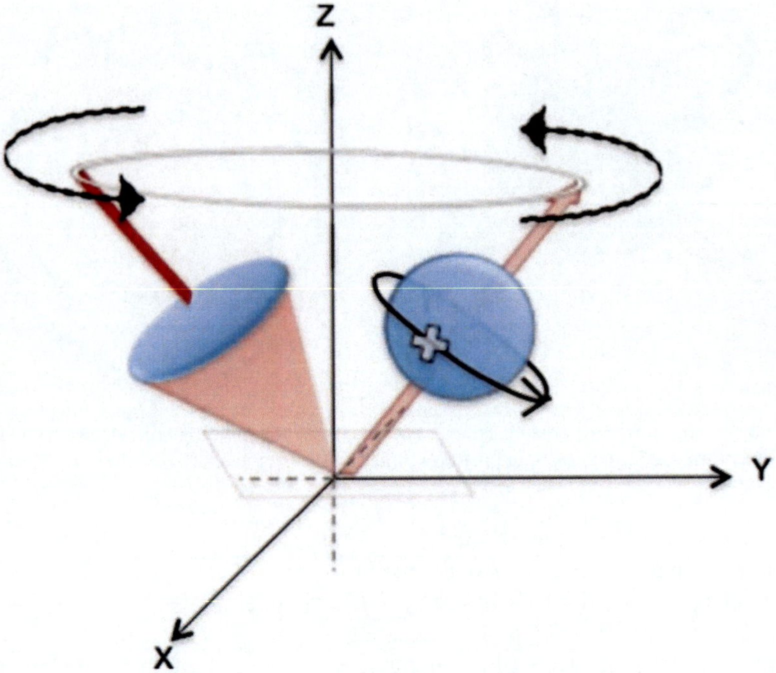

Fig. 7.7 Precession of protons when exposed to an external magnetic field

$$\omega_0 = \gamma B_0$$

γ Is a constant for a particular nuclear species (e.g., hydrogen) termed the gyromagnetic ratio. Its value for the proton is 42.6 MHz/T. The Larmor equation indicates that precession frequency is proportional to the strength of the magnetic field.

Longitudinal Magnetization [12]

Protons precessing parallel to B_0 begin to cancel each other out in all directions bar one: the direction of the z-axis, along B_0 (Fig. 7.8a).

In Fig. 7.8a, the magnetic moments of the protons precessing in the external magnetic field begin to cancel each other out. Opposing protons A, A_0, B, and B_0 'neutralize' each other leaving a net number of protons lying parallel to B_0. Protons precessing parallel to B_0 also begin to cancel each other out. C (pointing to the left) and C_0 (pointing to the right) oppose and neutralize each other. This occurs in all directions bar one: the direction of the z-axis, along B_0. The result is a sum magnetic field that is typically depicted as a vector in Fig. 7.8b.

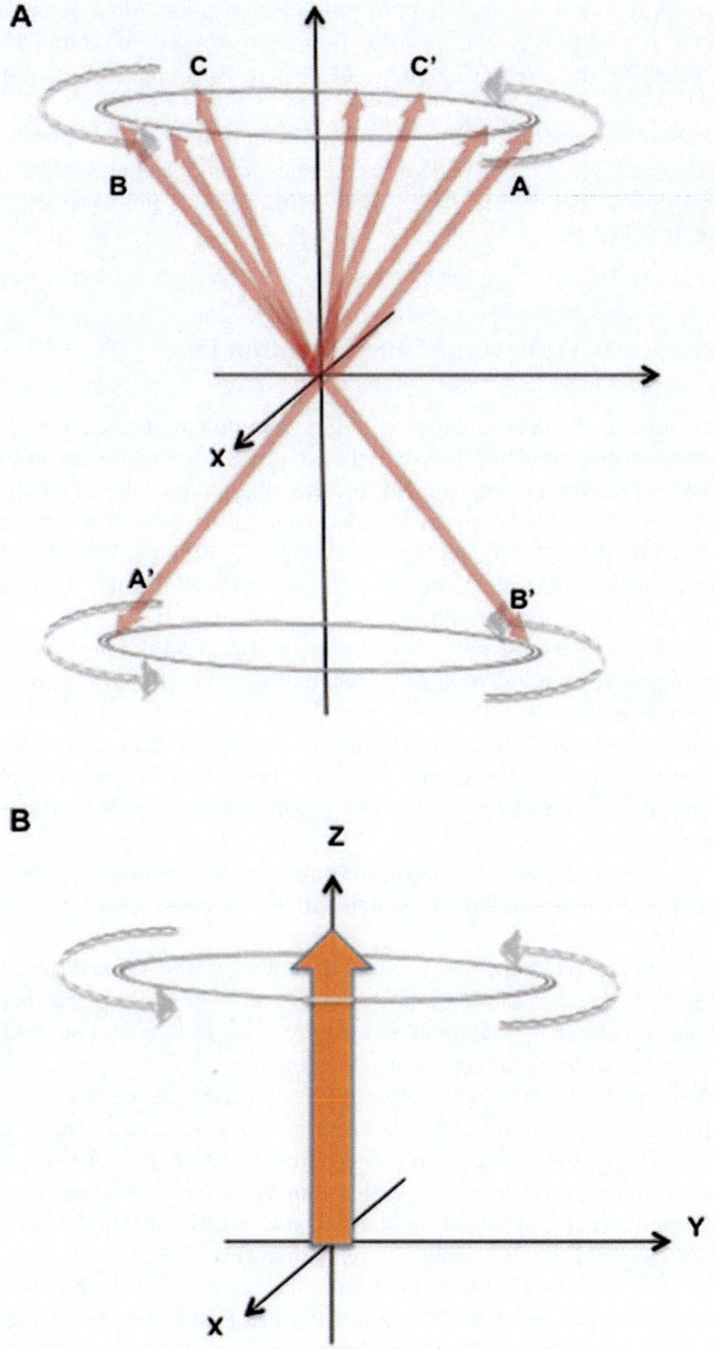

Fig. 7.8 Longitudinal magnetization. For simplicity, protons are now shown as vectors

The result is a sum magnetic field or sum magnetization, often given the symbol M, with the value M_0. It is characteristically shown as a vector. As this sum magnetization parallels the external magnetic field, it is also referred to as longitudinal magnetization (Fig. 7.8b).

The patient essentially becomes a magnet with a magnetic vector aligned with B_0. The magnetic force of the patient (as it stands) cannot be measured as it is in the same direction as the external field. A magnetization that lies at an angle to B_0 is, therefore, required [12].

RF Pulses and Transverse Magnetization [12]

With the patient in the magnet and possessing longitudinal magnetization, RF pulses are switched on and off. The purpose of the RF pulse is to disturb the protons so that they fall out of alignment with B_0. This disturbance occurs through the transference of energy from the RF pulse to the protons. This can only occur when the RF pulse has the same frequency as the precessional frequency of the protons, a phenomenon called resonance; hence the term magnetic resonance imaging. Accordingly, RF pulses are set at the Larmor frequency.

The activation of an RF pulse has two main effects on the protons. First, some protons gain energy and move to the higher energy state of being antiparallel to B_0. Consequently, opposing 'little bar magnets' (those parallel and antiparallel to B_0) once again cancel each other out, resulting in a reduction in overall longitudinal magnetization. Second, the RF pulse causes the protons to move in phase (that means, in the same direction, at the same time) with each other rather than in random directions.

The result is transverse magnetization in which a new magnetization vector is created in the x–y plane and moves in line with the precessing protons at the Larmor frequency [12].

The transverse magnetization vector is a moving magnetic field (rotating at the Larmor frequency), and, as such, if a conductive receiver coil is placed in proximity, an alternating voltage will be induced across it. This in turn generates an electrical current, which can be picked up (like an antenna would pick up radio waves) forming an MR signal. As soon as the RF pulse is switched off the protons start to fall out of phase with each other and also return to a lower energy state, that is, the protons relax. Relaxation occurs in two different ways [12]. Transverse magnetization begins to disappear, a process called transverse (or T2 ('Time' 2)) relaxation and the longitudinal magnetization starts to return to its original value, a process termed longitudinal (or T1 ('Time' 1)) relaxation (Fig. 7.9).

In Fig. 7.9a Protons aligned with B_0 produce a sum vector with the longitudinal magnetization. From Fig. 7.9b When an RF pulse is switched on, the longitudinal magnetization decreases and transverse magnetization propagates. Alternatively, it can be said that the sum magnetization 'tilts' to the side. An RF pulse that abolishes longitudinal magnetization to zero while inducing transverse magnetization is

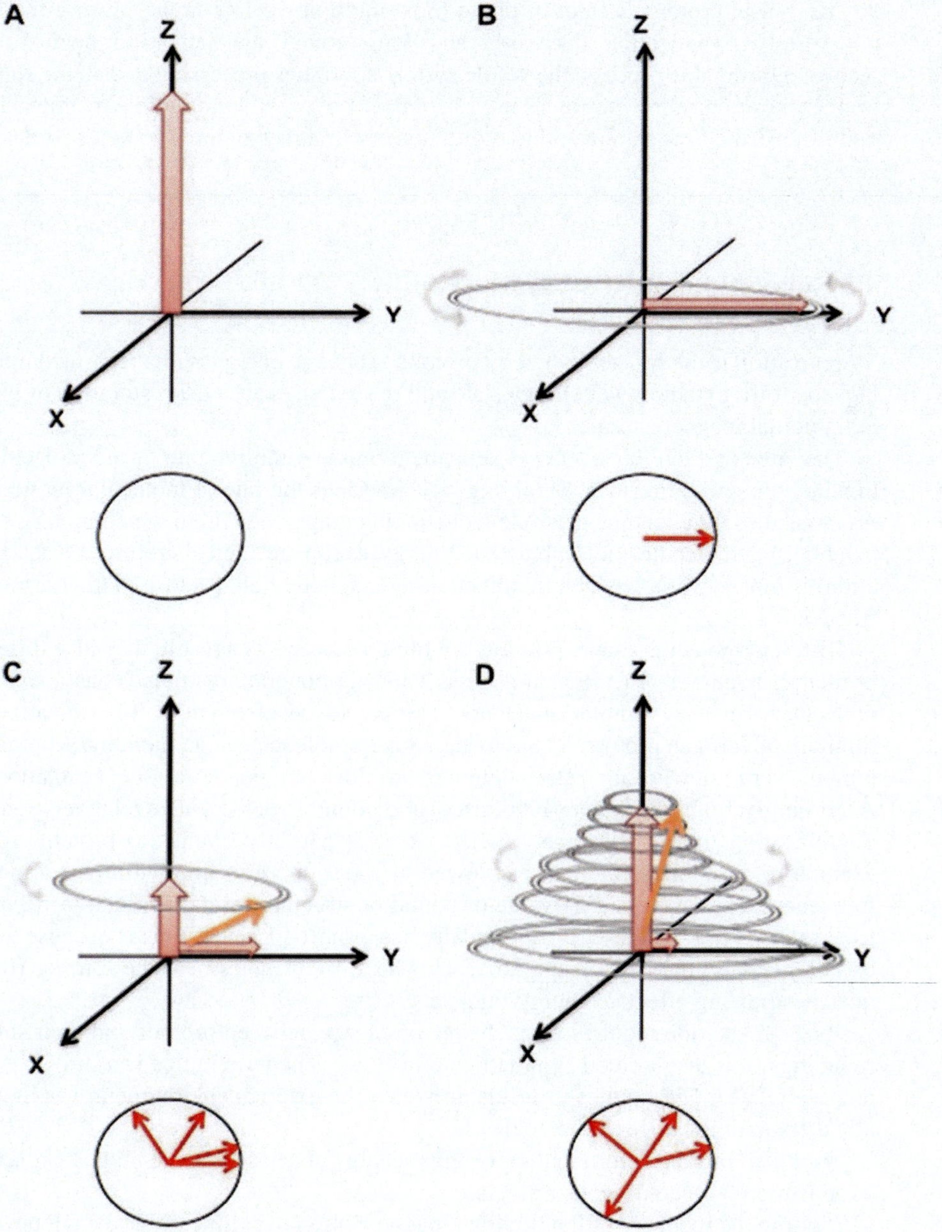

Fig. 7.9 Recovery of longitudinal magnetization following a 90° radiofrequency (RF) pulse

called a 90° (saturation) pulse as the sum magnetization vector is seen to tilt or flip 90°. Immediately following an RF pulse, protons precess in phase in the transverse plane, depicted by a single vector (arrow) in the lower circle. In Fig. 7.9c; After the

90° RF pulse, protons fall out of phase (now multiple vectors in the lower circle), transverse magnetization decreases and longitudinal magnetization begins to recover. During this process, the whole system continues precessing and so the sum vector takes a spiraling motion as in Fig. 7.9d. Recovery of longitudinal magnetization is termed T1 relaxation and loss of transverse magnetization is called T2 relaxation [12].

T1 Relaxation and T1 Values [12]

T1 relaxation is the process whereby protons exchange energy with their surroundings to return to their lower energy state and in so doing cause the restoration of the longitudinal magnetization.

The rate at which this occurs is dependent on the tumbling rate of the molecule in which the proton resides. Tumbling rate describes the rate of molecular motion. As molecules tumble, they generate a fluctuating magnetic field to which protons in neighboring molecules are subjected. Energy exchange (and therefore T1 relaxation) is more favorable when this fluctuating magnetic field is close to the Larmor frequency.

Different molecules have different tumbling rates and as a result, they also differ in their efficiency at T1 relaxation. Free water (unbound/unrestricted) has a small molecular size and tumbles much too quickly to be effective at T1 relaxation. Similarly, hydrogen protons bound to large macromolecules (e.g., membrane lipids) tumble very slowly and also demonstrate low efficiency at T1 relaxation. Accordingly, both free water (4000 msec) and bound hydrogen have relatively long T1 relaxation times. Conversely, when water is partially bound (to proteins for example) its tumbling rate can be slowed to a rate more in line with the Larmor frequency. As a result, the T1 value of bound or structured water is much less than free water (~400–800 msec). Fat typically has a short T1 value. This is because the carbon bonds at the ends of the fatty acids have frequencies near the Larmor frequency, allowing effective energy transfer.

As T1 relaxation requires an exchange of energy between protons and their surroundings, it is also termed spin–lattice relaxation. The term lattice is a throwback to early nuclear MR studies in solids, in which the external environment was literally a crystalline lattice of molecules.

As not all protons return to their original energy state at the same time, T1 relaxation is more of a continuous process.

Plotting the recovery of longitudinal magnetization over time (after the RF pulse is switched off) produces an exponential curve, called the T1 curve. It is difficult to exactly pinpoint the end of the longitudinal relaxation and so T1 (and similarly for T2) is not defined as the time of completion of the longitudinal relaxation.

Rather, T1 is a time constant that is used to describe how fast the process of T1 relaxation takes. Specifically, T1 is the time taken for the longitudinal magnetization to regrow from zero to $(1 - e{-}1)$, or approximately 63% of its final value. 1/T1 defines the longitudinal relaxation rate (Fig. 7.10).

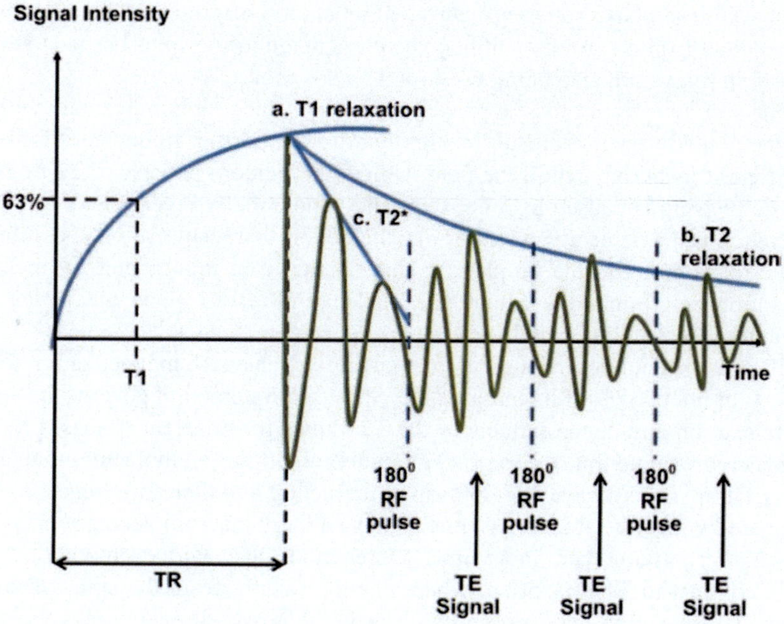

Fig. 7.10 T1, T2, and T2* relaxation

From Fig. 7.10, (a) the T1 curve: Plotting the recovery of the longitudinal magnetization over time following the switching off of an radiofrequency (RF) pulse results in aT1 curve. (b) T2 curve: A 180° refocusing pulse acts to 'combat' the effects of the external magnetic field inhomogeneity by rephasing the protons. This results in a temporary gain in signal intensity at time echo time (TE) termed spin echo. A sequence of 180° pulses results in a chain of spin echoes. Each subsequent echo will be of lower intensity due to T2 effects. A curve connecting the spin echo intensities is the T2 curve. (c) T2* curve: This curve results when 180° refocusing pulses are not used. The signal decays much faster due to T2* effects. TR indicates the repetition time.

T2 and T2* Relaxation [12]

Transverse relaxation describes the process whereby protons fall out of phase in the x–y plane and transverse magnetization decreases and disappears. There are two causes for this loss of phase coherence. The first, T2 relaxation results from slowly fluctuating magnetic field variations (inhomogeneity) within the local tissue. That is, the magnetic spin of protons is influenced by small magnetic fields from neighboring nuclei. This results in the random fluctuation of the Larmor frequency of individual protons causing an exchange of energy between proton spins, which leads to loss of phase coherence across a population of protons. The internal inhomogeneity of spins (protons) influencing other neighboring spins has led to the term spin–spin relaxation also being used for T2 relaxation.

The second cause of loss of phase coherence is due to inhomogeneity within B_0. Magnetic field variations result in slightly different Larmor frequencies for protons at different locations within the field. Unlike the random process of T2 relaxation, this de-phasing is caused by a constant and is potentially reversible. T2* (T2 star) relaxation is the name given to describe the effects that result from the combination of T2 relaxation and the de-phasing that results from inhomogeneity in B_0. T2* relaxation determines the actual rate of decay observed when measuring a free induction decay (FID) signal.

T2 is a constant describing the time taken for transverse magnetization to decay to e−1, or about 37% of its initial value. Spin–spin interaction governs the speed of T2 relaxation and hence influences the T2 values for different tissues. Free water (e.g., cerebrospinal fluid), comprising rapidly moving small molecules that are relatively far apart, will have less spin–spin interaction and therefore longer T2 values compared with water-based tissues that have a large macromolecular content (e.g., grey matter), giving them more time to interact or 'chat' with each other.

In contrast to T1 relaxation, where energy transfer from the spin system must occur, T2 relaxation may proceed with or without overall energy loss. For human tissue transverse relaxation is typically a much faster process than longitudinal relaxation; hence, T2 values are always less than or equal to T1. For biological tissues, T1 is approximately 5–10 times longer than T2 (300–2000 msec vs 30–150 msec, respectively).

Free Induction Decay [12]

When an RF pulse is switched off, T1 and T2 relaxations occur simultaneously and independently. The protons continue to precess and the sum magnetization vector follows a spiraling path whereby its direction and magnitude are constantly changing. Hence, an electrical signal is generated in a suitable receiver coil. The MR signal generated from the spiraling sum magnetization vector is termed FID. It has its greatest magnitude immediately after the RF pulse is switched off and then decreases as both relaxation processes occur. It also has a constant frequency (resonant frequency) and consequently, the FID signal takes the form of a sine wave with a rapidly decaying envelope (Fig. 7.11).

The transverse magnetization (with all protons rotating in phase) is at its greatest following an initial 90° excitatory radiofrequency (RF) pulse. Its amplitude (along with the signal intensity) then decreases as the protons begin to lose phase coherence. The resultant decay signal is termed free induction decay.

An FID is most commonly depicted with a 90° pulse, but an RF pulse of any flip angle can create an FID, because some components of the longitudinal magnetization are always tipped into the transverse plane. Theoretically, a 180° RF pulse

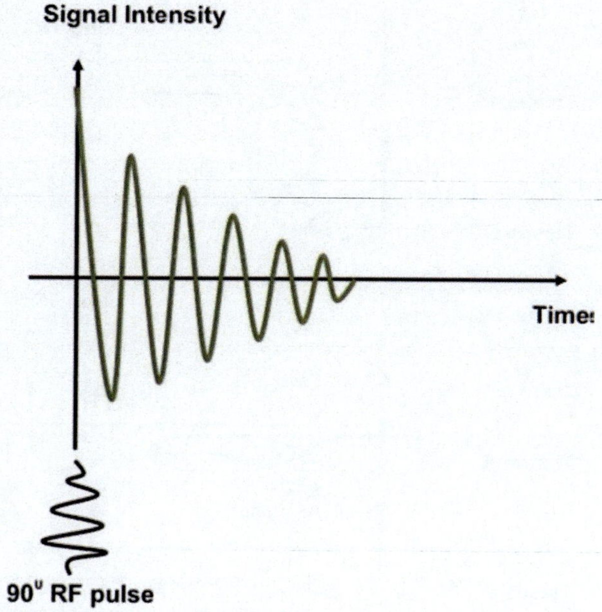

Fig. 7.11 Free induction decay

should refrain from generating an FID. However, in practice, all 180° pulses are imperfect and produce FID signals.

FID is subjected to further disruption (de-phasing) by the magnetic field gradients that are used to localize and encode the MR signal. Consequently, the signal generated by FID is not usually measured in MRI. Instead, it is common practice to generate and measure the MR signal in the form of an echo: typically, a spin echo (SE) or a gradient echo (GRE). Echoes can be appreciated by considering how T1- and T2-weighted images are formed.

T1-Weighted Images [12]

Contrast between tissues allows adjacent structures to be differentiated from one another. Contrast is determined by signal intensities, which in turn are governed (at least partly) by the T1 and T2 relaxation times of tissues within an image. An image in which the difference in signal intensity between tissues is predominantly due to differences in tissue T1 relaxation time is called a T1-weighted image. T1-weighted images are generated predominantly by manipulating the time between two RF excitation pulses, the so-called repetition time (TR) (Fig. 7.12).

Consider the following example in which two tissues, fat and fluid are being imaged where fat has shorter transverse and longitudinal relaxation times than fluid. An initial 90° radiofrequency (RF) pulse is switched on and off, time passes (TR), a second 90° RF pulse is switched on and off, MR signal is then sampled, and an

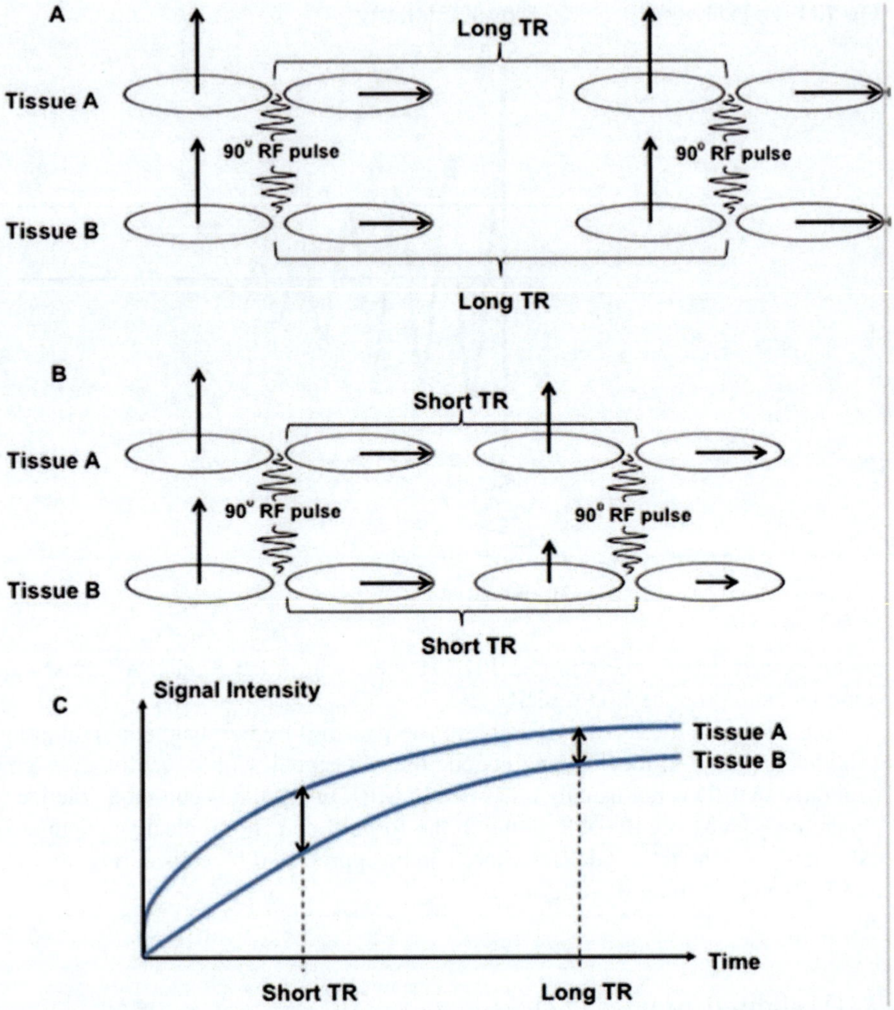

Fig. 7.12 Repetition time (TR) and T1-weighting

image is generated. As MR signal sampling occurs after the switching off of the second RF pulse, the signal intensity will be determined by the amount of transverse magnetization at that point. This in turn will be governed by the value of the longitudinal magnetization of the two tissues immediately before the second 90° RF pulse is switched on. Choosing a long TR (e.g., 1500 msec) would allow the tissues to recover their longitudinal magnetization fully with no appreciable difference in the value of longitudinal magnetization existing between the tissues. In this situation, the application of a second RF pulse with subsequent MR signal sampling would result in no discernible difference in MR signal between the tissues as the transverse magnetization values for the tissues are essentially equal (A) (i.e., no signal contrast). Choosing a short TR (20 msec) means that at the time of applying a second 90° RF pulse tissues with a long T1 (fluid) will show less recovery in the longitudinal magnetization than tissues with a short T1 (fat). The tissues will possess different transverse magnetization values (B), and thus, will generate different signal intensities and allow greater contrast between tissues (C).

This example (Fig. 7.12) was simplified for ease of understanding. It should be appreciated that many parameters influence signal intensity and hence tissue contrast, but in the example used, T1 had the greatest influence. Contrast in images obtained at long TR will not be influenced by T1 but instead may be influenced by differences in the T2 or proton density of the tissues in question.

T2-Weighted Images [12]

Recall that protons lose phase coherence following an RF pulse as a result of spin–spin interactions within the tissues (T2 relaxation) and because of inhomogeneity within the local static magnetic field (T2* relaxation). De-phasing caused by T2 relaxation is a random, irreversible process whereas the de-phasing caused by magnetic field inhomogeneity is potentially reversible.

Application of a 180° RF pulse following an initial 90° RF pulse rotates the protons through 180°, effectively making the protons 'turn around' and precess, still in the x–y plane, but in the opposite direction. Local field inhomogeneity remains and protons with a slightly faster Larmor frequency begin to catch up with slower protons. Eventually, the protons come back into phase, which results in an increase in the amplitude of the MR signal. Maximum signal amplitude is reached at the TE. To achieve maximal signal, the 180° RF pulse must be applied at time TE/2. As the protons (spins) bounce back or echo following the application of the 180° RF refocusing pulse, the signal obtained is given the name SE (Fig. 7.13).

Following a 90° excitatory radiofrequency (RF) pulse with consequent transverse magnetization, protons begin to de-phase (A–C) and signal amplitude decreases. Application of a 180° RF pulse at time TE/2 causes the protons to precess in the opposite direction. Protons that were initially 'in front' now find themselves at the back but begin to 'catch' the other protons, which are precessing at a slighter slower frequency (D–F). At time TE the protons regain phase, resulting in a stronger net transversal magnetization and thus a stronger signal. This signal re-emergence is termed spin echo.

Continued inhomogeneity in the static magnetic field means that protons will continue to lose phase coherence following a 180° RF refocusing pulse; however, it is possible to repeat the 180° RF refocusing pulse and obtain further SEs. Over time, the amplitude from each SE will decrease due to T2 effects. A curve connecting the SE intensities is called the T2 curve. T2* curve is the curve that results if 180° RF refocusing pulses are not used, that is, a curve that depicts the de-phasing caused by T2* effects Fig. 7.13b, c).

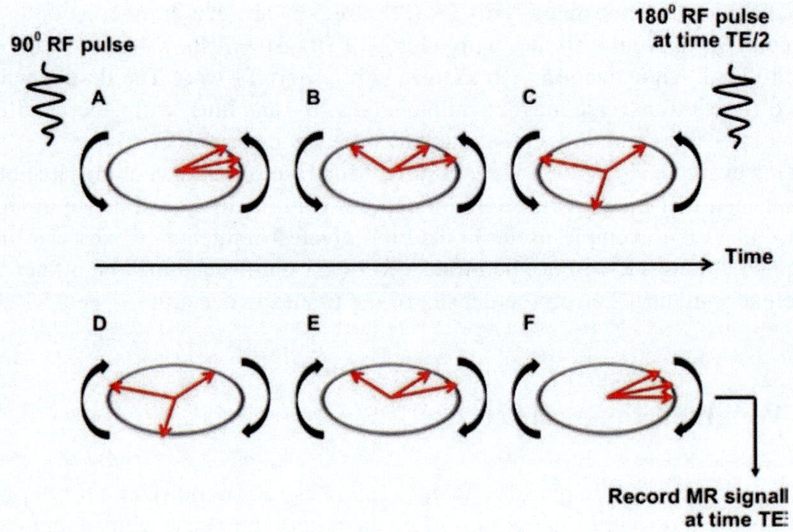

Fig. 7.13 Spin echo

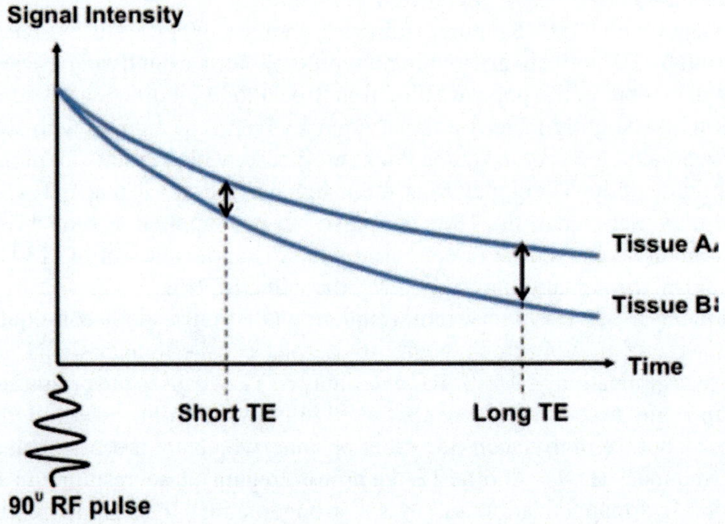

Fig. 7.14 Echo time (TE) and T2-weighting

Tissues have different T2 values. Brain, for example, has a shorter T2 than cerebrospinal fluid (CSF). TE is the time interval between the 90° RF pulse and the SE. It is one of the parameters whose value can be chosen by the operator of the MR machine in order to influence the signal intensities (hence contrast) between tissues. A much stronger signal is received when short TE (as opposed to long TE) is employed. However, at short TE (eg, <30 msec), differences in T2 have little influence on tissue contrast. Accordingly, T2-weighted images of the brain are obtained at long TE (e.g., 80 msec) (Fig. 7.14).

T2 curves for two different tissues are shown in Fig. 7.14. Tissue B has a shorter T2 relaxation time than tissue A. The difference in signal intensity between the tissues is more discernible at long TE than at short TE. RF depicts the radiofrequency.

A heavily T2-weighted image could be obtained at longer TE but loss of MR signal would impact on the signal to noise ratio making images potentially subdiagnostic.

Factors Influencing SE Contrast and Weighting [12]

Imaging parameters influencing the MR signal in the SE sequence include TE and TR. Values for both of these parameters are purposely chosen by the operator in order to influence the tissue weighting of the image. Short TR and short TE generate a T1-weighted image; short TR allows differences in longitudinal magnetization to develop before the next 90° excitation pulse while short TE limits the T2 effects. Long TR and long TE produce a predominantly T2-weighted image; long.

TR allows recovery of the longitudinal magnetization (thus limiting T1 effects) while at long TE, T2 effects become pronounced. A different type of image is produced at long TR and short TE. Long TR and short TE limit T1 and T2 effects, respectively. When this occurs, the signal is predominantly influenced by the proton density of the tissues (Fig. 7.15a–c).

From Fig. 7.15:

(A) The Sagittal T1-weighted image. Note that fat (around scalp and neck) is bright and cerebrospinal fluid (CSF) is dark.
(B) Illustrates the Axial proton density image. Note minimal contrast between grey matter and CSF.
(C) Illustrates the Axial T2-weighted image. Note that CSF is bright.
(D) Illustrates the Same image as (A) but following gadolinium administration. Note the dural venous sinuses appear bright due to T1 shortening.
(E) Illustrates the Axial fluid-attenuated inversion recovery image. Note suppression of the CSF signal. (F) illustrates the Axial gradient echo T2* image. Same patient as in (C). Note the increased conspicuity of multiple dark foci due to magnetic susceptibility of hemosiderin deposition in a patient with amyloid angiopathy.

It should be appreciated that the choice of imaging parameters for all MRI sequences can influence the sensitivity of the test for the pathology in question. For example, the sensitivity for the detection of lesions in multiple sclerosis is dependent on TE (lesion visibility decreases with increasing TE).

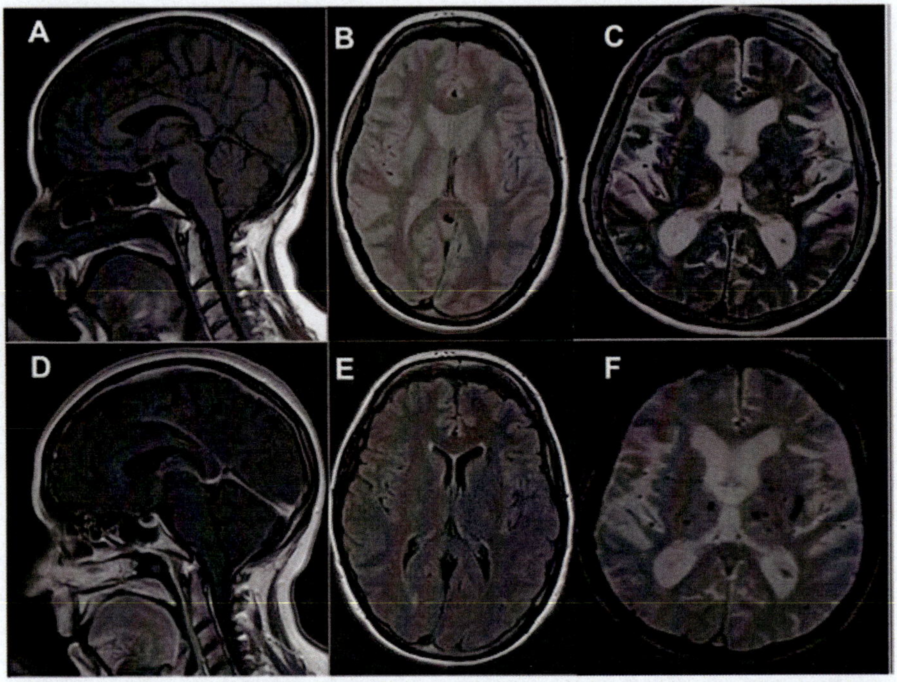

Fig. 7.15 Examples of typical clinical MRI

Gradient Echo [12]

SE sequences with their relatively long TR and TE are time-consuming. This limits the number of patients who can be scanned in a session and also risks movement artifact by a restless patient. Conversely, GRE sequences, which replace the 180° refocusing RF pulse with magnetic field gradients, are relatively short. However, they do produce different contrasts. Magnetic field gradients produce a change in field strength and thus a change in Larmor frequency along a particular direction. Application of a gradient pulse after an initial RF pulse causes protons to rapidly de-phase along the direction of the gradient resulting in rapid decline in the FID signal. This loss of phase coherence can be reversed by applying a second magnetic field gradient with a slope of equal amplitude but in opposite direction to the first. As a result, protons move back into phase and return a signal called GRE. TE is the time taken between the beginning of FID (i.e., the generation of transverse magnetization) following the initial RF pulse to the point at which the GRE reaches its maximum amplitude (Fig. 7.16).

An excitatory radiofrequency (RF) pulse causes transverse magnetization and initiation of a free induction decay signal. This signal rapidly de-phases following the application of a magnetic field gradient. Application of a second magnetic field gradient with a slope of equal amplitude but in opposite direction to the first causes

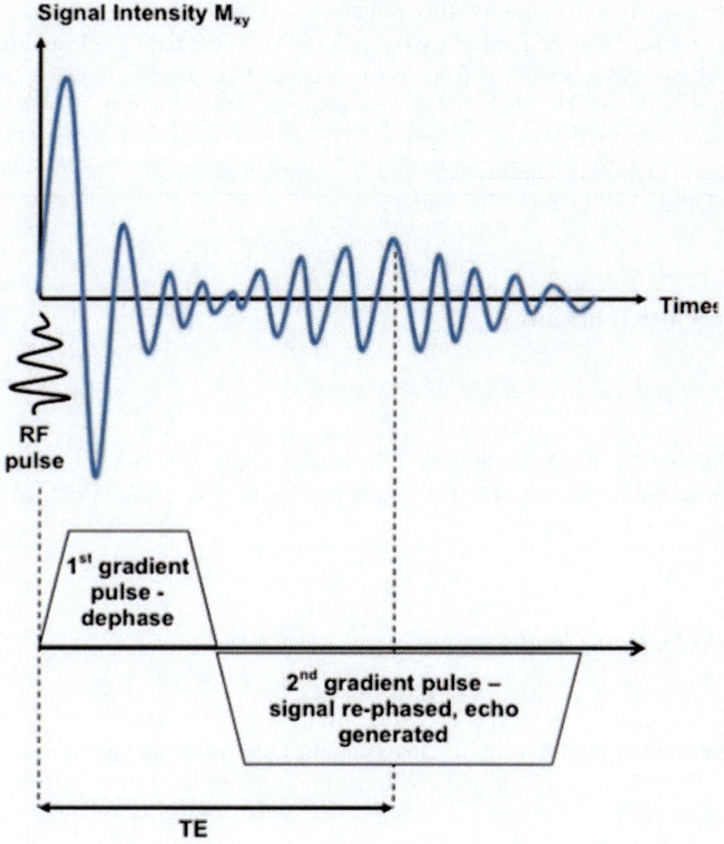

Fig. 7.16 Gradient echo

some re-phasing. The signal increases again at time TE to a maximal signal termed a gradient echo. The maximum amplitude of the gradient echo is dependent on the specified TE and the T2* relaxation rate of the tissues in question.

Differences Between SE and GRE [12]

Several differences exist between the two techniques. First, TE can be shorter than GRE as only one RF pulse is used. Second, GRE sequences are typically used with low flip angle excitations (e.g., 5–40°) as opposed to the 90° flip angle used in SE. As a consequence, longitudinal magnetization is not completely abolished with GRE and will provide reasonable signal even at very short TR. This combination of short TE and short TR permits rapid signal acquisition with GRE imaging. The price of short acquisition time in GRE, however, is a greater signal loss in the presence of magnetic susceptibility effects. Gradient reversal refocuses only those spins that have been de-phased by the action of the gradient itself and unlike SE, where phase shifts resulting from local magnetic field inhomogeneity are minimized through the 180° RF pulse, this inhomogeneity persists in GRE. Hence, image contrast in GRE is influenced by T2* effects and is not confined to true T2 relaxation. In GRE imaging, T1- and T2*-weighted images can be obtained by manipulating the flip angle, TE and TR.

Image Construction [12]

Localizing and Encoding MR Signals

The MR signal is localized in three dimensions using three separate magnetic field gradients termed (1) slice-selection gradient, (2) phase-encoding gradient (GP), and (3) frequency-encoding (GF).

Slice-Selection Gradient

Slice localization is achieved by using gradient coils to generate a gradient field orientated along a chosen axis. This gradient field alters the strength of B_0 in the chosen direction, so that protons within the gradient field have different Larmor frequencies (Fig. 7.17a).

From Fig. 7.17:

(A) Step 1: Slice selection. A slice-selecting gradient (GS) is applied at the same time as the excitatory radiofrequency (RF) pulse. In this example, the brain is being imaged and GS is applied along the z-axis, parallel with B_0. This means that different cross-sections of the brain experience magnetic fields of differing strength. Accordingly, protons will precess at different Larmor frequencies depending on their position along the gradient. Selecting an RF pulse (or range of RF pulse frequencies) that matches the Larmor frequency of the protons will determine the slice location.

(B) Step 2: Phase encoding. Protons are in phase after the RF pulse is applied. Applying a phase-encoding gradient along the y-axis causes the protons to increase their speed of precession relative to the strength of the magnetic field to which they are exposed. In this example, speed increases from top to bottom. This is depicted on the right of the schematic, which shows rows of protons with different precession speeds. When the gradient is switched off, all protons are once again exposed to the same magnetic field and as such they have the same frequency; only now, they are out of phase. (It may help to think of the protons giving off the same signal frequency but at different times.)

Fig. 7.17 Image construction

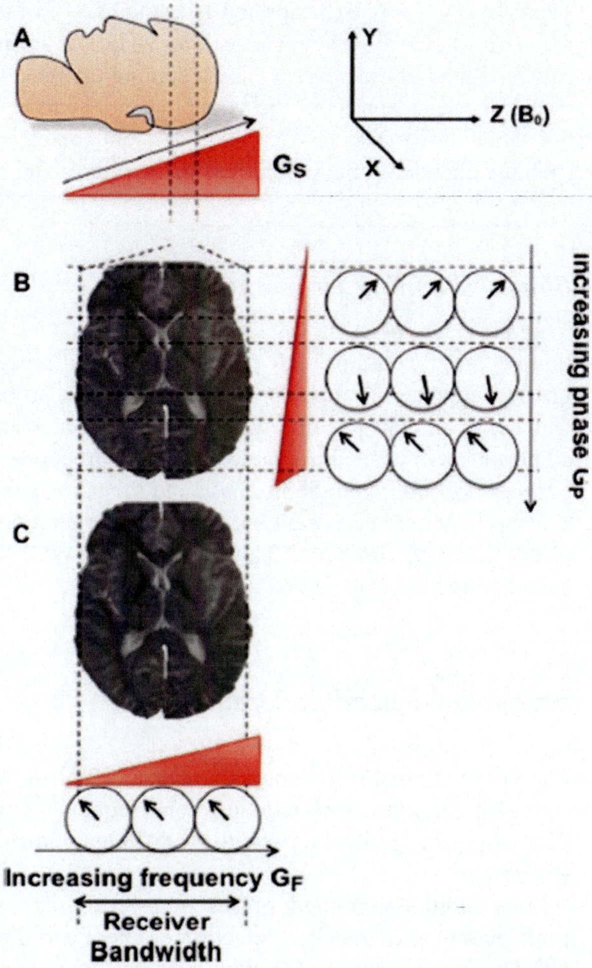

(C) Step 3: Frequency encoding. A further magnetic field frequency-encoding gradient is applied to help differentiate the signal from different protons by way of differing frequencies. In this example, the protons from the bottom row in step 2 (above) have been exposed to a gradient applied along the x-axis. While the protons are in phase, they now have different frequencies, allowing their differentiation in the third (x) plane. During the frequency-encoding step, the signal is measured (digitally sampled) at time TE, the point of maximal signal. The signal comprises a range of frequencies (or) that correspond to the Larmor frequencies of the proton magnetic moments at their different locations along the gradient. In this example, the orientation of the three gradients defined a slice perpendicular to the z-axis. Different slice orientations can be achieved by allocating the gradients to a different axis. Combining gradients along two or more axes for each localization task allows angled slices to be obtained.

Typically, the RF pulse is applied as a small range of frequencies (bandwidth) rather than as a single frequency. This excites a slice of a certain thickness. Slice thickness can be altered in two ways: (1) by changing the bandwidth of the RF pulse or (2) by changing the steepness of the gradient field. With a given slice selectively excited, the signals arising from each slice element (pixel) within that section need to be spatially encoded. This is achieved using phase- and frequency-encoding gradients.

Phase-Encoding Gradient

Following an excitatory RF pulse, the protons precess in phase. Applying a new gradient magnetic field will make some of the protons precess faster than others depending on their position within the gradient. When the gradient is switched off, all protons will express the same precession frequency and will emit the same signal but they will no longer be in phase and this allows the protons to be differentiated. The gradient applied is called the GP and the direction of application is termed the phase-encoding direction. Thus, spin phases will vary linearly over the phase-encode direction (Fig. 7.17b).

Frequency-Encoding Gradient

The GF is applied following and perpendicular to the phase-encoding gradient. It causes the protons to rotate at different frequencies according to their relative position along the gradient, permitting differentiation of the signal in a third plane (Fig. 7.17c).

Two additional gradient pulses are typically used in this three-step process, one immediately after the slice-selection gradient and the other immediately before the GF. These pulses are used to counteract any de-phasing of the transverse magnetization that may be caused by the imaging gradients and ensure that maximum echo (sampling signal) is achieved.

It should be emphasized that RF pulses excite all protons in a slice simultaneously and that a single echo signal is recorded from the entire slice for one phase-encoding step. Thus, to acquire sufficient phase-encoding information for a signal to be assigned to each location within the slice, the pulse sequence (comprising slice selection, frequency encoding, and phase encoding) is repeated many times. During each repetition, the same slice selection and frequency encoding are performed but the strength of the phase-encoding gradient is increased by equal increments. Each repetition of the phase-encoding step generates a signal echo that is digitized and stored in a raw data matrix called ‘k-space’. Data points in k-space represent the spatial frequencies content of an MRI. Data in k-space are converted into an image using a mathematical tool called a Fourier transform.

Image Acquisition Time

As each phase-encoding step requires a new pulse sequence, the total image acquisition time will depend on the product of TR (time interval between pulse sequences) and NP (number of phase-encoding steps). Conventional pulse sequences such as SE and GRE acquire only one phase-encoding step (one line of k-space) per TR, making the image acquisition time considerably long. This limitation is overcome by faster imaging techniques that acquire multiple lines of k-space per TR. The so-called Turbo (or fast) SE and GRE techniques are now commonplace in MRI.

Turbo SE

The turbo SE (TSE) sequence applies multiple 180° pulses after an initial 90° pulse (as opposed to the single 180° pulse used in a conventional SE sequence). Each 180° pulse generates an echo, which (after phase encoding) is used to fill a new line of k-space. The number of 180° refocusing pulses (and thus, the number of echoes generated) during one TR of the TSE sequence is termed the echo train length (ETL), echo factor, or turbo factor. ETL specifies the factor by which the pulse sequence is accelerated (e.g., an ETL of 16 means that the TSE sequence is 16 times faster than a conventional SE sequence).

The signal amplitude of each successive echo generated by the TSE sequence will be smaller than the last by way of T2 decay effects. The echoes will also have different TEs. This means that in the traditional sense, a single TE does not exist for the TSE sequence. Instead, an effective TE or pseudo TE is ascribed. The effective TE is defined as that of the echo, which is acquired closest to the center of k-space (with the smallest GP) as this is the echo that has the greatest influence on image contrast.

Echo-Planar Imaging

Echo-planar imaging (EPI) is a fast MRI technique in which the rapid oscillation of high-amplitude gradients is used to generate multiple GREs per TR. In single-shot EPI, all of the phase-encoding steps are acquired in a single TR. Multi-shot EPI uses a few TRs to acquire all phase-encoding steps.

3D Encoding

Although the selection of a set of slices encodes data in all three dimensions in space, the slice thickness (often several millimeters) is often much greater than the in-plane pixel resolution (often <1 mm). The initial slice-encoding step can be replaced by a further (termed secondary) phase-encoding iteration, along what was the slice-encoding direction. Instead of a set of slices (that may include slice gaps!), the inclusion of secondary phase encoding often produces a true 3D dataset comprised of voxels (3D pixels) whose dimensions can be <1 mm $\times$ 1 mm $\times$ 1 mm. This is often useful when user manipulation of data in all three perpendicular or oblique planes is necessary (for example, for intraoperative surgical guidance). Performing more phase-encoding steps has a cost, and takes time. As a result, the majority of true 3D sequences tend to be GREs with short TR or TSEs with high ETL factors.

Magnetic Susceptibility

Magnetic susceptibility describes the extent to which a substance becomes magnetized when placed in an external magnetic field. It results primarily from the interaction between electrons within the substance and the external magnetic field. More specifically, orbital and delocalized electrons within the substance produce circulating currents in response to the field. These currents generate internal magnetization (Mi) within the substance that either augments or opposes the local external magnetic field. The result is either positive susceptibility, whereby the magnetic field local to the substance is increased, or negative susceptibility in which the magnetic field local to the substance is decreased. Substances with negative susceptibility are termed diamagnetic; those with positive susceptibility are called paramagnetic.

Most biological tissues including water cause very weak diamagnetism and as such induce negligible susceptibility effects.

Conversely, matter such as iron, nickel, and cobalt show extreme positive susceptibility and are termed ferromagnetic.

Superparamagnetic is used to describe a substance with positive magnetic susceptibility that lies between paramagnetism and ferromagnetism (for example, small ferrous sulphate particles).

Intravenous Contrast

Gadolinium, a rare-earth metal and paramagnetic substance, is used as an intravenous MR contrast medium. Its strong paramagnetic properties result from the numerous unpaired electrons that exist within its inner shells. These unpaired electrons can interact with adjacent resonating protons and cause the protons to relax more rapidly. This results in a shortening of both longitudinal and transverse relaxation with a consequent reduction in the T1 and T2 values of the tissues in which it accumulates. This is depicted by an increase in signal in T1-weighted images and a decrease in signal in T2-weighted images. In practice, the signal increase detected in T1-weighted imaging is better appreciated than any corresponding signal decrease in T2-weighted imaging and makes T1-weighted imaging the method of choice for intravenous contrast studies.

Furthermore, as gadolinium-based contrast agents shorten T1, it is possible to also shorten TR, and hence, reduce overall scan time (Fig. 7.15d).

Exploiting Magnetic Susceptibility Using GRE Imaging [12]

Susceptibility differences among tissues and materials are more conspicuous on T2*-weighted GRE rather than SE imaging (recall that susceptibility differences promote magnetic field inhomogeneity that result in faster T2* relaxation, leading to signal intensity loss on GRE images; conversely, the 180° refocusing pulse in SE compensates for this). GRE sequences can be made more sensitive to T2* decay through the manipulation of imaging parameters such as TR, TE, and the flip angle. As T2* decay commences with the excitation pulse and progresses with time, a longer TE will result in greater signal loss.

A low flip angle reduces the influence T1 has on the subject matter (as longitudinal magnetization remains close to the fully relaxed state) and allows the T2* effects to dominate. Similarly, T1 effects are limited with a long TR. Hence, T2* GRE imaging is obtained at long TE, long TR, and low flip angle. T2* sensitivity also increases with field strength and voxel size. T2*-weighted sequences are particularly useful in the detection of hemorrhage and blood products (deoxyhemoglobin, methemoglobin, and hemosiderin are paramagnetic). Various pathologies may be detected using this imaging modality including arteriovenous malformations, cavernomas, superficial siderosis, intraparenchymal/intratumoral hemorrhage, and amyloid angiopathy (Fig. 7.15f). The literature also supports the use of T2*-weighted imaging for the differentiation of vestibular schwannomas from meningiomas as in most schwannomas, microhemorrhages can be detected.

Other Factors That May Influence Signal Intensity [12]

TI: The Inversion Recovery Sequence

The inversion recovery sequence uses a 180° pulse followed by a 90° pulse. The 180° pulse inverts longitudinal magnetization so that the sum magnetization aligns antiparallel to B_0.

Following the 180° pulse, the inverted longitudinal magnetization begins to shorten and return to its original position.

Application of a 90° pulse causes the longitudinal magnetization to flip to the transverse plane and allows a signal to be recorded. The strength of the signal is dependent on the magnitude of longitudinal magnetization just before the 90° pulse was applied.

Hence, the signal obtained by inversion recovery is dependent on the T1 properties of the tissues being examined. Inversion time (TI) is the time interval between the 180° and the 90° pulses (Fig. 7.18a).

From Fig. 7.18: (A) depicts the Inversion of the sum longitudinal magnetization vector following a 180° radiofrequency (RF) pulse. A further 90° RF pulse flips the recovering sum vector into the transverse plane allowing a signal to be obtained. The magnitude of the signal is dependent on the amount of T1 recovery that had occurred before the 90° pulse, and hence, on the inversion time (time between the 180° and 90° pulses). (B) depicts the T1 relaxation curves for fat and water demonstrating the recovery of longitudinal magnetization following a 180° RF pulse. Application of a 90° RF pulse at the null point leads to signal suppression of a chosen tissue.

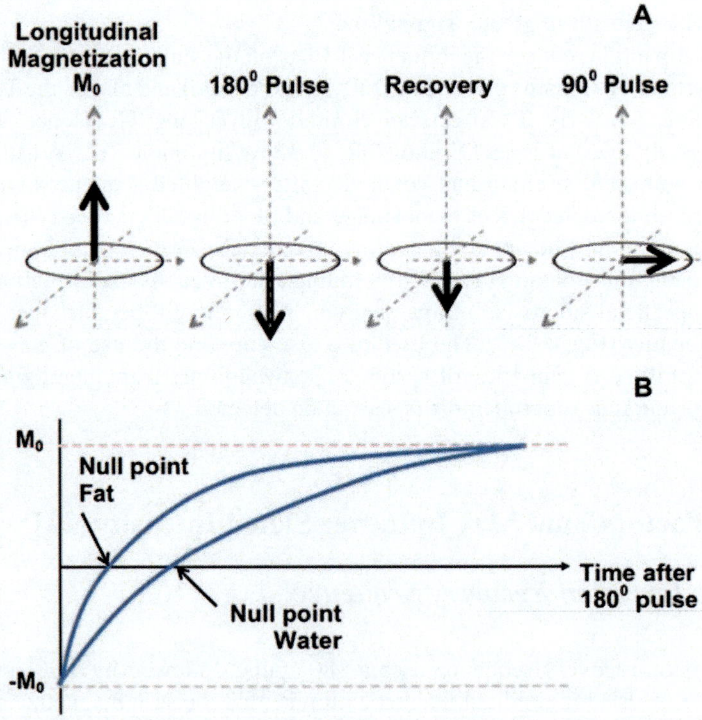

Fig. 7.18 Inversion recovery

Following the 180° pulse, there will be a time when longitudinal magnetization passes through the transverse plane.

Application of a 90° pulse at this point will result in a signal void (the protons will be flipped out of the orientation at which their signal can be recorded). Accordingly, this point is termed the null point. Knowledge of the null point for a particular tissue can be used to suppress the signal from that tissue (Fig. 7.18b). Fluid-attenuated inversion recovery (FLAIR) sequences are frequently used in neuroimaging to suppress the signal from cerebrospinal fluid (Fig. 7.15d). This has the particular advantage of making periventricular lesions much more conspicuous. As a further bonus, the combination of long TE and a TI can provide both T2-weighted contrast (aiding the delineation of, for example, ischemia or inflammation) and T1-weighted contrast (showing gadolinium-chelate deposition where there is a leaky blood–brain barrier). Gadolinium-enhanced FLAIR sequences are being used with increasing frequency in neuroscience centers. They can be particularly useful in the detection of subtle tumors or for assessing multiple sclerosis. Short TI inversion recovery is used commonly throughout radiology subspecialties to suppress fat signal. Other fat-suppression techniques are available.

Flow [12]

When imaging a slice through a vessel containing flowing blood, the blood within the slice at the time of an excitatory RF pulse will be influenced by the radio wave. However, by the time the MR signal is sampled, the blood influenced by the radio wave will have been replaced by blood flowing into the slice that has not undergone magnetization. Thus, no signal will be obtained from the vessel and it will appear black on the image: the so-called flow-void phenomenon.

This is not the only way in which flow influences signal intensity and merely constitutes an introduction as in reality, the subject is far more complex. For example, a signal may still be obtained from vessels that contain slow-flowing blood and this to the inexperienced looking at a single sequence of images may be interpreted as thrombosis.

Finally, a Word on Safety

Safety issues regarding clinical MRI can be broadly divided into those that result from the exposure to magnetic fields and those resulting from intravenous contrast administration.

Exposure to Magnetic Fields

Each of the three types of magnetic fields (static, gradient and RF) that the patient is exposed to during an examination carries a potential risk to patient safety. With regard to the static magnetic field (B_0), the main safety issue involves the attraction of ferromagnetic material towards the magnet. Material external to the patient (such as keys or coins) may become projectiles within the scanning room, putting the patient and staff (and machinery) at risk of being hit. Static magnetic fields can also exert mechanical forces on ferromagnetic components within implanted medical devices such as pacemakers, aneurysmal clips, and cardiac defibrillators. Devices may rotate or dislodge, potentially resulting in serious patient harm. Higher static magnetic fields lead to greater forces on ferromagnetic materials. Control of access to the magnet room is crucial in minimizing associated risks.

Excessive noise and potential auditory damage are the main concerns with gradient fields. Rapid alterations of currents within the gradient coils, in the presence of a strong magnetic field, generate significant force. This force acts upon the gradient coils and manifests as loud tapping or knocking sounds. Temporary hearing loss has been reported using conventional sequences. Due to the risk of damage caused by acoustic noise, it is common practice for the patient to wear protective ear-wear (often headphones and earplugs). Gradient magnetic fields also have the potential to induce voltages on pacing wires that may cause over-sensing and under-sensing. Excitation of peripheral nerves has also been reported.

RF pulses may lead to local tissue heating through the dissipation of energy. Typically, this is negligible (<1 °C) in clinical, but is potentially serious in patients with implantable devices. A pacemaker lead, for example, may act as an antenna and concentrate RF energy, resulting in dramatic local heating, tissue damage, and cardiac dysfunction.

A device must have undergone prior testing within MR electromagnetic fields for it to be deemed safe. Booklets listing MR-compatible objects/devices are available and are typically held in the MR scanner room. More and more MR-compatible medical devices and equipment are being generated. Most tertiary centers now have MR-safe anesthetic equipment and recent studies have reported favorably on the early outcome of MR-compatible pacemakers.

Intravenous Contrast Agents

Gadolinium-based contrast agents are generally well tolerated. A recent study suggests an adverse reaction rate of 0.46%. A rare but serious complication that may occur in patients with severe renal impairment is nephrogenic systemic fibrosis.

Although a systemic condition, cutaneous manifestations tend to dominate with a lower-limb cellulitic-like picture occurring early. Painful and pruritic, plaque-like 'woody' induration of the skin occurs later in the course of the disease with the legs more commonly affected than the arms and torso. The face is spared. These cutaneous manifestations may lead to contractures and limit mobility. Nephrogenic systemic fibrosis may also affect the liver, lungs, muscles, heart, and nerves. Death rate may be as high as 28%. Despite ongoing research, the exact mechanisms underlying the condition remain unclear. No universally effective treatment has been identified and prevention (mainly through limiting the use of high-dose gadolinium-based contrast agents and/or maximizing renal function in patients with renal failure) represents current recommendations.

Anaphylaxis resulting from gadolinium-based contrast agents is rare, with an estimated incidence of 0.02%. This compares favorably to the incidence with iodinated contrast media used with CT (0.03–0.16%). Patients with a pre-existing allergy to other contrast agents (such as iodine) may confer a greater risk of anaphylaxis to gadolium-based contrast agents, thus highlighting the need for a thorough screening questionnaire prior to any contrast administration.

Pregnancy and Breastfeeding

Large doses of gadolinium-based contrast agents have been shown to cause post-implantation fetal loss, retarded development, increased locomotive activity, and skeletal and visceral abnormalities in animal experiments. Any deleterious effects to humans are unconfirmed. However, gadolinium-based contrast agents can cross the placenta and appear in the fetal bladder from where they are excreted but enter the fetal system again through the swallowing of amniotic fluid. This makes it difficult to ascertain exact fetal-life measurements for these agents. Currently, it is recommended that gadolinium-based contrast should only be administered after a careful risk–benefit assessment.

The amount of contrast absorbed by the gastrointestinal tract of an infant, following breastfeeding, is believed to be negligible (<1% of the recommended intravenous infant dose) and as such no additional precautions are thought necessary for breastfeeding mothers. Some centers advise informed consent, however, with an option to stop breastfeeding (with continued expression and discarding of the milk) for 24 h following maternal administration.

Overview of Functional Magnetic Resonance Imaging [8]

Blood oxygen level-dependent (BOLD) functional magnetic resonance imaging (fMRI) is a noninvasive technique for studying brain activation. It depicts changes in deoxyhemoglobin concentration consequent to task-induced or spontaneous modulation of neural metabolism [57–59]. It measures changes in blood oxygenation and blood flow related to neuronal activity, providing the means to study human brain function in vivo, either in response to a certain task or when at rest [8].

Functional magnetic resonance imaging (fMRI demonstrates regional, time-varying changes in brain metabolism [63–65]. These metabolic changes can be consequent to task-induced cognitive state changes or the result of unregulated processes in the resting brain. It is used as a biomarker for disease, to monitor therapy, or for studying pharmacological efficacy.

Concept of Functional Magnetic Resonance Imaging (fMRI) [8]

Functional magnetic resonance imaging (fMRI) is based on magnetic resonance imaging (MRI), which in turn uses the basis of the nuclear magnetic resonance (NMR) coupled with gradients in magnetic field to create images that can incorporate many different types of contrast such as T1 weighting, T2 weighting, susceptibility, flow, etc. [8, 38, 67, 68]

The data acquired in an fMRI study consists of a sequence of 3-D magnetic resonance images (MRIs), each made up of a number of uniformly spaced volume elements, or voxels. The voxels partition the brain into a large number of equally sized cubes. A typical image may consist of roughly 100,000 voxels, where the image intensity value corresponding to each voxel represents the spatial distribution of the nuclear spin density, which relates to blood oxygenation and flow, in the local area. During an fMRI experiment, 100–1000 such 3-D images of the whole brain are acquired. In addition, a standard fMRI experiment consists of multiple subjects (e.g., 10–50), potentially brought in for multiple scanning sessions, each consisting of a number of replications of a certain experimental task [8, 26, 27, 30].

All the processes of neural signaling in the brain, including formation and propagation of action potentials, binding of vesicles to the pre-synaptic junction, the release of neurotransmitters across the synaptic gap, their reception, and regeneration of action potentials in the postsynaptic structures, scavenging of excess neurotransmitters, etc., require energy in the form of adenosine triphosphates (ATP). The nucleotide is produced principally by the mitochondria from glycolytic oxygenation of glucose, and its production results in carbon dioxide as a byproduct [8, 48, 50, 51]. When a region of the brain is upregulated (i.e., activated) by a cognitive task such as finger tapping, the additional neural firing, and other increased signaling processes result in a locally increased energy requirement, in turn resulting in upregulated cerebral metabolic rate of oxygen (CMR02) in the affected brain region. As the local stores of oxygen in tissues adjacent to capillaries are transiently consumed by glycolysis and waste products build up, various chemical signals (CO_2, NO, H+) cause a vasomotor reaction in arterial sphincters upstream of the capillary bed, which results in the dilation of these vessels. The increased blood flow acts to restore the local [O_2] level required to overcome the transient deficit [8, 32, 33]. However, for reasons that are still not fully understood, more oxygen is delivered than is needed to offset the increase in $CMRO_2$. As a result, neural upregulation results initially in a build-up of deoxygenated hemoglobin [Hb] and a decrease in deoxygenated hemoglobin [HbO_2] in the intra- and extravascular spaces, followed within a second or two by a vasodilatory response that reverses the situation to result in an increase in [HbO_2] and decrease in [Hb] over that in the resting condition. This sequence of processes is described as the hemodynamic response to the neural event [8, 52, 53].

Thus, there are two primary consequences of increased neural activity, and both can be detected by MRI. These consequences include increased local cerebral blood flow (CBF), and changes in oxygenation concentration (blood oxygen level dependent, or BOLD contrast). The change in CBF can be observed using an injected contrast agent and perfusion-weighted MRI, or noninvasively by arterial spin labeling (ASL). However, ASL suffers from reduction in sensitivity, increased acquisition time, and increased sensitivity to motion as compared with the BOLD contrast method, and its use has therefore centered on obtaining quantitative measurements of baseline cerebral blood flow (CBF) for studies involving the modeling of the neurobiological mechanisms of activation, or the calibration of the vasoreactivity, rather than in routine mapping of the brain function.

The second mechanism, termed blood oxygen level dependent (BOLD) contrast, was first demonstrated in rats and later in humans. It constitutes the contrast that is used in virtually all conventional fMRI experiments. BOLD contrast results from the change in magnetic field surrounding the red blood cells depending on the oxygen state of the hemoglobin [8, 37, 39, 41, 43]. When fully oxygenated, HbO_2 is diamagnetic and is magnetically indistinguishable from the brain tissue. However, fully deoxygenated Hb has four unpaired electrons and is highly paramagnetic. This paramagnetism results in local gradients in the magnetic field whose strength depends on the [Hb] concentration. These endogenous gradients in turn modulate the intra- and extra-vascular T2 and T2* relaxation times of the blood through diffusion and intra-voxel de-phasing, respectively. Using a gradient refocused echo (GRE) MRI pulse sequence, the acquisition is made sensitive to T2* and T2. At 1.5 T and 3 T, the T2* contrast is predominant and is largest in venules, while at higher field strength, the diffusion-weighted contrast of T2 relaxation becomes more important, and because signals are generated preferentially in capillaries and tissue with spin-echo acquisitions, provides greater spatial specificity. Since most fMRI is currently performed at 3.0 T or below, BOLD fMRI utilizes primarily GRE methods because of the increased T2* contrast.

Task activation fMRI studies seek to induce different neural states in the brain as the visual, auditory, or other stimulus is manipulated during the scan, and activation maps are obtained by comparing the signals recorded during the different states [42, 44, 60]. Therefore, it is important to collect each image in a snapshot mode to avoid head motion and physiological processes of respiration and cardiovascular functions from injecting noise signals unrelated to the neural processing being interrogated.

Generally, most fMRI is performed using an echo planar imaging (EPI) method, which can collect data for a two-dimensional image in approximately 60 ms at typical resolutions (3.4 × 3.4 × 4 mm3 voxel size). Typically, whole brain scans with approximately 32 2D slices are acquired with a repetition time (TR) of 2 s/volume. Each voxel in the resulting scan produces a time series that is subsequently analyzed in accordance with the task design.

Analysis Methods [8]

Once the images have been acquired, the time series data must be processed to obtain maps of the brain activation. Because the BOLD contrast is small (<1% in many studies of higher cognitive processes), simply averaging images over the experimental and control conditions and then subtracting is inadequate to reliably determine differences because noise will compete to render false positives and negatives. The noise results from thermal sources in the subject and electronics, bulk motion of the head, cardiac and respiratory-induced noise, and variations in baseline neural metabolism. Because the noise can sometimes be larger than the signal of interest, fMRI analyses compare the signal difference between the states using a statistical test. These tests result in an activation map that is a function of the probability that the brain states differ. The statistical test for activation can utilize a general linear model (GLM), cross-correlation with a modeled regressor, or one of several data-driven approaches such as independent components analysis (ICA). The models against which the acquired data are tested include the experimental design of interest as well as "nuisance regressors" of no interest such as signal drift, motion, and noise reflected in global or white-matter signals. In all cases, the activation testing is preceded by a series of preprocessing steps.

The steps in pre-processing may include all or some of the following; (1) time-slice correction, to eliminate differences between the time of acquisition of each slice in the volume; (2) motion co-registration, in which affine head motion is detected and the time series of the volumes is resampled to register each time frame to a reference frame, such as the first or middle time series point; (3) correction for physiological noise from breathing and cardiovascular function, low pass and/or high pass temporal filtering to improve the statistics while removing spectral components of no interest; (4) spatial smoothing to improve the signal to noise ratio (SNR) and improve the normality of the noise distribution; and (5) pre-whitening to correct for autocorrelation in the time series.

Comparisons with Other Functional Imaging Modalities [8]

Functional magnetic resonance imaging (fMRI) may be compared to other imaging methods used to obtain functional assessment of the brain metabolism in terms of spatial and temporal resolution and availability. The primary alternatives are positron emission tomography (PET), near infrared spectroscopy (NIRS), electroencephalography (EEG), and magnetoencephalography (MEG).

Spatial Resolution [8]

Resolution in fMRI is limited primarily by SNR because of the necessity for rapid acquisition of time series information. For MRI, where p is the pixel size, w the slice thickness, *Tacq* is the k-space readout time and N is the number of time frames. Thus, as *Tacq* is reduced for single-shot imaging (typically 20–30 ms) the pixel size must be increased over that for conventional anatomic imaging to maintain an acceptable SNR. Accordingly, the typical fMRI pixel size is 3–4 mm, although with higher field magnets (7 T) a pixel size of 500 microns or less may be readily achieved. The resolution of PET is limited by the size of the gamma ray detectors as well as the positron-electron annihilation range, and is typically either less or equal to 5–10 mm (≥5–10 mm). NIRS resolution is low (10–20 mm) and limited predominantly by the strong scatter and attenuation of IR photons (which also limits the depth of the cortex that can be imaged within a banana-shaped region connecting optodes), the modest density of optodes and the ill-conditioned inverse problem of reconstructing 3D maps of [Hb] from scalp recordings. The resolution in EEG and MEG is similarly limited to less than 10–20 mm (>10–20 mm) by the fact that a unique reconstruction of dipoles is not possible from the scalp-based measurements of electrical or magnetic distributions and models, and regularization must be employed for the model estimation. Unlike EEG, MEG does not have the confound that scalp recordings may be spatially distorted by heterogeneous electrical conduction paths within the brain/skull.

Temporal Resolution [8]

The temporal resolution of the fMRI is limited by the hemodynamic response time. Typically, the BOLD response has a width of approximately 3 s and a peak occurring between 5 and 6 s (~5–6 s) after the onset of a brief neural stimulus. This is much slower than the underlying neural processes, and temporal information is thereby heavily blurred. Nevertheless, by jittering event-related stimuli and using appropriate analysis methods, temporal inferences in the 100 ms resolution range may be achieved. PET scans require minutes to complete because of the low count rates of injected radio nuclides, so changes in neural processes may only be studied by repeated scanning. Like BOLD, NIRS reports changes in blood oxygenation and exacerbated by low SNR of NIR photons in the brain. It exhibits temporal limitations similar to those of fMRI. EEG and MEG, on the other hand, have millisecond temporal resolution and may easily capture the dynamics of the evoked responses that last a few milliseconds to several hundred milliseconds. Multimodal approaches combining fMRI and EEG use fMRI maps as spatial priors to reconstruct high temporal resolution electrophysiology, thereby gaining resolution in both dimensions.

Strengths and Weaknesses of fMRI [8]

From the discussion above, a primary strength of fMRI is its relatively high spatial resolution and availability. In addition, it is readily available to both clinical and academic researchers. It is noninvasive and can provide high-resolution anatomic scans in the same session to use for localization, vessel identification, or development of maps of white matter connectivity through the use of diffusion tensor imaging (DTI).

Because BOLD contrast derives from the sluggish hemodynamic response to metabolic changes, a significant weakness is its low temporal resolution. Another problem involves its signal dropout and/or spatial distortion in the frontal orbital and in the lateral parietal regions, caused by the less than 9 ppm difference in magnetic susceptibility at the interfaces between air and brain tissue. This can result in erroneous lack of BOLD signal in the ventral, temporal, and PFC regions that derive importance in many cognitive studies. Many methods have been developed to diminish these susceptibility losses, although most of them involve some tradeoff of SNR in magnetically uniform brain regions. Other weaknesses include the loud noise of the scanner associated with the switched magnetic fields, which may cause confounds in studies of audition and resting state networks. However, methods using interleaved scanning/stimulus delivery epochs may avoid these problems, albeit with some loss of flexibility in experimental design. Finally, the high magnetic fields require customized stimulus delivery and subject response systems, again limiting flexibility and complicating multimodal experiments such as concurrent EEG recording.

Future of fMRI [8]

For the most part, the MRI physics and technology development behind BOLD fMRI acquisitions are mature, and the tradeoffs between acquisition speed, resolution, SNR, signal dropout, and contrast are well understood. Over the years, a number of investigators have attempted to develop alternatives to BOLD contrast using direct neural current detection, although by now it is understood that the weak size of the neural current signal relative to physiological noise makes a breakthrough unlikely. Another alternative is the use of diffusion-weighted imaging to demonstrate activation-related changes in populations of bound versus free water distributions. A potential advantage is that such diffusion-related changes may have more rapid responses than BOLD methods. However, again the signals are weaker than BOLD contrast and their biophysical origin is still unclear. Others have reported the use of the spin echo rather than the gradient echo acquisitions of BOLD contrast, especially at higher fields where T2* is foreshortened.

While a modest research effort will continue in improving acquisition technology, the bulk of research in the development of fMRI has shifted to its application to answering more complex questions in cognitive neuroscience. One promising area is that of using activation maps as input to the classification and the state change algorithms to predict or classify cognitive behavior, such as predicting the brain states. Other emerging uses of fMRI include the development of quantitative measures, that means, biomarkers for disease or the monitoring of behavioral modifications such as reading disorders. A cautionary note, however, suggests that because of the small BOLD responses typical of cognitive processes, most studies are limited to employing group statistics to make inferences about populations rather than about individuals. Thus, the use of fMRI in quantifying individual characteristics may continue to be limited to those tasks for which relatively strong BOLD responses are observed, such as primary sensory systems. Resting state networks and their modification by disease conditions such as Alzheimer's, depression, and other psychiatric disorders are gaining attention. However, there is growing awareness that these networks may be much more complex in their spatiotemporal dynamics than previously thought, and much more work is indicated to understand their role and utility in predicting individual behavior/physiology. Finally, feedback derived from real-time fMRI has been shown to allow subjects to learn pain-reduction strategies, enhance sensorimotor control, and to control relevant brain regions in mood disorder experiments.

Conclusions [8]

Functional magnetic resonance imaging (fMRI) has enjoyed an exciting development course with an exponential growth in published studies since its inception in the early nineties. It has become commonplace for clinical uses such as pre-surgical planning, fundamental cognitive neuroscience investigations, behavior modification, and training. Informed by fMRI, more sophisticated modeling of the brain networks is certain to lead to new levels of understanding of the human brain.

Intervention in the Brain: Current Understanding and Technology Initiatives

Currently, just a few categories of technology initiatives towards interventions in the brain are known.

Noninvasive Brain Stimulation

Noninvasive brain stimulation has been seen more commonly in rehabilitation settings. It may be used for the treatment of stroke, spinal cord injury, traumatic brain injury, and multiple sclerosis, as well as for some diagnostic neurophysiological measurements. Two major modalities of noninvasive brain stimulation include the transcranial magnetic stimulation (TMS) and the transcranial direct current stimulation (tDCS). As an add-on therapy to the conventional rehabilitative treatments, the main goal of noninvasive brain stimulation is to create neuromodulation through the inhibition or activation of neural activity in the targeted cortical region. Indications for the therapeutic noninvasive brain stimulation in neurorehabilitation include motor recovery, aphasia, neglect, dysphagia, cognitive disorders, spasticity, and central pain [3].

Basic Principles of Noninvasive Brain Stimulation

The noninvasive brain stimulation consists of neuro-stimulation in a target area of the brain through the application of an electrical current directly or by creating an electrical field with magnetic induction on the scalp. Measurable outputs (motor-evoked potential, central transmission time, cortical silent period, intra-cortical inhibition, intra-cortical facilitation) are obtained with neuro-stimulation. In this way, it may be used for diagnostic purposes [3, 28, 29, 31]. In repeated applications, neuromodulation is made by increasing or decreasing the cortical excitability. Repeated application of noninvasive brain stimulation is used for therapeutic purposes.

The modalities used for noninvasive brain stimulation include:

- Transcranial brain stimulation (TBS)
- Transcranial magnetic stimulation (TMS)
- Transcranial direct current stimulation (tDCS)
- Transcranial alternating current stimulation (TACS)
- Deep brain stimulation (DBS)
- Brain-Computer Interfaces (BCIS)
- Neural stem cell therapy (NSCT)

Transcranial Brain Stimulation (TBS)

The transcranial brain stimulation (TBS) involves a noninvasive technology initiative in which the brain is stimulated either by inducing an electrical field using a magnetic coil placed against the head (transcranial magnetic stimulation, TMS) or by applying weak electrical currents via electrodes on the scalp (transcranial direct current stimulation, tDCS) and transcranial alternating current stimulation, (TACS).

Transcranial Magnetic Stimulation (TMS)

Transcranial magnetic stimulation (TMS) is a noninvasive tool for the electrical stimulation of the neural tissue, including the cerebral cortex, the spinal roots, and the cranial and the peripheral nerves [88–90].

Transcranial magnetic stimulation involves placing an electromagnetic coil on the scalp [45–47, 79]. The current flows briefly in the coil and generates a changing magnetic field that induces an electric current in the tissue, in the opposite direction (Fig. 7.19).

High-intensity current is rapidly turned on and off in the coil through the discharge of capacitors. This produces a time-varying magnetic field that lasts for about 100–200 microseconds. The magnetic field typically has a strength of about 2 T (40,000 times the magnetic field of the earth, or about the same intensity as the static magnetic field used in clinical magnetic resonance imaging). The proximity of the brain to the time-varying magnetic field results in current flow in the neural tissue.

A striking effect of TMS occurs when one places the coil on the scalp over the primary motor cortex [34–36, 89]. A single TMS pulse of sufficient intensity causes

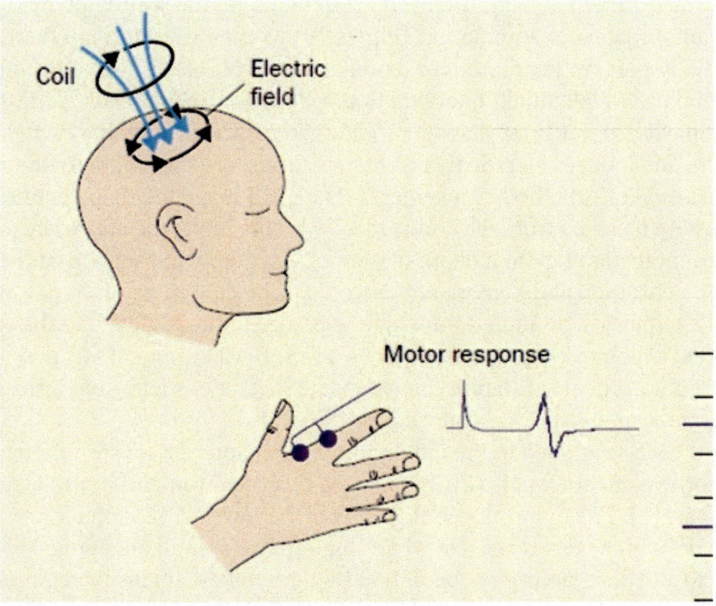

Fig. 7.19 Basic principle of transcranial magnetic stimulation

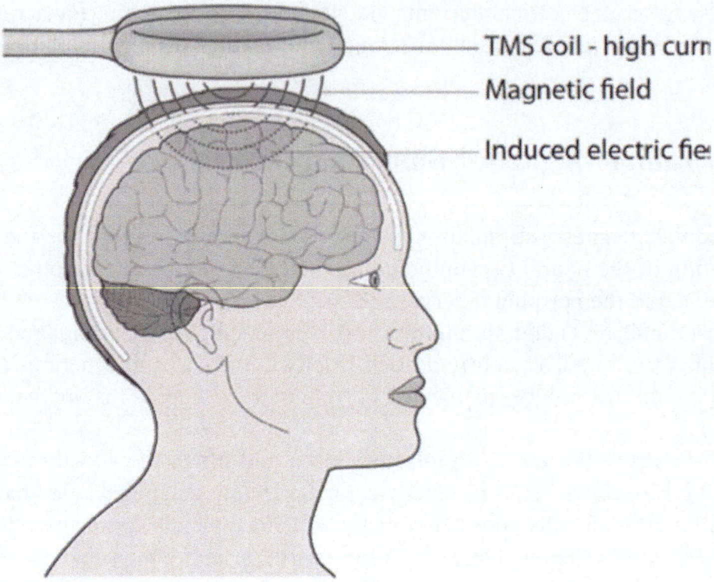

Fig. 7.20 Concept of the TMS

involuntary movement. The magnetic field intensity needed to produce motor movement varies considerably across individuals, and is known as the motor threshold. Placing the coil over different areas of the motor cortex causes contralateral movement in the different distal muscles, corresponding to the well-known homunculus. Transcranial magnetic stimulation (Fig. 7.20) may be used to map the representation of body parts in the motor cortex on an individual basis [40, 49]. Subjectively, this stimulation feels much like attention reflex movement. Thus, TMS pulse produces a powerful but brief magnetic field that passes through the skin, soft tissue, and skull, and induces electrical current in neurons, causing depolarization that then has behavioral effects (body movement). The TMS magnetic field declines logarithmically with distance from the coil [3, 54–56]. This limits the area of depolarization with current technology to a depth of about 2 cm below the surface of the brain.

Some neurological disorders may involve or be caused by an impairment of the cortical excitability or altered interactions between the cortical and the subcortical structures, which may be detected by the TMS. Furthermore, TMS may be used to modify intra-cortical excitability and activate distant cortical, subcortical, and spinal structures along specific connections [87, 89].

When TMS is applied to the motor cortex at an appropriate stimulation intensity, the motor-evoked potentials (MEPs) may be recorded from the contralateral extremity muscles [3]. The motor threshold refers to the lowest TMS intensity necessary to evoke MEPs in the target muscle when single-pulse stimuli are applied to the motor cortex. Motor threshold may be defined as the lowest intensity required to elicit

MEPs of more than 50 microvolts (50 μV) peak-to-peak amplitude in at least 50% of successive trials, in resting or activated (slightly contracted) target muscles. The motor threshold is believed to reflect membrane excitability of the corticospinal neurons and the interneurons projecting onto these neurons in the motor cortex, as well as the excitability of the motor neurons in the spinal cord, neuromuscular junctions, and muscles. In addition to the membrane excitability itself, the motor threshold must also relate to the activity of the neural inputs into the pyramidal cells that affect their membrane excitability (that means, tonic inhibitory and excitatory drives onto the cortical output neurons).

Transcranial magnetic stimulation (TMS) may be used to measure various parameters in the motor cortex, allowing the evaluation of different aspects of the cortical excitability [3]. Such measures are useful in understanding the changes in the brain physiology seen in the setting of the cortical plasticity and brain disorders.

Repetitive Transcranial Magnetic Stimulation (rTMS)

Repetitive transcranial magnetic stimulation (rTMS) is a noninvasive neurostimulation and neuromodulation technique that is delivered as a series of pulses, that means, a train. The most typical technical parameters of rTMS involve the frequency (high-frequency; 1) stimulation delivered greater than one pulse per second (>1 pulse per second), but generally is either greater or equal to 5 Hz (≥5 Hz) is applied as High-Frequency, or 2) low-frequency stimulation delivered either below or equal to one pulse per second (≤1 pulse per second), intensity (expressed as a percentage of the resting motor threshold generally set at 100–120%), train duration, inter-train interval, number of trains per session, and number of pulses per session, respectively.

Repetitive TMS (rTMS) is used to induce changes in the brain activity that may last beyond the stimulation period. Contrary to the single-pulse TMS, rTMS is able to change and modulate the cortical activity beyond the stimulation period, as a potential method for the treatment of neurological and psychiatric disorders [78, 81].

Transcranial Direct Current Stimulation (tDCS)

Transcranial direct current stimulation (tDCS) of Fig. 7.21, constitutes a noninvasive method, allowing for the reversible modulation of activity in particular brain regions. This has provided a valuable tool for establishing brain-behavior relationships across a variety of cognitive, motor, social, and affective domains, and in healthy populations, it has been shown to temporarily modify behavior, accelerate learning, and boost task performance. Transcranial direct current stimulation (tDCS) specifically targets areas of the brain through low levels of constant electric currents

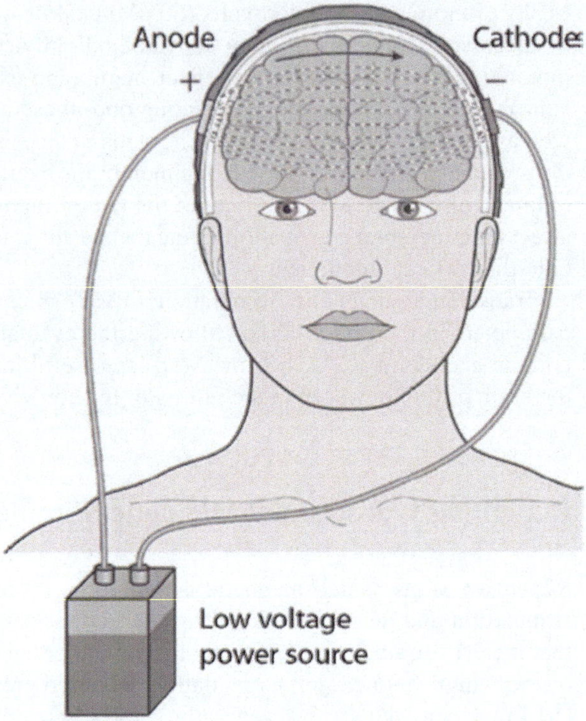

Fig. 7.21 Principle of the TDCS

[81, 82, 86, 94]. TDCS devices, as illustrated in Fig. 7.21, deliver cathodal (negative) stimulation or anodal (positive) stimulation via small electrodes placed on different regions of the scalp and are thought to influence the rate of neuronal activity [91–93, 95, 96].

Transcranial Alternating Current Stimulation (TACS)

Transcranial alternating current stimulation (tACS) is a unique form of noninvasive brain stimulation. Sinusoidal alternating electric currents are delivered to the scalp to affect mostly the cortical neurons. The transcranial alternating current stimulation (tACS) seems likely to open a new era in the field of noninvasive electrical stimulation of the human brain by directly interfering with the cortical rhythms [97, 100]. It is expected to synchronize (by one single resonance frequency) or desynchronize (for example, by the application of several frequencies) the cortical oscillations. If applied long enough, it may cause neuro-plastic effects. In the theta range, it may improve cognition when applied in phase. Alpha rhythms could improve the motor performance, whereas the beta intrusion may deteriorate them [81, 83, 84]. TACS with both alpha and beta frequencies has a high likelihood to induce retinal phosphenes. Gamma intrusion may possibly interfere with attention. Stimulation in the "ripple" range induces intensity-dependent inhibition or excitation in the motor cortex (M1) most likely by the entrainment of the neuronal networks, whereas stimulation in the low kHz range induces excitation by the neuronal membrane interference. TACS in the 200 kHz range may have potential in oncology.

Transcranial alternating current stimulation (TACS) devices deliver neurostimulation to specifically targeted areas of the brain with alternating electoral currents. The alternating currents are thought to change the rate of abnormal neuronal activity that may be related to a variety of behavioral and cognitive conditions.

Transcranial alternating current stimulation (tACS) produces acute perceptual, motor, and cognitive changes in humans [83–85].

TDCS and TACS are noninvasive, painless treatments that are being evaluated for a variety of conditions including, chronic pain, brain and spinal cord injury, Parkinson's disease, and psychiatric conditions such as major depressive disorder, substance use disorders, and anxiety [97–99, 101].

Deep Brain Stimulation (DBS)

Deep brain stimulation (DBS) (Fig. 7.22) is the implantation of electrodes into deep brain regions for the purpose of modulating neural function in order to treat neurological/psychiatric conditions [3]. The electrodes are implanted using stereotactic techniques and are attached to an implantable pulse generator (IPG), which is

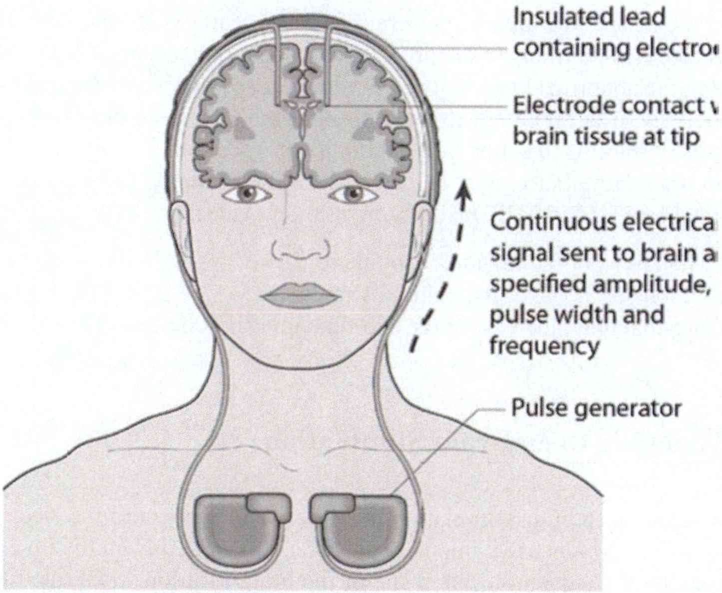

Fig. 7.22 Principle of the DBS

typically placed sub-dermally below the clavicle. DBS consists of implanted electrodes that provide continuous tiny electrical currents into deep structures of the brain [102–104]. This DBS therapeutic procedure involves the application of low-intensity electrical impulses typically at or near a frequency of 130 Hz, to critical sites in the brain through permanently embedded electrodes. These impulses are transiently activated by nearby axons. The distant effects of the stimulation are dependent on the function of certain neurons stimulation [78, 80, 102, 104]. There are three components involved in the deep brain stimulation device. These include:

- The electrode
- The pulse generator
- The extension

The electrode constitutes a small tip-shaped device that is placed deep into the deep region of the disease site of the brain. Its surface consists of four metal pads that generate the electric pulse. The electric pulse generated by the electrode only stimulates the brain tissue within close range of the electrode.

The pulse generator (also called the stimulator) constitutes a small, box-shaped device that generates the electrical signals for the electrode. Usually, this pulse generator is embedded under the skin of the chest of the patient. It consists of a battery with a lifespan of 2–7 years. The electrical patterns are generated at quite high frequencies in fluctuating pulses usually over 100 times/s.

The extension component consists of an insulated cable that carries the signals from the pulse generator to the implanted electrode in the brain.

Deep brain stimulation (DBS) is a surgical therapy used for the treatment of Parkinson's disease (PD). During DBS surgery, a special wire, called a lead, is inserted into a specific area of the brain [3]. The lead, which has four electrodes, delivers electrical currents to the precise brain locations responsible for movement, regulating the abnormal brain cell activity that causes symptoms such as tremor and gait problems. It is important to keep in mind that DBS can only help relieve the symptoms of Parkinson's, not cure or stop its progression.

Deep brain stimulation may be categorized into three types based on the target site in the brain [102, 104]. The categorization includes:

- The Thalamus, or the thalamic stimulation
- The Globus pallidus, or the pallidal stimulation
- The sub-thalamic nucleus, or the sub-thalamic stimulation

The Thalamic Deep Brain Stimulation

The Thalamic deep brain stimulation involves the placement of the electrode down into the thalamus. Electrical impulses are then transmitted from the tip of the electrode into the thalamus to locate a site of the brain location where the tremor may be affected or stopped. Monitoring is simultaneously performed to determine any possible side effects during the stimulation, such as numbness of the face, mouth, heaviness or weakness of the limb, change in speech, etc. If results are satisfactory, the electrode is left in place and joined to a plastic clip that is attached to the skull opening and the wound is closed [102, 104].

In the next stage, a small incision is given below the collarbone to create a small pouch underneath the skin that will hold the pulse generator or battery. Again a small incision is given behind the ear and a cable passed from the chest incision up to the head and then this cable is connected to the electrode coming out of the brain. All the entire system remains underneath the skin.

Sub-Thalamic Deep Brain Stimulation

After the creation of the skull opening, a microelectrode (very small metal wire) is inserted into the brain towards the thalamus and sub-thalamic region. Identification of specific brain cells in these regions is performed. The reason is to map out the area for the optimized placement of the electrode. During this time period, the room is kept dark. The patient must be made comfortable as the brain is assessed. When the particular area is identified, test stimulation is performed to check that the electrode is in safe position that will not disrupt the function of the brain. After that, the electrode is placed in safe area and clipped to the skull bone-fastening device [102, 104].

When surgical intervention on both sides of the brain is to be performed in the same setting, a second incision is made on the other side and the procedure is repeated.

In the first stage of the surgery, the electrodes are placed into the brain in 1 day. The scalp incisions are closed and the patient is asked to return 3–7 days later for the second stage of the procedure. The second stage is also performed under general anesthesia. Cables and batteries are inserted into the neck and chest area. Once the device is inserted, the neurologist activates the stimulators.

Globus Pallidus Deep Brain Stimulation

Exceptionally, the procedure is similar to the thalamic and the sub-thalamic stimulations. The electrodes are placed into the brain using MRI stereotactic guidance in dystonia patient [104]. Both electrode insertions are performed under local anesthesia, and the cable and pulse generator placement is done under general anesthesia on the same day.

Deep brain stimulation is indicated in treatment of motor symptoms of Parkinson's disease (PD), essential tremor, dystonia, and a wide variety of neurological and psychiatric conditions including epilepsy, obsessive-compulsive disorder (OCD), and major depression.

Brain-Computer Interface System (BCIS)

A Brain-Computer-Interface system (BCIS) (Fig. 7.23) uses mental activity, voluntarily produced by the patient, to control a computer or an embedded system via electroencephalogram (EEG) signals which allow communication or interaction with the surrounding environment [3, 106]. Like any communication or control system, a BCI has input (e.g., electrophysiological activity from the user), output (e.g., device commands), components that translate input into output, and a protocol that determines the onset, offset, and timing of operation.

The success of a BCI system depends as much on the system itself as on the ability of the user to produce distinctive EEG activity. BCI systems may be divided into two groups according to the placement of the electrodes used to detect and measure the neurons firing in the brain [3, 104]. These groups include; (1) the invasive systems, electrodes are inserted directly into the cortex are used for single cell or multi-unit recording, the electrocorticography (EcoG), electrodes are placed on the surface of the cortex (or dura); and (2) the noninvasive systems, they are placed on the scalp and use electroencephalography (EEG) or magnetoencephalography (MEG) to detect neuron activity.

Paradigm of the Brain-Computer Interface, BCI

Any natural form of communication or control requires peripheral nerves and muscles. The process begins with the intent of the user. This intent triggers a complex process in which certain brain areas are activated, and hence, signals are sent via the

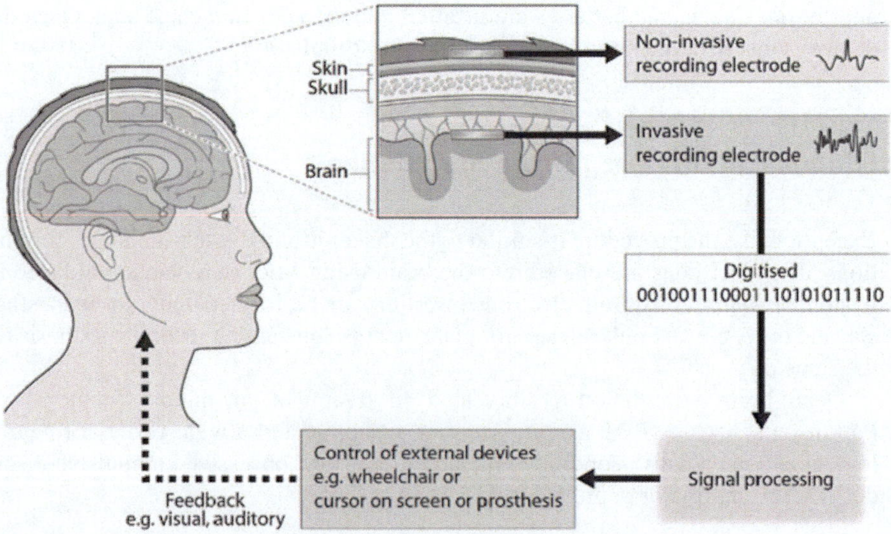

Fig. 7.23 Principle of the BCIS

peripheral nervous system (specifically, the motor pathways) to the corresponding muscles, which in turn perform the movement necessary for the communication or the control task. The activity resulting from this process is often called motor output or efferent output. Efferent means conveying impulses from the central to the peripheral nervous system and further to an effector (muscle). Afferent, in contrast, describes communication in the other direction, from the sensory receptors to the central nervous system [106–108]. For motion control, the motor (efferent) pathway is essential. The sensory (afferent) pathway is particularly important for learning motor skills and dexterous tasks, such as typing or playing a musical instrument.

A BCI offers an alternative to natural communication and control. It is an artificial system that bypasses the normal efferent pathways of the body, which are the neuromuscular output channels. Figure 7.24 illustrates this functionality [105, 109].

Instead of depending on peripheral nerves and muscles, a BCI directly measures brain activity associated with the user's intent and translates the recorded brain activity into corresponding control signals for BCI applications. This translation involves signal processing and pattern recognition, which is typically performed by a computer. Since the measured activity originates directly from the brain and not from the peripheral systems or muscles, the system is called a Brain–Computer Interface.

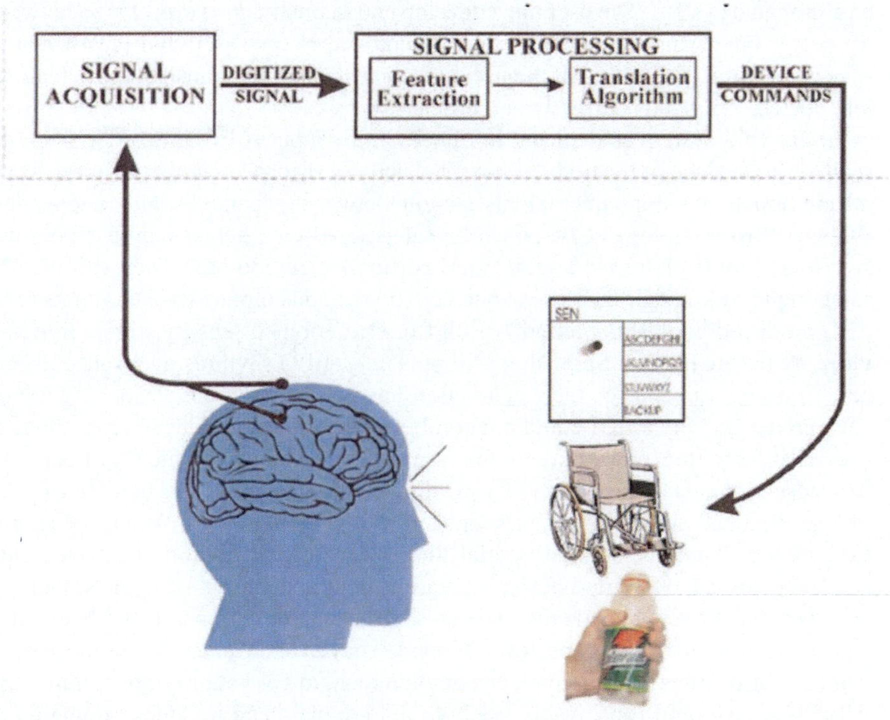

Fig. 7.24 Principle of the BCI system

A BCI must have four components. It must record activity directly from the brain (invasively or noninvasively). It must provide feedback to the user, and must do so in real time. Finally, the system must rely on intentional control [105, 106]. That is, the user must choose to perform a mental task whenever s/he wants to accomplish a goal with the BCI. Devices that only passively detect changes in the brain activity that occur without any intent, such as EEG activity associated with workload, arousal, or sleep, are not BCIs.

The BCIS applies electrodes that may either be implanted in the brain, or resting on the scalp to record the brain signals of the patient. These brain signals may then be translated into commands to operate computer-controlled devices. By actively producing brain signals, users can control these devices. BCIs could in principle assist users to communicate, control prostheses or wheelchairs, support rehabilitation, or facilitate detection of consciousness—making these technologies potentially useful to those with paralysis [3].

Figure 7.24 shows the block diagram to illustrate the basic design and operation of any BCI system [105]. Signals from the brain are acquired by electrodes on the scalp or in the head and processed to extract specific signal features (e.g., amplitudes of evoked potentials or sensorimotor cortex rhythms, firing rates of cortical neurons) that reflect the intent of the user. These features are translated into commands that operate a device (e.g., a simple word processing program, a wheelchair, or a neuro-prosthesis). Success depends on the interaction of two adaptive controllers, user, and system. The user must develop and maintain good correlation between his or her intent and the signal features employed by the BCI; and the BCI must select and extract features that the user can control and must translate those features into device commands correctly and efficiently.

In this BCI system as depicted in Fig. 7.24, the input is EEG recorded from the scalp or the surface of the brain or neuronal activity recorded within the brain. Thus, in addition to the fundamental distinction between dependent and independent BCIs, electrophysiological BCIs can be categorized by whether they use noninvasive (e.g., EEG) or invasive (e.g., intra-cortical) methodology. They can also be categorized by whether they use evoked or spontaneous inputs. Evoked inputs (e.g., EEG produced by flashing letters) result from stereotyped sensory stimulation provided by the BCI [105]. Spontaneous inputs (e.g., EEG rhythms over sensorimotor cortex) do not depend for their generation on such stimulation. There is, presumably, no reason why a BCI could not combine noninvasive and invasive methods or evoked and spontaneous inputs. In the signal-acquisition part of the BCI operation, the chosen input is acquired by the recording electrodes, amplified, and digitized.

The digitized signals are then subjected to one or more of a variety of feature extraction procedures, such as spatial filtering, voltage amplitude measurements, spectral analyses, or single-neuron separation. This analysis extracts the signal features that (hopefully) encode the messages of the user or commands. BCIs may use signal features that are in the time domain (e.g., evoked potential amplitudes or neuronal firing rates) or the frequency domain (e.g., mu or beta-rhythm amplitudes). A BCI system could conceivably use both time-domain and frequency-domain signal features, and might thereby improve performance [105].

In general, the signal features used in present-day BCIs reflect identifiable brain events like the firing of a specific cortical neuron or the synchronized and rhythmic synaptic activation in the sensorimotor cortex that produces a mu rhythm. Knowledge of these events may help guide BCI development. The location, size, and function of the cortical area generating a rhythm or an evoked potential may indicate how it should be recorded, how users might best learn to control its amplitude, and how to recognize and eliminate the effects of the non-CNS artifacts. It is also possible for a BCI to use signal features, like sets of autoregressive parameters, that correlate with the intent of the user but do not necessarily reflect specific brain events. In such cases, it is particularly important (and maybe more difficult) to ensure that the chosen features are not contaminated by EMG, electrooculography (EOG), or other non-CNS artifacts [105].

The first part of signal processing simply extracts specific signal features. The next stage, the translation algorithm, translates these signal features into device commands-orders that carry out the user's intent. This algorithm might use linear methods (e.g., classical statistical analyses, or nonlinear methods (e.g., neural networks). Whatever its nature, each algorithm changes independent variables (that means, signal features) into dependent variables (i.e., device control commands).

Effective algorithms adapt to each user on three levels. First, when a new user first accesses the BCI the algorithm adapts to the signal features of that user [105]. If the signal feature is mur-hythm amplitude, the algorithm adjusts to the range of mu-rhythm amplitudes of the user. If the feature is P300 amplitude, it adjusts to the characteristic P300 amplitude of the user, and if the feature is the firing rate of a single cortical neuron, it adjusts to the characteristic range of firing rates of the neuron. A BCI that possesses only this first level of adaptation, that means, that adjusts to the user initially and never again, will continue to be effective only if the performance of the user is very stable. However, EEG and other electrophysiological signals typically display short- and long-term variations linked to time of day, hormonal levels, immediate environment, recent events, fatigue, illness, and other factors. Thus, effective BCIs need a second level of adaptation. These include periodic online adjustments to reduce the impact of such spontaneous variations. A good translation algorithm may adjust to these variations so as to match as closely as possible the user's current range of signal feature values to the available range of device command values.

While they are clearly important, neither of these first two levels of adaptation addresses the central fact of effective BCI operation, its dependence on the effective interaction of two adaptive controllers, the BCI and the brain of the user. The third level of adaptation accommodates and engages the adaptive capacities of the brain [105]. When an electrophysiological signal feature that is normally merely a reflection of the brain function becomes the end product of that function, that is, when it becomes an output that carries the intent of the user to the outside world, it engages the adaptive capacities of the brain. Like activity in the conventional neuromuscular communication of the brain and the control channels, BCI signal features may be affected by the device commands they are translated into. The results of BCI operation will affect future BCI input. In the most desirable (and hopefully typical) case, the brain may modify the signal features so as to improve the BCI operation. If, for example, the feature is mu-rhythm amplitude, the correlation between that amplitude and the intent of the user will hopefully increase over time. An algorithm that incorporates the third level could respond to this increase by rewarding the user with faster communication. It would thereby recognize and encourage the user's development of greater skill in this new form of communication. On the other hand, excessive or inappropriate adaptation could impair performance or discourage further skill development [105]. Proper design of this third level of adaptation is likely to prove crucial for BCI development. Because this level involves the interaction of two adaptive controllers, the brain of the user, and the BCI system. Its design may rank among the most difficult problems confronting BCI research.

The Output Device

For most current BCIs, the output device is a computer screen and the output is the selection of targets, letters, or icons presented on it. Selection is indicated in various ways (e.g., the letter flashes). Some BCIs also provide additional, interim output, such as cursor movement toward the item prior to its selection. In addition to being the intended product of BCI operation, this output is the feedback that the brain uses to maintain and improve the accuracy and speed of the communication [105, 106].

Neural Stem Cell Therapy (NSCT)

Neural stem cell therapy (Fig. 7.25) involves the injection of stem cells into the brain in order to repair damage caused by acute events such as stroke or neurodegenerative conditions such as Alzheimer's disease, etc. [3].

Stem cells are the essential building blocks of multicellular organisms. They have two defining properties; (1) they can self-renew to produce more stem cells and (2) they can differentiate to generate specialized cell types. Stem cells come in different varieties, depending on when and where they are produced during development and how versatile they are [110–112].

Totipotent stem cells can give rise to a full organism once implanted in the uterus of a living animal. Both embryonic stem (ES) cells (produced in culture from the epiblast cells of a blastocyst) and embryonic germ (EG) cells (produced in culture from primordial germ cells of an early embryo) are pluripotent, capable of giving rise to every cell type of an organism, except the trophoblasts of the placenta. Somatic stem cells and germ-line stem cells are produced later in the development from the pluripotent stem cells. Hematopoietic stem cells are probably the best-characterized somatic stem cells and are able to generate all the cell types of the blood and the immune system [110–113]. Stem cells are also responsible for the generation of other somatic systems during development, such as the central

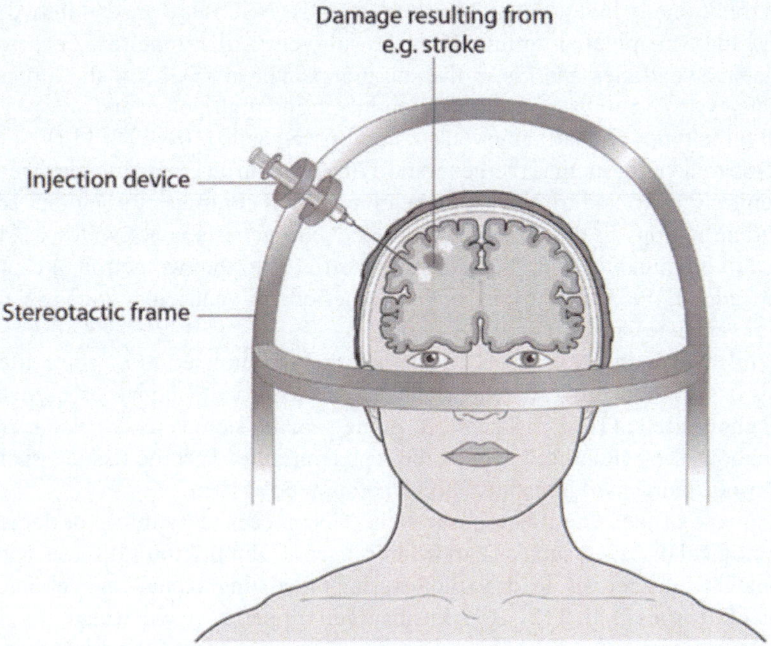

Fig. 7.25 Concept of the neural stem cell therapy

nervous system (CNS). Neural stem cells (NSCs) in the central nervous system (CNS) have tri-lineage ability, capable of generating neurons, oligodendrocytes, and astrocytes [112, 113].

Neural stem cells are present not only in the developing nervous systems, but also in the adult central nervous system of mammals, including humans. The mature central nervous system has been traditionally regarded as an unfavorable environment for the regeneration of damaged axons of mature neurons and the generation of new neurons. In the adult central nervous system, however, newly generated neurons from the adult neural stem cells in specific regions exhibit a striking ability to migrate, send out long axonal and dendritic projections, integrate into pre-existing neuronal circuits, and contribute to normal brain functions [101, 110, 112].

Rapid advances in stem cell biology have raised exciting possibilities of replacing damaged or lost neurons by activation of endogenous neural stem cells and/or transplantation of in vitro-expanded stem cells and/or their neuronal progeny. Before the full potential of adult stem cells may be realized for regenerative medicine, it may be necessary to identify the sources of the stem cells, in order to understand the mechanisms regulating their proliferation, fate specification, and, most importantly in the case of neuronal lineages, to characterize their functional properties. Equally important, we need to understand the neural development processes in the normal and diseased adult central nervous system environment, which is quite different from the embryonic central nervous system, where neural development has been traditionally investigated. Neural stem cells (NSCs) are present in two regions of the adult mammalian brain; (1) in the sub-ventricular zone (SVZ) surrounding the lateral ventricles and (2) in the sub-granular layer (SGL) of the hippocampal dentate gyrus. Neural stem cells are clonogenic and multipotent and able to give rise to all three types of central nervous system (CNS) cells [101, 110, 112].

Research suggests that the generation of new neurons, namely neurogenesis, is not entirely restricted to prenatal development, but continues throughout adult life in certain regions of the mammalian brain. Neurogenesis is not widespread within the adult mammalian brain but is largely restricted to the two germinal centers, the hippocampal, the dentate gyrus, and the anterior sub-ventricular zone (SVZ) of the lateral ventricles, respectively.

Whilst diversification of cell types is largely completed at or soon after birth, many tissues undergo self-renewal in response to acute injury or normal usage throughout life [111, 112]. Therefore, some somatic stem cells are also maintained throughout the whole life span for the regeneration of specific tissues, such as the epidermis, hair, small intestine, and hematopoietic system.

Damage of the central nervous system (CNS), caused by injury or degenerative neurological diseases, such as Parkinson's disease, Huntington's disease, and amyotrophic lateral sclerosis, leads to the severing of existing axonal connections and the death of neurons [111, 112]. Both intrinsic cell properties and extrinsic environmental factors contribute to the lack of regenerating capacity of the adult mammalian central nervous system (CNS).

References

1. Taylor, B., & Marsiske, M. (2019, October). *Central nervous system.* https://doi.org/10.1007/978-3-319-69892-2-668-1.
2. Society of Science (2002) A primer on the brain and nervous system; The Society for Neuroscience, Fourth edition.
3. Nuffield Council on Bioethics. (2013). *Novel neurotechnologies: Intervening in the brain.* Nuffield Council on Bioethics ISBN: 978-1-904384-29-8
4. WHO. *Neurological disorders: Public health challenges.* World Health Organization; ISBN: 978-92-4-156336-9.
5. Yin, J., & Yuan, Q. (2015, January). Structural homeostasis in the nervous system: A balancing act for wiring plasticity and stability. *Frontiers in Cellular Neuroscience, 8*, Article 439.
6. Goldstein, D. S., & Kopin, I. J. (2017, December). Homeostatic systems, biocybernetics, and autonomic neuroscience. *Autonomic Neuroscience, 208*, 15–28; HHS Public Access.
7. Nishikawa, K., Biewener, A. A., Aerts, P., Ahn, A. N., Chiel, H. J., Daley, M. A., Daniel, T. L., Full, R. J., Hale, M. E., Hedricks, T. L., Kristopher Lappin, A., Richard Nichols, T., Quinn, R. D., Satterlie, R. A., & Azymik, B. (2006, January 4–8). Neuromechanics: An integrative approach for understanding motor control. In *From the symposium 'Biomechanics and neuromuscular control' presented at the annual meeting of the Society for Integrative and Comparative Biology*, at Orlando, Florida.
8. Glover, G. H. (2011, April). Overview of functional magnetic resonance imaging. *Neurosurgery Clinics of North America, 22*(2), 133–139.

9. James, T. L. (1998). *Chapter 1: "Fundamentals of NMR" Department of Pharmaceutical Chemistry*. University of California.
10. Edwards, J. C. (2009). *Principles of NMR*. Process NMR Associates LLC.
11. Wei, Y., Yang, C., Jiang, H., Li, Q., Che, F., Wan, S., Yao, S., Gao, F., Zhang, T., Wang, J., & Song, B. (2022). Multi-nuclear magnetic resonance spectroscopy: State of the art and future directions. *Insights Into Imaging, 13*, 135; Springer Open Access.
12. Currie, S., Hoggard, N., Craven, I. J., Hadjivassiliou, M., & Wilkinson, I. D. (2013). Understanding MRI: Basic physics for physicians. *Postgraduate Medical Journal, 2013*(89), 209–223.
13. Plewes, D. B., & Kucharczyk, W. (2012). Physics of MRI: A primer. *Journal of Magnetic Resonance Imaging, 35*, 1038–1054.
14. Basser, P. J., Pajevic, S., Pierpaoli, C., et al. (2000). In vivo fiber tractography using DT-MRI data. *Magnetic Resonance in Medicine, 44*, 625. [PubMed: 11025519].
15. Belliveau, J. W., Kennedy, D. J., McKinstry, R. C., et al. (1991). Functional mapping of the human visual cortex by magnetic resonance imaging. *Science, 254*, 716. [PubMed: 1948051].
16. Bernstein, M. A., King, K. F., & Zhou, X. J. (2004). *Handbook of MRI pulse sequences*. Elsevier Press.
17. Birn, R. M., Smith, M. A., Jones, T. B., et al. (2008). The respiration response function: The temporal dynamics of fMRI signal fluctuations related to changes in respiration. *NeuroImage, 40*, 644. [PubMed: 18234517].
18. Bandettini, P. A. (2009). What's new in neuroimaging methods? *Annals of the New York Academy of Sciences, 1156*, 260. [PubMed: 19338512].
19. Buxton, R. B., Wong, E. C., & Frank, L. R. (1998). Dynamics of blood flow and oxygenation changes during brain activation: The balloon model. *Magnetic Resonance in Medicine, 39*, 855. [PubMed: 9621908].
20. Calhoun, V. D., Adali, T., Pearlson, G. D., et al. (2001). A method for making group inferences from functional MRI data using independent component analysis. *Human Brain Mapping, 14*, 140. [PubMed: 11559959].
21. Chang, C., Cunningham, J. P., & Glover, G. H. (2009). Influence of heart rate on the BOLD signal: The cardiac response function. *NeuroImage, 44*, 857. [PubMed: 18951982].
22. Chang, C., & Glover, G. H. (2009). Relationship between respiration, end-tidal CO(2), and BOLD signals in resting-state fMRI. *NeuroImage, 47*(4), 1381–1393.
23. Weisskoff, R. M., Zuo, C. S., Boxerman, J. L., et al. (1994). Microscopic susceptibility variation and transverse relaxation: Theory and experiment. *Magnetic Resonance in Medicine, 31*, 601. [PubMed: 8057812].
24. Wise, R. G., & Preston, C. (2010). What is the value of human FMRI in CNS drug development? *Drug Discovery Today, 15*, 973–980.
25. Worsley, K. J., Liao, C. H., Aston, J., et al. (2002). A general statistical analysis for fMRI data. *NeuroImage, 15*, 1. [PubMed: 11771969].
26. Yacoub, E., Van De Moortele, P. F., Shmuel, A., et al. (2005). Signal and noise characteristics of Hahn SE and GE BOLD fMRI at 7 T in humans. *NeuroImage, 24*, 738. [PubMed: 15652309].
27. Yang, Q. X., Dardzinski, B. J., Li, S., et al. (1997). Multi-gradient echo with susceptibility inhomogeneity compensation (MGESIC): Demonstration of fMRI in the olfactory cortex at 3.0 T. *Magnetic Resonance in Medicine, 37*, 331. [PubMed: 9055220].
28. Bandettini, P. A., Jesmanowicz, A., Wong, E. C., et al. (1993). Processing strategies for time-course data sets in functional MRI of the human brain. *Magnetic Resonance in Medicine, 30*, 161. [PubMed: 8366797].
29. Bandettini, P. A., Wong, E. C., Hinks, R. S., et al. (1992). Time course EPI of human brain function during task activation. *Magnetic Resonance in Medicine, 25*, 390. [PubMed: 1614324].

30. Bangen, K. J., Restom, K., Liu, T. T., et al. (2009). Differential age effects on cerebral blood flow and BOLD response to encoding: Associations with cognition and stroke risk. *Neurobiology of Aging, 30*, 1276. [PubMed: 18160181].
31. Buckner, R. L., Bandettini, P. A., O'Craven, K. M., et al. (1996). Detection of cortical activation during averaged single trials of a cognitive task using functional magnetic resonance imaging. *Proceedings of the National Academy of Sciences of the United States of America, 93*, 14878. [see comments]. [PubMed: 8962149].
32. Buxton, R., & Frank, L. (1997). A model for the coupling between cerebral blood flow and oxyen metabolism during neural stimulation. *Journal of Cerebral Blood Flow and Metabolism, 17*, 64. [PubMed: 8978388].
33. Chang, C., & Glover, G. H. (2010). Time-frequency dynamics of resting-state brain connectivity measured with fMRI. *NeuroImage, 50*, 81. [PubMed: 20006716].
34. Cho, Z. H., & Ro, Y. M. (1992). Reduction of susceptibility artifact in gradient-echo imaging. *Magnetic Resonance in Medicine, 23*, 193. [PubMed: 1734179].
35. Constable, R., & Spencer, D. (1999). Composite image formation in Z-shimmed functional MR imaging. *Magnetic Resonance in Medicine, 42*, 110. [PubMed: 10398956].
36. Cui, X., Bray, S., Bryant, D. M., et al. (2010). A quantitative comparison of NIRS and fMRI across multiple cognitive tasks. *NeuroImage, 54*, 2808.
37. Stenger, V. A., Boada, F. E., & Noll, D. C. (2000). Three-dimensional tailored RF pulses for the reduction of susceptibility artifacts in T2*-weighted functional MRI. *Magnetic Resonance in Medicine, 44*, 525. [PubMed: 11025507].
38. Thulborn, K. R., Waterton, J. C., Matthews, P. M., et al. (1982). Oxygenation dependence of the transverse relaxation time of water protons in whole blood at high field. *Biochimica et Biophysica Acta, 714*, 265. [PubMed: 6275909].
39. Weiger, M., Pruessmann, K. P., Osterbauer, R., et al. (2002). Sensitivity-encoded single-shot spiral imaging for reduced susceptibility artifacts in BOLD fMRI. *Magnetic Resonance in Medicine, 48*, 860. [PubMed: 12418001].
40. Bodurka, J., & Bandettini, P. A. (2002). Toward direct mapping of neuronal activity: MRI detection of ultraweak, transient magnetic field changes. *Magnetic Resonance in Medicine, 47*, 1052. [PubMed: 12111950].
41. Boxerman, J. L., Bandettini, P. A., Kwong, K. K., et al. (1995). The intravascular contribution to fMRI signal change: Monte Carlo modeling and diffusion-weighted studies in vivo. *Magnetic Resonance in Medicine, 34*, 4. [PubMed: 7674897].
42. Davis, T. L., Kwong, K. K., Weisskoff, R. M., et al. (1998). Calibrated functional MRI: Mapping the dynamics of oxidative metabolism. *Proceedings of the National Academy of Sciences of the United States of America, 95*, 1834. [PubMed: 9465103].
43. deCharms, R. C., Christoff, K., Glover, G. H., et al. (2004). Learned regulation of spatially localized brain activation using real-time fMRI. *NeuroImage, 21*, 436. [PubMed: 14741680].
44. deCharms, R. C., Maeda, F., Glover, G. H., et al. (2005). Control over brain activation and pain learned by using real-time functional MRI. *Proceedings of the National Academy of Sciences of the United States of America, 102*, 18626. [PubMed: 16352728].
45. Detre, J. A., Leigh, J. S., Williams, D. S., et al. (1992). Perfusion imaging. *Magnetic Resonance in Medicine, 23*, 37. [PubMed: 1734182].
46. Edmister, W. B., Talavage, T. M., Ledden, P. J., et al. (1999). Improved auditory cortex imaging using clustered volume acquisitions. *Human Brain Mapping, 7*, 89. [PubMed: 9950066].
47. Dale, A. M., & Halgren, E. (2001). Spatiotemporal mapping of brain activity by integration of multiple imaging modalities. *Current Opinion in Neurobiology, 11*, 202. [PubMed: 11301240].
48. Richards, T. L., & Berninger, V. W. (2008). Abnormal fMRI connectivity in children with dyslexia during a phoneme task: Before but not after treatment 1. *Journal of Neurolinguistics, 21*, 294. [PubMed: 19079567].
49. Roland, P. E. (1993). *Brain activation*. Wiley.
50. Sarty, G. E. (2007). *Computing brain activity maps from fMRI Teime series images*. Cambridge University Press.

51. Shmuel, A., Yacoub, E., Chaimow, D., et al. (2007). Spatio-temporal point-spread function of fMRI signal in human gray matter at 7 Tesla. *NeuroImage, 35*, 539. [PubMed: 17306989].
52. Ogawa, S., Lee, T. M., Stepnoski, R., et al. (2000). An approach to probe some neural systems interaction by functional MRI at neural time scale down to milliseconds. *Proceedings of the National Academy of Sciences of the United States of America, 97*, 11026. [PubMed: 11005873].
53. Ogawa, S., Menon, R. S., Tank, D. W., et al. (1993). Functional brain mapping by blood oxygenation leveldependent contrast magnetic resonance imaging. A comparison of signal characteristics with a biophysical model. *Biophysical Journal, 64*, 803. [PubMed: 8386018].
54. Phan, K. L., Fitzgerald, D. A., Gao, K., et al. (2004). Real-time fMRI of cortico-limbic brain activity during emotional processing. *Neuroreport, 15*, 527. [PubMed: 15094517].
55. Ogawa, S., Lee, T. M., Kay, A. R., et al. (1990). Brain magnetic resonance imaging with contrast dependent on blood oxygenation. *Proceedings of the National Academy of Sciences of the United States of America, 87*, 9868. [PubMed: 2124706].
56. Ogawa, S., Lee, T. M., Nayak, A. S., et al. (1990). Oxygenation-sensitive contrast in magnetic resonance image of rodent brain at high magnetic fields. *Magnetic Resonance in Medicine, 14*, 68. [PubMed: 2161986].
57. Menon, R. S. (2002). Postacquisition suppression of large-vessel BOLD signals in high-resolution fMRI. *Magnetic Resonance in Medicine, 47*, 1. [PubMed: 11754436].
58. Miller, K. L., Bulte, D. P., Devlin, H., et al. (2007). Evidence for a vascular contribution to diffusion FMRI at high b value. *Proceedings of the National Academy of Sciences of the United States of America, 104*, 20967. [PubMed: 18093924].
59. Norman, K. A., Polyn, S. M., Detre, G. J., et al. (2006). Beyond mind-reading: Multi-voxel pattern analysis of fMRI data. *Trends in Cognitive Sciences, 10*, 424. [PubMed: 16899397].
60. Luo, Q., & Gao, J. H. (2010). Modeling magnitude and phase neuronal current MRI signal dependence on echo time. *Magnetic Resonance in Medicine, 64*(6), 1832–1837.
61. Lauterbur, P. C. (1973). Image formation by induced local interactions. Examples employing nuclear magnetic resonance. *Nature, 242*, 190.
62. Le Bihan, D., Urayama, S., Aso, T., et al. (2006). Direct and fast detection of neuronal activation in the human brain with diffusion MRI. *Proceedings of the National Academy of Sciences of the United States of America, 103*, 8263. [PubMed: 16702549].
63. Friston, K. J., Holmes, A. P., Poline, J. B., et al. (1995). Analysis of fMRI time-series revisited. *NeuroImage, 2*, 45. [PubMed: 9343589].
64. Gaab, N., Gabrieli, J. D., & Glover, G. H. (2007). Assessing the influence of scanner background noise on auditory processing. I. An fMRI study comparing three experimental designs with varying degrees of scanner noise. *Human Brain Mapping, 28*, 703. [PubMed: 17080440].
65. Gaab, N., Gabrieli, J. D., & Glover, G. H. (2007). Assessing the influence of scanner background noise on auditory processing. II. An fMRI study comparing auditory processing in the absence and presence of recorded scanner noise using a sparse design. *Human Brain Mapping, 28*, 721. [PubMed: 17089376].
66. Gaab, N., Gabrieli, J. D., & Glover, G. H. (2008). Resting in peace or noise: Scanner background noise suppresses default-mode network. *Human Brain Mapping, 29*, 858. [PubMed: 18454447].
67. Glover, G. H., & Law, C. S. (2001). Spiral-in/out BOLD fMRI for increased SNR and reduced susceptibility artifacts. *Magnetic Resonance in Medicine, 46*, 515. [PubMed: 11550244].
68. Liu, T. T., & Frank, L. R. (2004). Efficiency, power, and entropy in event-related FMRI with multiple trial types. Part I: Theory. *NeuroImage, 21*, 387. [PubMed: 14741676].
69. Greicius, M. (2008). Resting-state functional connectivity in neuropsychiatric disorders. *Current Opinion in Neurology, 21*, 424. [PubMed: 18607202].
70. Greicius, M. D., Srivastava, G., Reiss, A. L., et al. (2004). Default-mode network activity distinguishes Alzheimer's disease from healthy aging: Evidence from functional MRI. *Proceedings of the National Academy of Sciences of the United States of America, 101*, 4637. [PubMed: 15070770].

71. Martinez-Ramon, M., Koltchinskii, V., Heileman, G. L., et al. (2006). fMRI pattern classification using neuroanatomically constrained boosting. *NeuroImage, 31*, 1129. [PubMed: 16529955].
72. Hsu, J. J., & Glover, G. H. (2005). Mitigation of susceptibility-induced signal loss in neuroimaging using localized shim coils. *Magnetic Resonance in Medicine, 53*, 243. [PubMed: 15678531].
73. Mansfield, P. (1977). Multi-planar image formation using NMR spin echoes. *Journal of Physical Chemistry C, 10*, L55.
74. Kwong, K. K., Belliveau, J. W., Chesler, D. A., et al. (1992). Dynamic magnetic resonance imaging of human brain activity during primary sensory stimulation. *Proceedings of the National Academy of Sciences of the United States of America, 89*, 5675. [PubMed: 1608978].
75. Li, T., & Song, A. W. (2003). Fast functional brain signal changes detected by diffusion weighted fMRI. *Magnetic Resonance Imaging, 21*, 829. [PubMed: 14599532].
76. Kim, D. I., Sui, J., Rachakonda, S., et al. (2010). Identification of imaging biomarkers in schizophrenia: A coefficient-constrained independent component analysis of the mind multisite SchizophreniaStudy. *Neuroinformatics, 8*(4), 213–229.
77. Huettel, S. A., Song, A. W., & McCarthy, G. (2004). *Functional magnetic resonance imaging*. Sinauer Associates, Inc.
78. Siebner, H. R., et al. (2004). Preconditioning of low-frequency repetitive transcranial magnetic stimulation with transcranial direct current stimulation: Evidence for homeostatic plasticity in the human motor cortex. *The Journal of Neuroscience, 24*, 3379–3385.
79. Schwarzkopf, D. S., et al. (2011). Stochastic resonance effects reveal the neural mechanisms of transcranial magnetic stimulation. *The Journal of Neuroscience, 31*, 3143–3147.
80. Thut, G., et al. (2011). Entrainment of perceptually relevant brain oscillations by non-invasive rhythmic stimulation of the human brain. *Frontiers in Psychology, 2*, 170.
81. Silvanto, J., et al. (2007). Neural adaptation reveals state-dependent effects of transcranial magnetic stimulation. *European Journal of Neuroscience, 25*(6), 1874–1881.
82. Fröhlich, F. (2015). Experiments and models of cortical oscillations as a target for noninvasive brain stimulation. *Progress in Brain Research, 222*, 41–73.
83. Helfrich, R. F., et al. (2014). Entrainment of brain oscillations by transcranial alternating current stimulation. *Current Biology, 24*, 333–339.
84. Witkowski, M., et al. (2015). Mapping entrained brain oscillations during transcranial alternating current stimulation (tACS). *NeuroImage*. https://doi.org/10.1016/j.neuroimage.2015.10.024
85. Stefan, K., Kunesch, E., Cohen, L. G., et al. (2000). Induction of plasticity in the human motor cortex by paired associative stimulation. *Brain, 123*, 572–584.
86. Opitz, A., Paulus, W., Will, S., et al. (2015). Determinants of the electric field during transcranial direct current stimulation. *NeuroImage, 109*, 140–150.
87. George, M. S., Lisanby, S. H., & Sackeim, H. (1999, April). Transcranial magnetic stimulation: Applications in neuropsychiatry. *Archives of General Psychiatry, 56*(4), 300–311.
88. Hallett, M. (2000) *Transcranial magnetic stimulation and the the human brain*. Human motor control section, National Institutes of Neurological Disorders and Stroke, NIH Building, Bethesda, Maryland; Nature 2000.
89. Valero-Cabre, A., Amengual, J. L., Stengel, C., Pascual-Leone, A., & Coubard, O. A. (2017). Transcranial magnetic stimulation in basic and clinical neuroscience: A comprehensive review of fundamental principles and novel insights. *Neuroscience and Biobehavioral Reviews, 83*(2017), 381–404. Elsevier.
90. Pascual-Leone, A., Bartres-Faz, D., & Keenan, J. P. (1999). *Transcranial magnetic stimulation: Studying the brain-behavior relationship by induction of 'virtual lesions'*. The Royal Society, Laboratory for Magnetic Brain Stimulation, Department of Neurology, Beth Israel Deaconess Medical Center, Harvard Medical School.

91. Thair, H., Holloway, A. L., Newport, R., & Smith, A. D. (2017). Transcranial direct current stimulation (tDCS): A beginner's guide for design and implementation. *Frontiers in Neuroscience, 11*, Article 641.
92. Agarwal, S. M., Shivakumar, V., Bose, A., Subramaniam, A., Nawani, H., Chhabra, H., et al. (2013). Transcranial direct current stimulation in schizophrenia. *Clinical Psychopharmacology and Neuroscience, 11*, 118–125. https://doi.org/10.9758/cpn.2013.11.3.118
93. DaSilva, A. F., Truong, D. Q., DosSantos, M. F., Toback, R. L., Datta, A., & Bikson, M. (2015). State-of-art neuroanatomical target analysis of high definition and conventional tDCS montages used for migraine and pain control. *Frontiers in Neuroanatomy, 9*, 1–12. https://doi.org/10.3389/fnana.2015.00089
94. Bikson, M., Datta, A., & Elwassif, M. (2009). Establishing safety limits for transcranial direct current stimulation. *Clinical Neurophysiology, 120*, 1033–1034. https://doi.org/10.1016/j.clinph.2009.03.018
95. Dieckhöfer, A., Waberski, T. D., Nitsche, M., Paulus, W., Buchner, H., & Gobbelé, R. (2006). Transcranial direct current stimulation applied over the somatosensory cortex -differential effect on low and high frequency SEPs. *Clinical Neurophysiology, 117*, 2221–2227. https://doi.org/10.1016/j.clinph.2006.07.136
96. Utz, K. S., Dimova, V., Oppenländer, K., & Kerkhoff, G. (2010). Electrified minds: Transcranial direct current stimulation (tDCS) and Galvanic Vestibular Stimulation (GVS) as methods of non-invasive brain stimulation in neuropsychology—A review of current data and future implications. *Neuropsychologia, 48*, 2789–2810. https://doi.org/10.1016/j.neuropsychologia.2010.06.002
97. Tavakoli, A. V., & Yun, K. (2017). Transcranial alternating current stimulation (tACS) mechanisms and protocols. *Frontiers in Cellular Neuroscience, 11*, Article 214.
98. Amengual, J. L., Vernet, M., Adam, C., & Valero-Cabre, A. (2017). Local entrainment of oscillatory activity induced by direct brain stimulation in humans. *Scientific Reports, 7*, 41908. https://doi.org/10.1038/srep41908
99. Bolognini, N., Pascual-Leone, A., & Fregni, F. (2009). Using non-invasive brain stimulation to augment motor training-induced plasticity. *Journal of Neuroengineering and Rehabilitation, 6*, 8. https://doi.org/10.1186/1743-0003-6-8
100. Fröhlich, F. (2016). Mechanism of transcranial alternating current stimulation (tACS). In *Transcranial electrical stimulation: Mechanisms, technology and therapeutic applications Webinar NIMH*. [Google Scholar]
101. Giovannucci, A., Badura, A., Deverett, B., Najafi, F., Pereira, T. D., Gao, Z., et al. (2017). Cerebellar granule cells acquire a widespread predictive feedback signal during motor learning. *Nature Neuroscience, 20*, 727–734. https://doi.org/10.1038/nn.4531
102. Gautam, A., Rai, M., Srivastava, R., Jyoti, B., & Priyadarshi, P. K. (2018). Deep brain simulation-a novel approach for relieving neurological disorders: A review. *International Journal of Contemporary Medicine Surgery and Radiology, 3*(3), C86–C92.
103. Pycroft, L., Stein, J., & Aziz, T. (2018). Deep brain stimulation: An overview of history, methods, and future developments. *Brain and Neuroscience Advances, 2*, 1–6.
104. Zangen, A., Roth, Y., Voller, B., & Hallett, M. (2005). Transcranial magnetic stimulation of deep brain regions: Evidence for efficacy of the H-Coil. *Clinical Neurophysiology, 116*(2005), 775–779.
105. Wolpaw, J. R., Birbaumer, N., McFarland, D. J., Pfurtscheller, G., & Vaughan, T. M. (2002). Brain-computer interfaces for communication and control. *Clinical Neurophysiology, 113*, 767–791. Elsevier.
106. Grainmann, B., Allison, B., & Pfurtscheller, G. (2010). Brain-computer interfaces: A gentle introduction. *Brain-Computer Interfaces, The Frontiers Collection*. https://doi.org/10.1007/978-3-642-02091_1. Springer Verlag Berlin Heidelberg.
107. Sonam, Y. S. (2015). A review paper on brain computer interface. *International Journal of Engineering Research & Technology (IJERT)*; ISSN: 2278–0181; NCETEMS 2015 Conference Proceedings.

108. Capogrosso, M., Milekovic, T., Borton, D., et al. (2016). A brain spine interface alleviating gait deficits after spinal cord injury in primates. *Nature, 539*, 284–288.
109. Abdulkader, S. N., Atia, A., & Mostafa-Sami, M. (2015). Brain computer interfacing: Applications and challenges. *Egyptian Information Journal, 16*(2), 213–230.
110. Shanmuganathan, D., & Sivakumaran, N. (2018, March). Review: The development of neural stem biology and technology in regenerative medicine. *International Research Journal of Natural and Applied Sciences, 5*(3); Article: January 2018.
111. Valenzuela, M., Sidhu, K., Dean, S., & Sachdev, P. (2007). Neural stem cell therapy for neuropsychiatric disorders. *Acta Neuropsychiatrica, 2007*(19), 11–26.
112. Okano, H. (2010). Neural stem cells and strategies for the regeneration of the central nervous system. *Proceedings of the Japan Academy. Series B, Physical and Biological Sciences, 86*, 438–450.
113. Glass, J. D., Boulis, N. M., Johe, K., Rutkove, S. B., Federici, T., Polak, M., Kelly, C., & Feldman, E. L. (2012). Lumbar intraspinal injection of neural stem cells in patients with amyotrophic lateral sclerosis: Results of a phase I Trial in 12 Patients. *Stem Cells Regenerative Medicine, 30*, 1144–1151.

Chapter 8
Artificial Intelligence in Human Locomotor System

The human exhibits the ability and capability to behave adaptively both in diverse and complex/unstructured environments. This adaptive behavior emerges from the interplay among the descending output from the central nervous system (CNS), sensory input from the body and the environment, the dynamics of the muscles, and the complex dynamics of the entire human body [1, 4, 5].

The nervous system is the most complex body system, and exhibits the prime communication of the body and coordination network. The brain and spinal cord constitute the central nervous system (CNS), whereas the nerves and sensory organs make up the peripheral nervous system (PNS). Together, the CNS and the PNS transmit and process sensory information and coordinate bodily functions. The nervous system controls the functioning of the human body both internally and externally. It consists of the brain, spinal cord, cranial nerves, and spinal nerves. Figure 8.1 illustrates the spinal nerves [6, 8].

The thirty-one pairs of peripheral spinal nerves emerge from the spinal cord through spaces between the vertebrae. Each nerve divides and subdivides into a number of branches; the dorsal branches serve the rear portion of the body, while the ventral serves the front and sides. The branches of one spinal nerve may join with other nerves to form meshes called plexuses where information is shared [7, 8]. The plexuses send signals along secondary nerve branches to areas of complex function or movement. Figure 8.1 depicts an illustration of the spinal nerves.

Along with the endocrine system, it is also responsible for regulating and maintaining body homeostasis. The nervous system processes and interprets the sensory input and makes decisions about what should be done at each moment (integration). The nervous system then causes a response by activating effector organs such as muscles or glands in a response called motor output.

The nervous system has a special status in human life that distinguishes it from other organs. Its healthy functioning plays a central role in the operation of our bodies, our capacities for autonomous agency, our conceptions of ourselves and our relationships with others, and thus in our abilities to lead fulfilling lives. Hence, the

© The Author(s), under exclusive license to Springer Nature Switzerland AG 2023

T. Sarkodie-Gyan, H. Yu, *The Human Locomotor System*,
https://doi.org/10.1007/978-3-031-32781-0_8

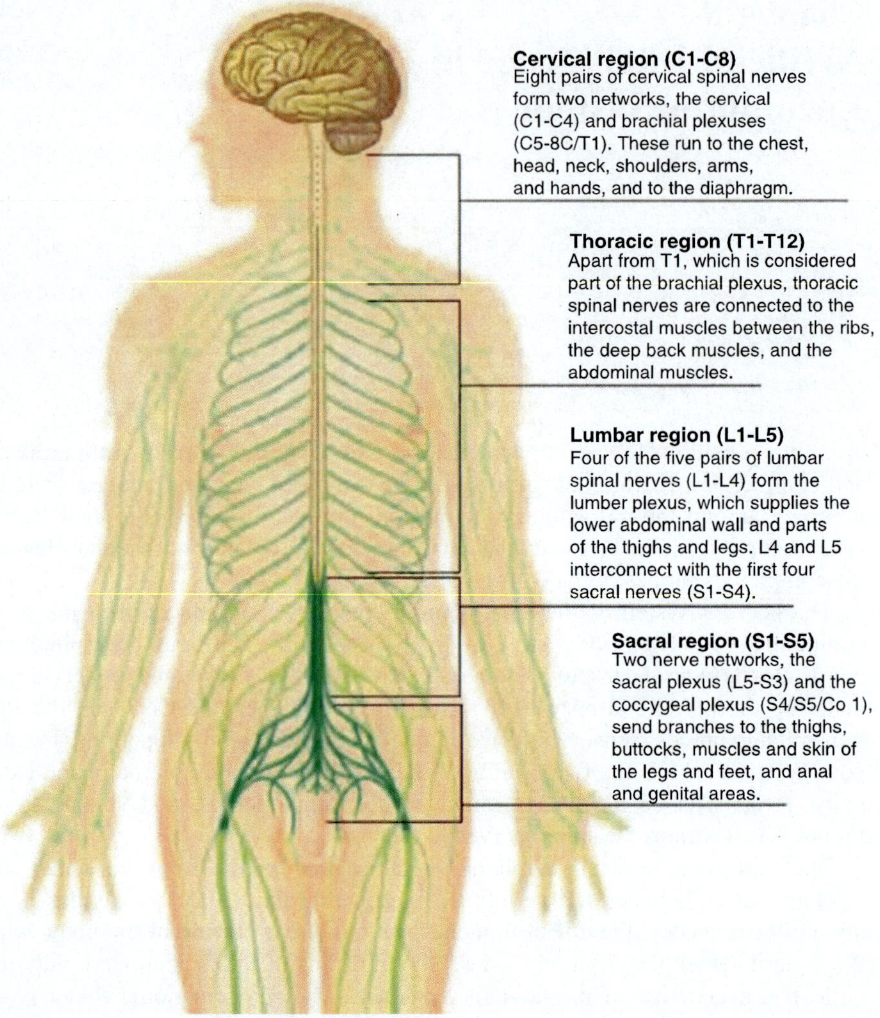

Fig. 8.1 Illustration of spinal nerves

brain plays an irreducible role in the maintenance and performance of the body, including controlled movements and autonomic functions, as well as thoughts, emotions, memories, and behavior [6].

It is known that the function of such adaptive behaviors is disturbed in patients who exhibit intractable medical conditions, and also patients who exhibit injuries, infections, or diseases through damages to the brain or spinal cord. Some of these types of damage are quite common, and together they are responsible for some of the most intractable medical conditions. The most serious of the intractable diseases include traumatic brain injury (TBI), which follows an impact to the skull; stroke, where damage results from the interruption of the blood supply to the brain; and a range of neurodegenerative disorders, such as Parkinson's disease or Alzheimer's disease, where populations of brain cells undergo degenerative change. Whatever the cause of the brain damage, there are two particular consequences [6, 8]. First, neurons are lost. This can be a diffuse loss, such as happens in Alzheimer's disease, or the complete loss of a particular area of brain, which can occur following a stroke. Specific populations of neurons may also be targeted by disease processes, such as the midbrain dopaminergic neurons lost in Parkinson's disease, the spinal and cortical motor neurons in motor neuron disease, or spinal inhibitory neurons in neuropathic pain [98, 101, 119]. In other conditions, for example Batten's disease, almost every nerve cell can be affected. Similarly, the loss can be acute as in TBI or stroke, or chronic and progressive as in Alzheimer's disease. Whatever the specifics, however, lost neurons cannot generally be replaced. The second consequence of damage is that the connections between brain cells are lost [8]. A central example of this is spinal cord injury (SCI), where typically a fall or a motor vehicle accident causes damage to the spinal column. Consequently, the connection between the brain and the spinal cord is partly or completely severed, and both sensation and control of muscles are lost or reduced significantly. Many of these intractable conditions and/or types of damages lead to serious disorders that affect memory, cognition, movement, or consciousness. They cause conditions such as chronic pain, and many of the most intractable medical conditions [3, 29].

Space Exploration

Exposure to microgravity and the space environment during short-and long-duration space missions provokes physiological impairments in astronauts. These impairments include changes in perception, balance, locomotion, and gaze control; nausea and spatial disorientation; problems with eye-head coordination; bone loss; muscle atrophy; orthostatic hypotension; and increased cancer risk from space radiation [9, 11, 12, 22, 24]. Optimal speed and quality of spaceflight recovery are likely dependent upon improved characterization and understanding of the flight-related changes in neuro-vestibular, sensorimotor and other motor control systems.

Spaceflight significantly affects the neuromuscular system, causing a loss of muscle strength and functionality. The microgravity environment induces atrophy in the muscles, leading to declines in muscle mass, volume, and strength due to the reduced mechanical loading on the body [11–13, 23]. The decreased muscle and joint loading also lead to decreases in neuromuscular activation. Animal studies [9, 11, 25, 26] have shown that the detrimental effects of microgravity are more extensive in slow-twitch muscle fibers than fast-twitch and in extensors more than flexors. This may significantly impair the movement of the astronaut since slow-twitch muscle fibers are the first to be activated in all movements, jeopardizing the ability of the astronaut to respond and the success of the mission.

Alterations in the sensorimotor behavior tasks include head-trunk coordination, lower limb muscle activity patterns, kinematics, and the ability to coordinate effective landing strategies during jump tasks. To counter these effects, astronauts regularly perform weight-bearing exercises; however, these tasks do not thwart loss of muscle strength and functionality during long space flights [18–21]. These changes include losses of muscle collagen concentration, mature actin filament cross-links, and non-collagenous muscle protein. Recovery from the changes in muscle structure and metabolism due to spaceflight may take months or even years [15, 16].

In addition to the structural changes in the muscles, there are detrimental changes in the nervous system as well. This sophisticated sensory-motor system provides the proper balance and control of movement by using the motion-sensing capabilities of the auditory, visual, vestibular, and somatosensory systems. The visual system is the primary system for spatial orientation; and an impairment of gaze and head stabilization from spaceflight may lead to perceptual distortion, disorientation, and performance changes [15–17, 27]. The vestibular system functions as an ultra-sensitive accelerometer to detect head motion and direction, whereas the somatosensory system assists in body spatial awareness and perception. Astronauts rely heavily on these systems for movement control and balance, not just in flight, but post-flight as well. Thus, any changes induced during flight must be countered to ensure the health and safety of the astronaut and the mission.

Increasing attention is being paid to physiologic challenges faced in human spaceflight. These challenges require the acquisition of knowledge about the hostile environment that astronauts encounter during spaceflight, an environment that is characterized by extreme variations in temperature, the absence of atmospheric pressure, solar and galactic cosmic radiation, and zero gravity. These physiological challenges will require either partial or complete solutions to allow for sustainable human exploration beyond the protective environment of Earth. Key physiological issues have been examined in the context of a human expedition to Mars, and several possible countermeasures have been evaluated [15–17, 28]. However, most or even all of these measures have had only marginal results. Despite extensive in-flight exercise, most astronauts experience difficulties in standing, walking, and adjusting for several days after landing. This poses a serious problem in the case of an emergency landing on Earth or a landing on Mars after a long-duration space flight.

Automatic postural responses, which are mediated in the brain stem and subcortical area, coordinate movements across joints. Like reflexes, automatic responses are activated by external stimuli. These responses may be thought of as over-learned, "long-loop" reflexes that rapidly respond by resisting disturbances. Automatic responses depend on movement strategy, which in turn is dependent on the experience of the person and on surface conditions. Automatic reactions are also adaptable to specific balance demands. Unlike reflexes and automatic responses, voluntary postural movements, mediated by the brainstem and cortical area, may be initiated in response to an external stimulus, or they may be self-initiated and generate purposeful movements and behavior [14–16].

Postural adjustments associated with voluntary movements are organized on the basis of internal representation. In order to know when and how to apply restoring forces to keep the center of mass (COM) within the base of support (BOS), the CNS must have an accurate picture of where the body is in space and whether it is stationary or in motion [10, 12, 14]. Therefore, internal representations (provided by the postural body scheme) of body geometry, body dynamics (support conditions), and body orientation through sensory inputs are essential for the mapping of sensation to action.

Countermeasures are a means to (i) maintain physiological balance, health, physical fitness, and mission performance and (ii) reduce the health risks and improve the safety of human spaceflight. The countermeasures typically aim at preventing, mitigating, or minimizing the effect of adverse or harmful risks on the crew [13, 14]. From a neurophysiological/biomechanical perspective, countermeasures involve the interaction of different levels of balance control mechanisms, to maintain or return the center of mass of the body within the limits of stability as determined by the base of support (i.e., the area of the feet). Balance is related to the inertial forces acting on the body and the inertial characteristics of the body segments. Furthermore, the limits of stability are the boundaries of an area of space in which the body may maintain its position without changing the base of support [16, 17].

Application of Human Locomotor System in Surveillance

The physiological and technological advances inherent in the human locomotor system have given birth to an emerging technology for the identification of suspicious/malicious behavior in mass transit [30, 32–34]. This new technology applies multi-disciplines ranging from machine vision, neuroscience and the evolutionary theories, progress in brain and behavioral research, physiology, and psychology, towards the understanding of mechanisms, functions, and emotional processes.

The main thrust of this emerging technology involves the introduction of visual automated systems for surveillance at airports and places of mass transit. It may consequently involve motion detection and tracking, behavior and activity analysis and understanding, and the identification of person(s) [34, 35].

There is an acute need for development of intelligent systems that would enable both the verification and identification of humans for the main purpose of identifying malicious and suspicious behaviors in mass transit. It is urgent to invest substantial effort into detecting and thwarting terrorist and insurgent attacks by observing suspicious behaviors of individuals at transportation checkpoints and elsewhere [33–35, 37]. Related technologies and methodologies abound, but their volume and diversity have sometimes been overwhelming. Also, effectiveness claims sometimes lack a clear basis in science and technology.

Recognition systems based on biometrics originate from real-life criminal and forensic applications. Biometrics may be defined as the science and technology of measuring and analyzing biological aspects of human being [37, 38, 41]. Biometrics Technologies are classified into two categories shown in Fig. 8.2.

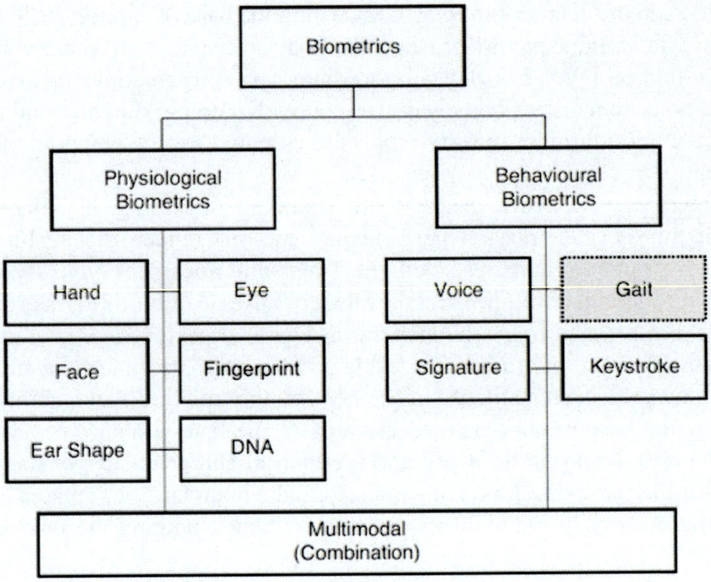

Fig. 8.2 Classification of biometrics technologies

Physiological biometrics are those biometrics derived from direct measurements of a part of the human body. The most common and successful types of such measures to date are fingerprints, face recognition, iris scans, and hand scans. Behavioral Biometrics involves those characteristics based on an action performed by an individual [37, 38, 41].

According to theories from neurophysiology/biomechanics, gait dynamics have meaning and may be useful in providing insight into the neural control of locomotion of the human (neuromusculoskeletal system).

New technologies involving the application of electroencephalograms (EEGs) allow some physiological features to be observed without "wiring up" individuals, sometimes at a distance, and sometimes covertly or surreptitiously, as with using heat-sensitive cameras to detect capillary dilation and blood flow to the face and head. There is some evidence of unique value in indicating deception or imminent action by an individual if baseline information is available for that specific individual ahead of time or if credible intelligence about a possible attack is available [43, 44, 46].

Most of the technologies are in a relatively early stage of development, but there does seem to be potential. Measurement of physiological signals closer to the central nervous system (i.e., the brain) holds the most promise for detecting guilt and behavioral intent [45, 46]. Another approach may seek to automate recognition of potentially threatening body postures or poses.

Recognition of emotional expressions based on automated algorithms and computation is still in its infancy, but this is an active field of development, and improved algorithms are likely to yield greater accuracy and robustness. Furthermore, as with many pre-incident indicators of attack, emotion-recognition algorithms that fuse multiple parameters may exhibit better performance than inferring emotional state simply from facial expressions alone. This visual surveillance technologies, CCD cameras, thermal cameras, and night vision devices, count among the three most widely used devices in the visual surveillance market.

Current and existing technologies have always applied a single source of information data for the detection of suspicious and/or malicious behaviors in mass transit. It becomes more apparent that false alarms take control over the efficiency of such systems because these single-sensor single algorithms are severely limited in their ability to resolve ambiguities in a complex system involving the detection and analysis of complex human behavior [40, 41, 46]. These capabilities definitely require the application of the principles of computational intelligence to integrate information from different sensory sources, to coordinate and guide the observations made by each sensor, and to improve the reliability of automated assessment by reducing the uncertainty and ambiguity inherent in making decisions based solely on a single type of information source.

Fusion of multiple sensor data is critical to achieving efficiency when identifying and differentiating aberrations in behavior patterns, and to achieve objectivity when making a determination of a situation assessment [1, 3]. There is the urgent need of a system that will automate the visual surveillance system including motion detection and tracking, behavior and activity analysis and understanding, and person identification.

It is urgent to be able to identify a suspicious, malicious behavior of an individual in mass transit and to conduct a continuous conditioned monitoring of this individual in "tag and track" manner in real-time. Disciplines including, but not limited to, neurophysiological/biomechanics (three-dimensional movement analysis of the lower limb, the stride length/sec) data, electrocardiogram (ECG) for monitoring the heart rate (HR), heat-sensitive cameras for detection of capillary dilation and blood flow, and the understanding of mechanisms, functions, and evolutionary processes play a significant role in the determination of emotions The information from the multiple sources; gait data, heart rate variability, blood flow, and emotional expressions need to be fused together using the methods of computational intelligence.

According to Steimer [32], emotions have often been considered as unique to human beings, and studies basically treated emotions from a philosophical perspective. In fact, a major goal of modern neuroscience is the progressive introduction of the study of emotions into the field of biology, and the understanding of mechanisms, functions, and evolutionary significance of emotional processes. This notion suggests the progressively introduction of evolutionary theories and progress in brain and behavioral research, physiology, and the psychology of emotions into the field of biology [32, 35, 37].

The modern era of emotion research has progressively led to view emotions as being experienced or expressed at three different, but closely interrelated levels: the mental or psychological level, the (neuro) physiological level, and the behavioral level. These three complementary aspects are present in even the most basic emotions, such as fear.

The recommended practice of the APTA SS-SRM-RP-009-09 entitled "Identifying Suspicious Behavior in Mass Transit" clearly contends that the perpetrators may be nervous, and nervous tendencies are also indications of suspicious behavior [35]. Therefore, transit personnel can assess passengers by looking for nervous trends. Table 8.1 shows the nervous tendencies suspects may demonstrate.

Table 8.1 offers the opportunity to monitor such parameters like trembling, pacing, and other physical reactions, as significant features for the detection of suspicious behaviors [35].

Many methods exist for understanding and categorizing emotions. Emotion, one crucial aspect of human behavior often overlooked by researchers, operators, and policymakers who often view it as too "soft" for serious consideration or research, serves a crucial purpose in understanding any individual or group behavior. For the individual, emotions are evolved information-processing systems that aid in survival. These transient, fleeting reactions to events may impact a person's welfare and require immediate response. Emotions prime behaviors by initiating unique physiological signatures and mental structures, aid in bonding memories and cognitions, and, most important, serve as a motivator of human behavior [33, 35, 37, 38].

Humans may also have standard ways of nonverbally communicating to others threats of impending violence.

Identifying the nonverbal cues humans use to warn others of their growing aggression levels could be important for research and practice in criminal justice and psychology. Such knowledge could help criminal justice and mental health professionals recognize when a client is posing a potential threat. Such knowledge could better inform research and therapy with chronically violent persons, helping them identify their own behavioral precursors of violence. It could also be useful in educational programs for those with autism, or brain injuries to the frontal lobe, who have difficulty naturally interpreting the social meanings behind the nonverbal cues of others.

Hausdorff et al. [53] performed studies for simultaneous monitoring of the electrocardiogram (ECG) and walking stride rate in order to study the relationship between ECG changes (e.g., arrhythmias, ischemia, mean heart rate, and heart rate variability) and physical activity (e.g., supine, standing, walking, and walking rate) on a beat-by-beat and stride-by-stride basis.

Table 8.1 Nervous tendencies, Apta Standards Development Program 2009 [12]

Trembling	Continuously scanning the area	Aggressively biting nails
Unusual perspiration	Heavy breathing	Eyes wide open
Fidgeting	Rubbing hands	Pacing
Clock watching	Repetitively touching face	Exaggerated yawning
Avoiding uniformed officers	Exaggerated emotions such as crying or laughing to self	Other physical reactions

Many investigations had shown that measures of heart rate dynamics and heart rate variability, an indicator of cardiovascular health, are sensitive to alterations in the autonomic nervous system's regulation of heart rate [44, 45].

However, little was known about how gait affected these heart rate dynamics. Among other things, initial use of this technique revealed an inverse relationship between heart rate variability and stride rate (cadence) variability: healthier cardiovascular dynamics were associated with healthier gait dynamics. These findings suggested and later studies confirmed that, at least in certain populations, gait instability is related to, and perhaps a function of, cardiovascular health and also demonstrate that stride variability reflects more than just musculoskeletal mechanics. More generally, this initial work suggested that just like there is much to be gained by investigating heart rate dynamics, above and beyond the study of the average heart rate, similar investigations of gait dynamics may provide insight into locomotor control [37, 39, 41].

These findings also demonstrate that measures of variability are sensitive markers of physiological function and amenable to change with the appropriate intervention.

The overview of the computational intelligence scheme for the detection of suspicious and/or malicious persons is illustrated in Fig. 8.3. The multi-modality data acquired by the RGBD camera stands for Red, Green and Blue camera that provides both depth and color data (RGBD) cameras and proprioceptive sensors are fed into

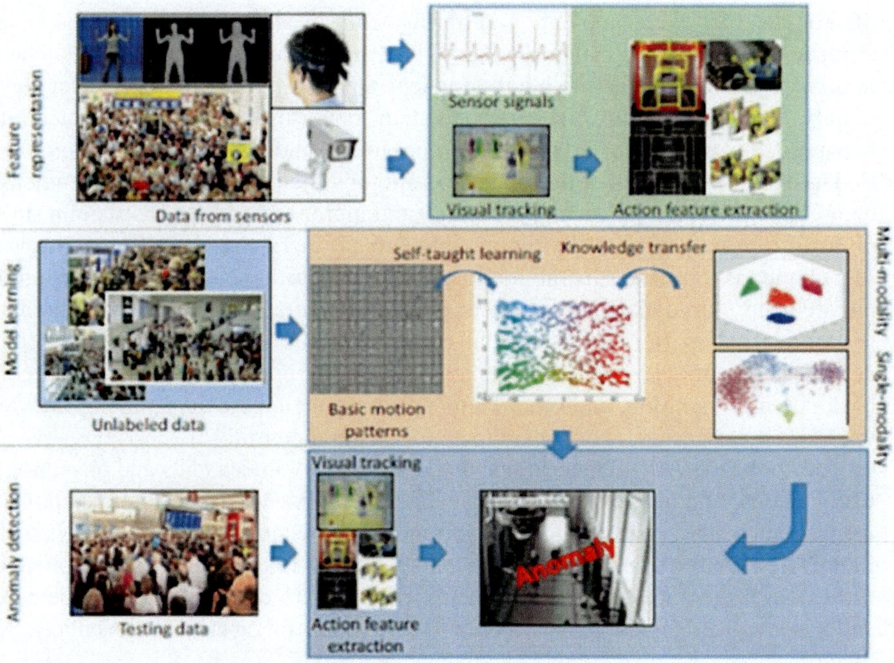

Fig. 8.3 Principle of automated detection system

the system. Signals such as ECG signals from proprioceptive sensors and visual tracks from RGBD videos are extracted from the input data. Action features are then extracted from the input videos. Data collected from real-world scenarios are also used to enrich the training data. The relationships between single-modality visual data and internal mental state may be learned by projecting the data onto a subspace, where their reconstruction error is minimized. Visual representative clusters are also learned from single-modality visual data. A certain amount of unlabeled data is introduced into the data pool in order to further enrich the training data, and learn the basic motion patterns using the self-taught learning approach.

Given real-world testing data, the anomaly may be detected by transferring the learned knowledge and using the learned appearance clusters [38, 41].

The methodology takes advantage of both normal and abnormal activity data. These data may be clustered into several representative clusters, some of which are normal activity data and some of which are abnormal data. Since the abnormal data are significantly sparse than the normal data, a data mining algorithm may be applied to discover the most representative normal data as the templates of normal data. The training data and testing data may be captured in different scenarios.

Emerging Technologies

Technological initiatives are currently at a pivotal point to improve the quality of life for the very large population suffering from disabilities, injuries, and other intractable diseases and handicaps due to impairment or aging [53–55, 70]. These initiatives may lead to facilitating their independent living, promoting greater community/social integration, and enhance their opportunities for communication, transportation, education and training, employment, and leisure.

The dynamic behavior of the human locomotor system in space exhibits immense complexity involving physical, perceptual, and motor aspects. It consists of a stack of superimposed segments involving a kinematic hierarchy whereby each lower segment tends to transfer its motion to all superimposed ones [1, 4]. The system is hierarchically structured and consists of linked references that are anchored in gravito-inertial space. The perceptual notion of space is derived mainly through vestibular sensors [1, 4]. Biomechanics, multi-body dynamics, multimodal feedback control, cognition, and other factors obscure the interactions among the vestibular, somatosensory, and visual inputs for postural control [31, 36, 42].

The application of engineering methods, biology, biomechanics and physiology, mathematics, and artificial and computational intelligence in an interdisciplinary manner may offer insight into new and emerging technologies that may help to illuminate our understanding of neuro- and pathophysiology, analyze physiological processes in the musculoskeletal system, as well as to develop methods, devices, and systems to diagnose, and monitor systems and therapeutic interventions [1, 30, 46].

This would lead to the improvement of the quality of diagnosis, prognosis and follow-up for individual patients suffering from neurological disorders/impairments, or intractable diseases.

By virtue of the immense complexity inherent in the human locomotor system, the application of systems analysis approach, psychophysics, measurements of the complex motor behavior, as well as artificial and computational intelligence may offer insight into our understanding of the neural processes inherent in the musculoskeletal system [52, 58].

Artificial Intelligence (AI)

Artificial intelligence (AI) refers to systems that display intelligent behavior by perceiving, reasoning, and analyzing their environment and taking action, with some degree of autonomy, to achieve specific goals, or the automation of intelligent behavior.

Artificial intelligence constitutes the science and engineering discipline that makes intelligent machines (sensors, actuators, controllers, robots, etc.), especially intelligent computer programs. It constitutes a way of making a computer, a computer-controlled robot (or other systems), or a software think intelligently, in a similar manner the intelligent humans think [57, 59, 60]. AI is accomplished by studying how human brain thinks and how humans learn, decide, and work while trying to solve a problem, and then using the outcomes of this study as a basis of developing intelligent software and systems (sensors, actuators, controllers, robots, etc.).

The notion of intelligent behavior may be depicted as the ability to perceive one's environment (through sensors), to act in complex environments, to learn and understand from experience, to reason towards solving problems and to discover hidden knowledge, to apply knowledge successfully in new situations, to think abstractly using analogies, to communicate with others, and more [60].

Intelligent Measurement and Instrumentation

The science of measurement embraces both experimental and theoretical determinations at any level of uncertainty in any field of science and technology [63, 64, 66]. It is, therefore, at the core of all practical scientific endeavors. It exhibits significance because almost all of everyday life, including but not limited to practical science, technology, engineering, and medicine, involves measurements that we rely on for our health, commercial prosperity, quality of life, and the protection of the environment. It constitutes the structure that ensures that these measurements are stable, comparable, and accurate, providing confidence in measurement at a stated level (usually by quoting a measurement uncertainty). When these qualities are associated with a measurement, it reduces waste, allows trade, enables infrastructure to function, technology to advance, economy to prosper, encourages global agreement, collaboration, and trade, and ensures our ongoing health and safety and quality of life. In fact, measurement science generates systems and frameworks for quantification, and through these, underpins consistency and assurance in all measurements.

Improvements in science and technology enable advances in measurement science. There is an utmost requirement for measurement more than ever before to ensure sustainable progress in science and in society [59, 60]. Good measurement enables societal progress and economic growth via an agreed quality infrastructure. Measurement sits at the top of this quality infrastructure providing the stable definitions of measurement units and realization of primary standards for these, from which traceability chains can flow; and improving these realizations to continue to reduce uncertainties for the end user.

Intelligent instrumentation constitutes the key enabler to enhance productivity and to ensure consistent quality through the integration of sensors, information processing, actuators, and control systems into neuro-devices. Through this integration, the /machines/systems acquire intelligence (Machine Intelligence Quotient, MIQ).

Modern health care relies heavily on a large array of sophisticated medical instrumentation to diagnose health problems, to monitor patient condition, and administer therapeutic treatments, most often in a noninvasive or minimally invasive manner. Some examples of bio instrumentation include:

- Electrocardiographs (a device that measures the electrical activity of the heart).
- Electroencephalograph (a device that measures the electrical activities of the brain).
- Electromyography (a device that measures the electrical activities of muscles).
- Laser systems for eye surgery.
- Image scanners (nuclear cameras, CAT, MRI).
- Drug infusion and insulin pumps.
- Anesthesia machine.
- Kidney dialysis machine.
- Electrosurgical equipment.
- Cardiac pacemaker.

Case Study I: Measurement Science in the Human Locomotor System

In the Laboratory for Human Motion Analysis and NeuroRehabilitation, Murad Alaqtash and Huiying Yu designed an experiment to extract the kinematic, kinetic, and the electrical activities of the muscles (electromyography) during walking tasks [67–69].

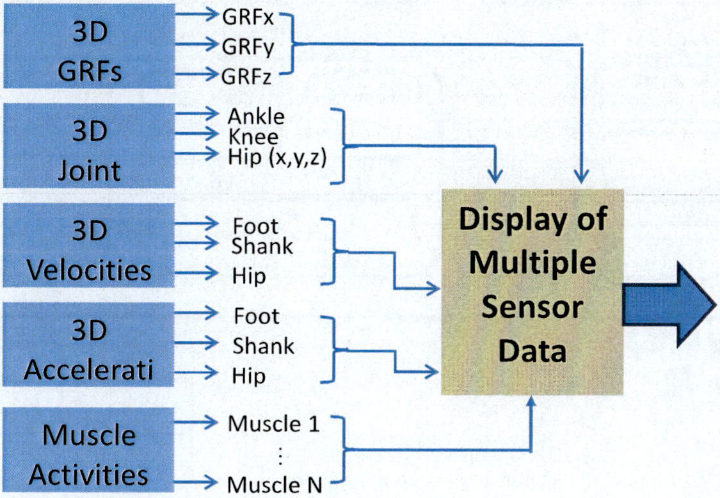

Fig. 8.4 Concept of the measurement design

The concept of the measurement (data acquisition) is illustrated in Fig. 8.4.

The data acquisition system consists of an instrumented treadmill for the ground reaction forces (GRFs), surface electromyography (EMG) system for monitoring the electrical activities of the muscles, and wearable inertial sensors for the gait kinematics.

Psychophysics was applied whereby each subject walked on the instrumented treadmill (Bertec Corporation, USA) at a self-selected natural speed continuously for 3 min (this constitutes the input stimulus). The self-selected natural speed was a subjective speed selection (comfortable speed) by the subjects.

The analysis of the acquired kinetic, kinematics, and electromyographic data was performed using intelligent systems involving the methods of feature extraction, multi-sensor data fusion, knowledge base system, and fuzzy-based inference system, Fig. 8.5.

The responses from the psychophysics input (stimuli) were recorded, processed, and displayed. These responses include the kinetic, the kinematics (from the wearable sensors, the IMUs), and the electromyographic (EMG) data, and are depicted in Fig. 8.5.

It may be noted that the responses on the left side refer to the averaged responses of able-bodied subjects (reference patterns), whereas the responses on the left depict the "blinded" input patterns, respectively.

From Fig. 8.6a, the curves on the left side illustrate the muscle activities in averaged able-bodied subjects. The curves on the right depict the muscle activities of a subject suffering from Cerebral Palsy.

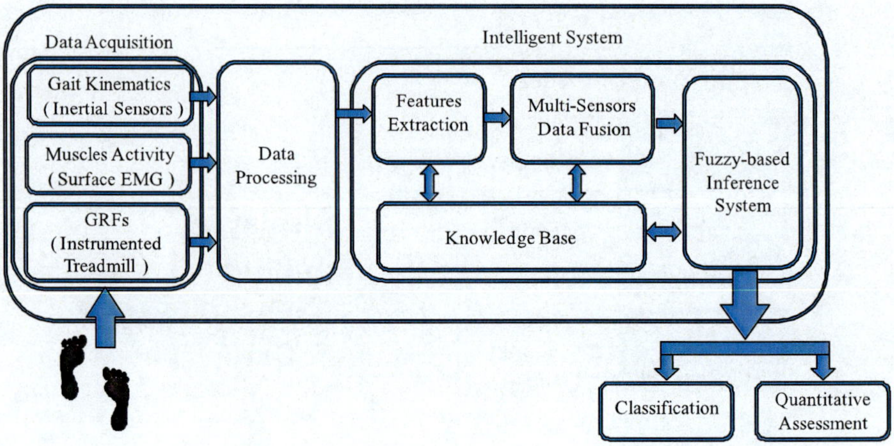

Fig. 8.5 Concept of the intelligent measurement system

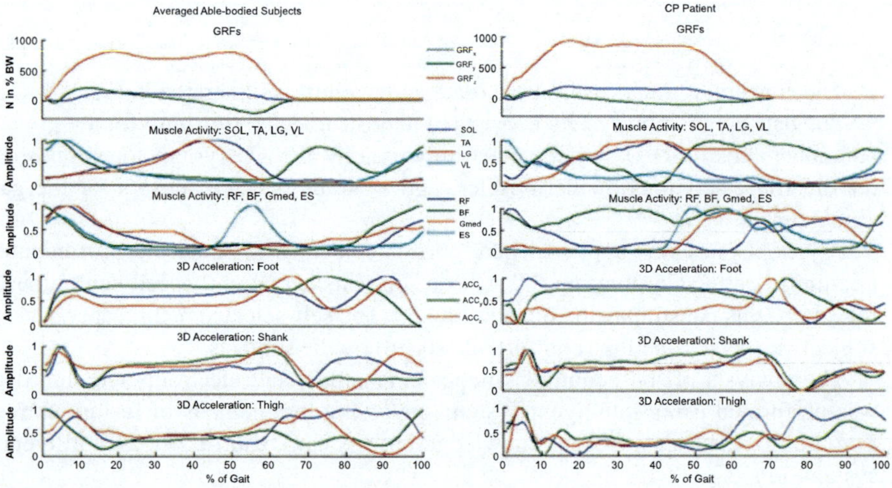

Fig. 8.6a Illustration of muscle activities and kinematics data

From Fig. 8.6b, the curves on the left side illustrate the muscle activities in averaged able-bodied subjects. The curves on the right depict the muscle activities of a subject suffering from Multiple Sclerosis.

The further processes depicted in Figs. 8.3 and 8.4 will be tackled in the coming section (on Computational Intelligence).

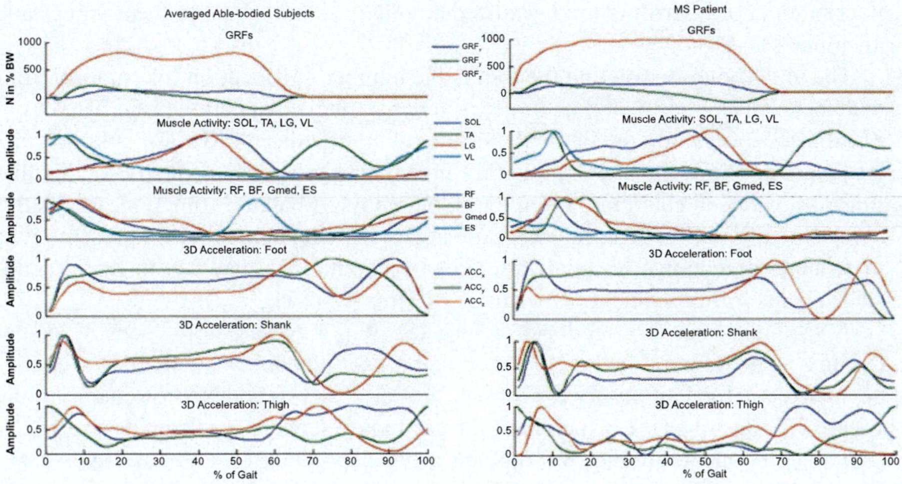

Fig. 8.6b Illustration of muscle activities and kinematics data

Case Study II: Measurement Science in the Human Locomotor System (Conceptualized and Performed in the Laboratory for Human Motion Analysis and NeuroRehabilitation)

About 60–70% of people with diabetes have mild to severe forms of nervous system damage, causing peripheral neuropathy including impaired sensation or pain in the feet or hands, foot deformity and callus formation resulting in focal areas of high pressure, Carpal Tunnel Syndrome, poor glucose control leading to impaired wound healing, and other nerve problems [79–82]. In diabetes, peripheral neuropathy is an anatomically diffuse process primarily affecting sensory and autonomic fibers; however, distal motor findings may be present in advanced cases. Patients with diabetes mellitus and peripheral neuropathy develop loss of protective sensation (LOPS), whereby a person is unable to feel minor trauma from mechanical, thermal, or chemical sources. When foot lesions are present (foot ulceration), the reduction in autonomic nerve functions may also inhibit wound healing, which leads to amputation. About 15% of people with diabetes develop at least one foot ulcer in their lifetime. The risk of lower extremity amputation is 15–46 times higher in diabetics than in persons who do not have diabetes mellitus. More than 60% of nontraumatic lower-limb amputations in the United States occur among people with diabetic neuropathy [8, 41, 83, 99]. Patients with diabetes mellitus with long-term peripheral neuropathy eventually develop loss of neural sensory feedback. Furthermore, limited joint mobility is common in patients with diabetes mellitus. Peripheral neuropathy progresses in long-standing diabetes, the joints are unable to recognize the forces put across them and the relative positions of the various joints, sustaining microtrauma or microfracture because the body does not adjust to these forces and positions [84–87].

The underlying reason that the foot is the number one location for comorbidities in diabetes places peripheral neuropathy and its sequela that include LOPS and Loss of Proprioception atop the list with peripheral arterial disease (PAD) ranking a distant second. These risk factors magnify preexisting biomechanical risk factors that function in closed chain into often overlooked loss of balance, function, productivity, and quality of life that precede or are adjacent to the more obvious presentations of wounds, infections, hospitalizations, and amputations that are, in reality, end-stage events for more and more diabetic patients [88–91].

A secondary sensory feedback device (SSFC), a novel ectopic sensory center that may substitute for the loss of the somatosensory feedback on the foot, was conceptualized in the Laboratory for Human Motion Analysis and NeuroRehabilitation. Figure 8.7 illustrates the principle of the secondary sensory feedback device.

The components of Fig. 8.7 include a novel polymer-based capacitive sensor (mechanoreceptors) as the in-sole plantar pressure sensor for the acquisition of the plantar pressure data; a processing unit for the display of the acquired data using a sensor validation scheme; a transducer for the conversion of the in-sole plantar pressure into a proportional output-voltage, and a corresponding visual display of these values on the subject's thigh (this will make the subject aware of the pressures emanating from which area of the foot); a rapid and reliable data analysis system capable of identifying, differentiating and quantifying clinically relevant variations in the output plantar pressures; and a final intelligent polymer-material to convert the

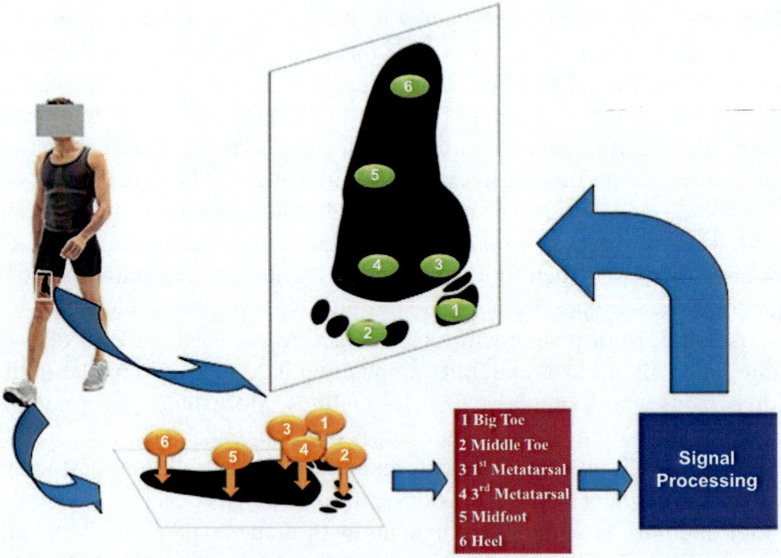

Fig. 8.7 Principle of the secondary feedback device

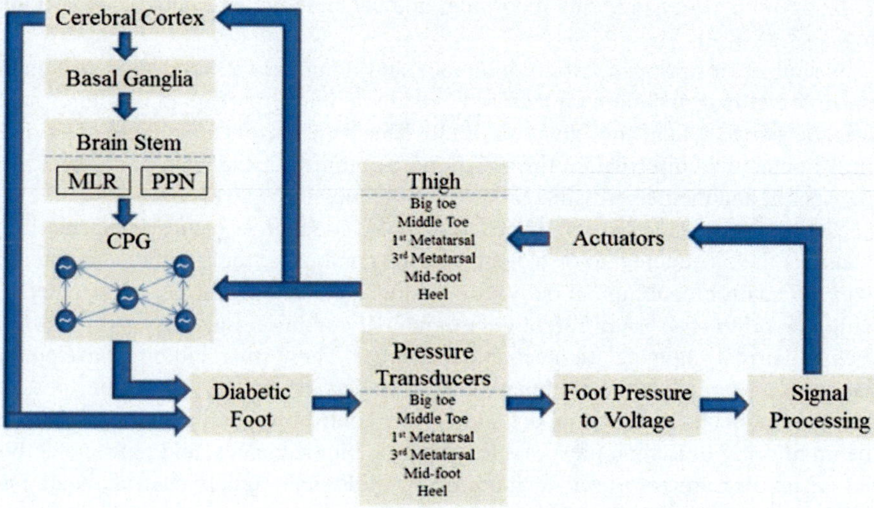

Fig. 8.8 Block diagram illustrating the neurophysiological concept of the SSFC

output signal into some sensation on the bearer's thigh (similar to an actuator to activate the sensations on the subject's thigh).

The mechanorecptors provide the central nervous system with information including the muscle spindles, joint afferents, and Golgi tendon organs. These receptors are also sensitive to ankle rotation and will give information regarding balance [78, 94, 99].

The secondary sensory feedback device will transmit signals of the foot plantar pressure to an alternative site of the body. The sensory feedback device will eventually develop a secondary sensory feedback center (SSFC) that will help patients with peripheral neuropathy to recognize the high pressures applied on the areas of their foot plantar, and therefore to reduce the chance of traumatic injury, infection, and foot amputation.

Over time, the brain of the patient undergoes a process of cerebral plasticity and begins to identify the remote or ectopic sites as the plantar aspect of the foot [92, 94–96]. In addition, the SSFC will also help patients with prosthetic devices below the knee to provide proprioceptive sensory feedback of the foot through alternate segments of the body to achieve optimal posture and locomotion. Figure 8.8 provides the basic block diagram for understanding the neurophysiological principle of the SSFC.

Biosensors

Wearable biosensors exhibit the potential in the monitoring of human health and personalized medicine. They are portable electronic devices that integrate sensors into/or with the human body in various forms; gloves, clothing, and implants, realizing in vivo sensing, data recording, and calculation using mobile or portable devices [29, 118].

Wearable biosensors exhibit continuous and noninvasive monitoring of human physiology through dynamic measurements of chemical markers in bio-fluids such as tears, sweat, saliva, and interstitial fluid. These biosensors are able to offer real-time biochemical information towards a more comprehensive view of the health of the wearer, and the performance or stress at the molecular level.

Smart clothing consists of fibers, filaments, conductive devices, and clothing materials that is attached to or woven with the conductive tools, which can interact with the environment/human body. It constitutes the intelligent or smart materials with the ability to sense different environmental conditions and respond to stimuli such as thermal, chemical, or mechanical changes. These smart clothing have potential applications in healthcare, fitness, health, and safety. Usually, E-textiles possess three components; a sensor, an actuator, and a controlling unit, and are involved in the monitoring of human physiological signals, biomechanics, and physical activities. E-textiles are also used to monitor physiological signals such as heart rate (HR), temperature, and breathing rate.

The lachrymal gland secretes the lachrymal fluid or the human tears. They contain both low and high-molecular-weight compounds that diffuse directly from the blood, and exhibit close tear-blood concentration correlations. They are also less complex than blood and are part of the antifouling mechanism of the eye. These characteristics make human tears an attractive noninvasive diagnostic bio-fluid for healthcare monitoring applications.

They contain both low and high-molecular-weight compounds that diffuse directly from the blood, and exhibit a close correlation in concentration between tear and blood concentration. Tears are also less complex than blood and are part of the antifouling mechanism of the eye. These characteristics make human tears an attractive noninvasive diagnostic bio-fluid for healthcare monitoring applications. In fact, tear glucose concentrations have been shown to exhibit good correlation with blood glucose levels, reflecting the diffusion from the lachrymal artery. Reflex tears, generated during emotional or mechanical stimulation, have different compositions than basal tears, which make up the protective tear film covering the eye surface at all times. These variations and challenges highlight the need for developing wearable noninvasive tears-sensing platforms without eye irritation.

Tongue-Computer Interface

There are several individuals who have severe disability/impairment due to causes that range from traumatic brain injury (TBI) and spinal cord injury (SCI) to amyotrophic lateral sclerosis (ALS) and stroke. This population of people finds it extremely difficult to carry out activities of daily living [97, 100, 102, 103].

The cranial nerves that are protected by the skull, often escape injuries even in severe cases of the spinal cord injury. Many of these nerves are not afflicted by neuromuscular diseases either. Therefore, the parts of the body that are innervated by the cranial motor nerves, such as eyes, jaws, facial muscles, and tongue, emerge as good candidates for discerning the intentions of the user as intermediate links to the brain.

Both the tongue and the mouth occupy an amount of sensory and motor cortex in the human brain. Therefore, they are inherently capable of sophisticated motor control and manipulation tasks with many degrees of freedom. The tongue muscle is similar to the heart muscle in that its rate of perceived exertion is very low and it does not fatigue easily [51, 71, 72, 104]. The tongue is noninvasively accessible and is not influenced by the position of the rest of the body, which can be adjusted for maximum user comfort. Figure 8.9 illustrates the inductive tongue control system. It consists of:

(a) The activation unit.
(b) The palatal plate.
(c) The inductors.

The tongue activates the sensors by placing the tongue-mounted activation unit at or inside a coil [73–75, 105].

The principle depicted in Fig. 8.9 is based on Faraday's law of induction for a coil, and uses variable inductance techniques. It is aimed at changing the inductance of an air-cored inductance coil, by moving a ferromagnetic material, attached to the tongue, into the core of the coils [51, 73, 75].

As the ferromagnetic material is placed in the coil, the core becomes a combination of air and ferromagnetic material, and the relative magnetic permeability of the

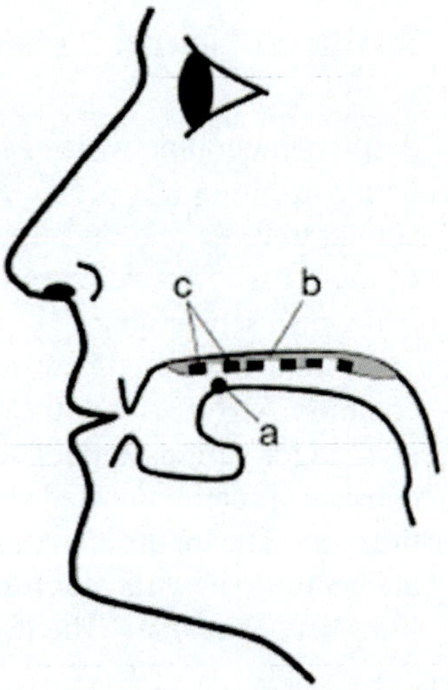

Fig. 8.9 Principle of the tongue-computer interface

core material changes according to the magnetic permeability of the ferromagnetic material.

When a sine-wave current of constant peak-to-peak amplitude is applied, a constant amplitude voltage drop is obtained across the coil. The introduction of the ferromagnetic material into the air gap of the coil results in an increase in the induction, which is sustained until the material is removed [75, 105, 106]. This will be utilized for activation of a command in the inductive tongue control system. The method resembles the known techniques used for displacement sensors.

The sensor outputs are wirelessly transmitted to a smartphone or personal digital assistant, also worn by the user. A sensor signal processing algorithm running on the personal digital assistant classifies the sensor signals and converts them into user control commands, which are then wirelessly communicated to the targeted devices in the environment of the user.

Magnetoelastic Wireless Biosensor

The magnetoelastic wireless detection sensor is a biosensor that is capable of detecting biological warfare agents (Bacillus anthracts spores). This biosensor is affinity-based and it utilizes a phage-derived diagnostic probe as the biomolecular recognition element for the multivalent capturing of the target agents. The sensor consists of a combination of the phage display technique and a magnetoelastic wireless detection platform. Its function is based on the resonance frequency of the magnetoelastic component. When the target agent is bonded onto the surface of the sensor, the resonance frequency of the magnetoelastic biosensor decreases due to the deposition of additional mass of the target agent [74, 107–109].

Artificial Neural Networks

Artificial neural networks (ANNs) are models of biological neural networks which derive inspiration to produce artificial systems that exhibit intelligent computations similar to those that the human brain routinely performs. The neural network is one of the most effective learning methods for the learning and interpreting complex real-world sensor data.

A multi-layer feedforward neural network, shown in Fig. 8.10, is one of the most common ANNs. A multi-layer feedforward neural network has been a standard method for various applications including the analysis of human locomotion [68]. It consists of interconnected set of simple nodes, neurons, distributed in a hidden layer or layers. The inputs are mapped to the nodes through an input layer and the outputs are controlled by transfer function within each node and by adjusting the weights of links between nodes. The learning process in neural networks consists of adjusting

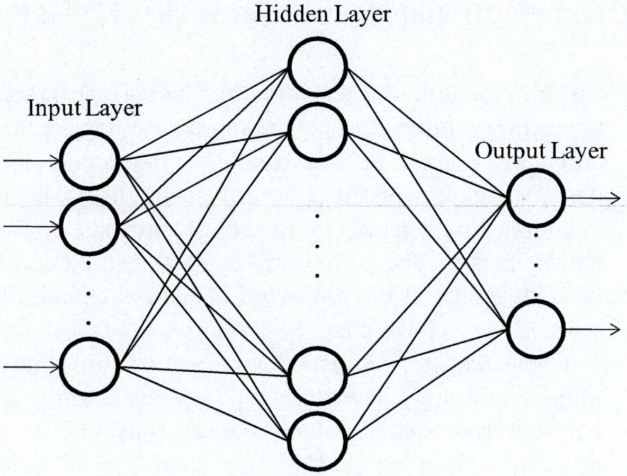

Fig. 8.10 Principle of a multi-layer neural network

the weights through training in which the actual output approximately matches the desired output.

Artificial neural networks (Fig. 8.10) constitute networks that mimic the way the brain works; computer programs that actually learn patterns; forecasting without having to know statistics.

Neural networks that are combined together, constitute deep learning networks. These techniques have enabled significant progress in the fields of sound and image processing, including facial recognition, speech recognition, computer vision, automatic language processing, text classification, etc.

A set of methods that allows a machine to be fed with raw data and to automatically discover the representations needed for detection or classification is known as Representation Learning.

Deep-learning methods are representation-learning methods with multiple levels of representation. They are obtained by composing simple but nonlinear modules that each transform the representation at one level (starting with the raw input) into a representation at a higher, slightly more abstract level [61].

The composition of an appropriate amount of such transformations, very complex functions may be learned. For classification tasks, higher layers of representation amplify aspects of the input that are important for discrimination and suppress irrelevant variations.

An image, for example, comes in the form of an array of pixel values, and the learned features in the first layer of representation typically represent the presence or absence of edges at particular orientations and locations in the image. The second layer typically detects motifs by spotting particular arrangements of edges, regardless of small variations in the edge positions. The third layer may assemble motifs into larger combinations that correspond to parts of familiar objects, and subsequent layers would detect objects as combinations of these parts. The key aspect of deep learning is that these layers of features are not designed by human engineers: they are learned from data using a general-purpose learning procedure.

Convolutional Neural Networks (CNNs)

Convolutional neural networks (CNNs) are designed to process data that come in the form of multiple arrays. Examples include a color image composed of three 2D arrays that contain pixel intensities in the three-color channels. Many data modalities are in the form of multiple arrays; one-dimension (1D) for signals and sequences, including language; 2D for images or audio spectrograms; and 3D for video or volumetric images. There are four key ideas behind convolutional neural networks that take advantage of the properties of natural signals. They include (1) local connections, (2) shared weights, (3) pooling, and (4) the use of many layers [76, 120, 121].

CNNs are a type of deep learning algorithms that are used to process data with a grid-like topology. They are a type of deep learning algorithm that is used to process data that has a spatial or temporal relationship [121, 122]. They are similar to other neural networks, but they have an added layer of complexity due to the fact that they use a series of convolutional layers. Convolutional layers are an essential component of convolutional neural networks (CNNs). Fig. 8.11 represents a typical CNN architecture.

A typical CNN consists of three types of operation layers: (1) convolution layer, (2) pooling layer, and (3) classifier layer.

A convolution layer reads input feature maps or images from previous layers or data input ports, and transforms these inputs using several groups of kernels into a series of output feature maps. Each convolution layer is usually followed by a pool layer that subsamples the features from previous convolution layer. Finally, the classifier layer generates the probability of each category implied in the initial input data.

The convolutional layers constitute a set of filters (also called kernels) that are applied to an input image. The output of the convolutional layer is a feature map. This feature map is a representation of the input image with the filters applied. Convolutional layers may be stacked to create more complex models that may be able to learn more intricate features from the images [61, 120, 121].

The Pooling layers constitute a type of convolutional layer used in deep learning. They reduce the spatial size of the input, thus making it easier for processing and requiring less memory. Pooling also helps to reduce the number of parameters and

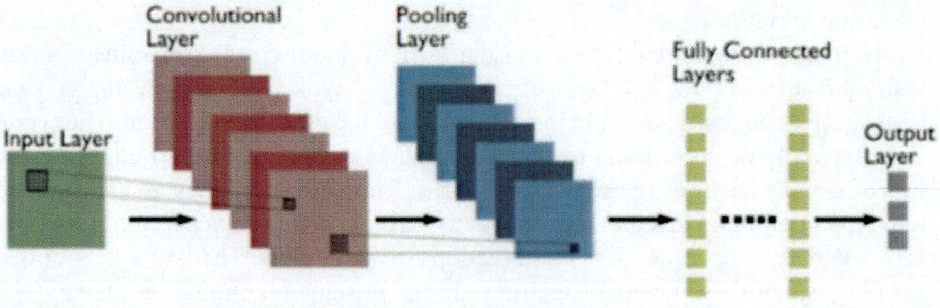

Fig. 8.11 A typical CNN architecture

it makes training faster. There are two main types of pooling; (1) the max pooling and (2) the average pooling, respectively. The Max pooling takes the maximum value from each feature map, whereas the average pooling takes the average value. Pooling layers are typically used after convolutional layers in order to reduce the size of the input before it is fed into a fully connected layer [61].

The fully connected layers constitute one of the most basic types of layers in a convolutional neural network (CNN). As the name already suggests, each neuron in a fully connected layer is fully connected to every other neuron in the previous layer. Fully connected layers are typically used towards the end of a CNN, when the goal involves taking the features learned by the previous layers and using them to make predictions.

Computational Intelligence

Lotfi Zadeh proposed a different view about machine intelligence in which he made a distinction between hard and soft computing techniques, respectively. Hard computing is based on artificial intelligence whereas soft computing is based on computational intelligence. Hard computing exhibits imprecision and uncertainty features of a system whereas these are the foremost features in soft computing. Soft computing may be viewed as a large subset of computational intelligence [62]. The features depicted in these intelligent systems are their human-like capability to make decisions based on information with imprecision and uncertainty.

An intelligent system exhibits learning and adaptation capabilities. Adaptation may be viewed as the ability of a system to change or evolve its parameters or structure in order to better meet its goal [55, 56, 110–112].

The intelligent system has the ability to learn and/or deal with new or unknown situations and is able to make predictions or decisions about future events.

Computational intelligence may be a combination of soft computing and numerical processing, and is, in fact, interdisciplinary, Fig. 8.12. It includes disciplines involving modern adaptation control, optimal control, learning theory, reinforcement learning, fuzzy logic, neural networks, and evolutionary computations [62, 63, 113–115].

Application of Computational Intelligence in Human Locomotor System

The process of neuro-diagnosis presupposes the reliable and efficient acquisition and evaluation of pertinent information at various intermittent stages [1]. The outcome of diagnosis is a significant determinant towards the identification and explanation of the ailments, disorders, and diseases present in a particular subject.

Fig. 8.12 Conceptual scheme of computational intelligence

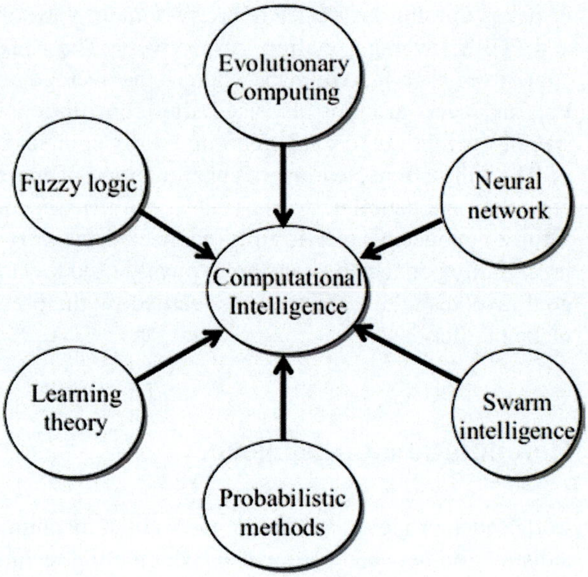

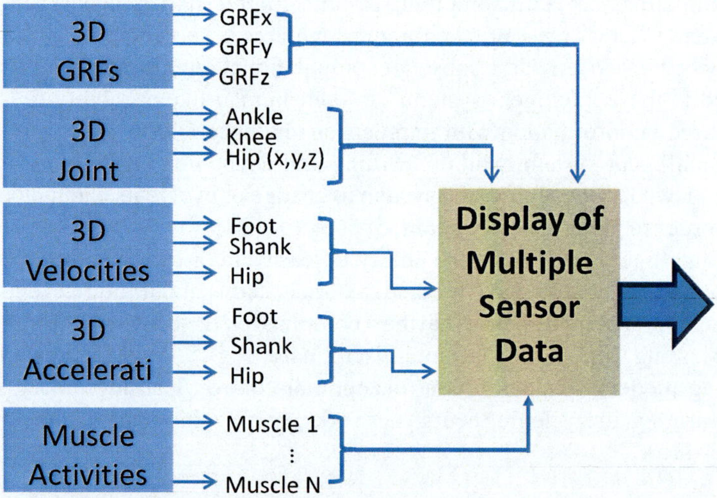

Fig. 8.13 Setup for extraction of biomechanical parameters

Figure 8.13 illustrates the experimental design for the acquisition of the kinetic, kinematics, and electromyographic data in 3D space of the human subject (Courtesy of the Laboratory for Human Motion Analysis and NeuroRehabilitation).

In the study as depicted in Fig. 8.13, masses of data relating to the kinetic, kinematics, and muscle activities were acquired using the inertial sensor array, the instrumented treadmill, and the EMG device during normal walking tasks. The fusion of the multiple complex data streams may provide insights into the neural control of the locomotion [1, 67, 68, 70]. This may provide improved clinical diagnosis, prognosis, and follow-up procedures.

One way of dealing with the functional impairments in locomotor systems may be qualitative, nonnumerical, and imprecise in nature. Therefore, the ratings of the evaluator may be influenced by qualitative factors such as emotion, perception, shifting, and imprecision. One way of unscrambling the data or partitioning the weighted evaluation points into subsets is via the approach of cluster analysis [2, 68, 116, 117]. The subsets may be distinct or fuzzy. However, in situations such as this experimental study, it appears instructive to analyze the data via fuzzy clustering techniques. The general objective of cluster analysis is to group similar data or variables into clusters. This is usually a sequential process, involving pair-wise grouping of variables or clusters of variables. A cluster may be defined as a fuzzy subset of an item's space or graph. This clustering can continue until all data or variables are contained in the same overall cluster or can be designated to stop at a certain number of clusters. Thus, data clustering techniques offer a simple way of finding the relationship between data and grouping data together.

Intelligent reasoning, especially in the presence of ambiguity, confusion, missing information, deception, and so forth, are major attributes of human problem-solving capabilities. Therefore, the main task for this system is to perform intelligent decision-making through inference and extraction of conclusions from available data, an inferential reasoning system that can more closely emulate human performance [2, 68].

The acquired data (from the described experiment as depicted in Fig. 8.13) are treated as aggregate information granules that enable the efficient partition of input space and more rapid analysis. This means that we deal with the relationships of the kinetic, kinematics, and muscle activity functions. The relationships depict the attributive features of the human dynamic behavior and are expressed in an implication table, giving rise to a fuzzy relational matrix, established between the dynamic activities during the walking tasks [2]. In this case study, three relationships were established as follows:

1. The acquired kinematics data as a function of the seven gait phases.
2. The acquired kinetic data as a function of the seven gait phases.
3. The acquired muscle activity data as a function of the seven gait phases.

Since data is always conveniently expressed in a tabular form, it may also be appropriate to consider this table as a matrix. Therefore, fuzzy relational matrix may be applied to establish the strength of association or interaction amongst the elements of the human dynamic functions [1, 2, 67, 68]. This association also elucidates a rule base that may be used to provide a model, a model of a feature matrix.

The dynamic analysis of the locomotor system involves a large amount of data and a large number of characteristics. In fact, such complex analysis requires a large number of measurements for a reliable assessment of the situation. In this context, data analysis may be viewed as the search for structure in data by extracting the specific attributes of the subject in order to gain structural information about these data. This may be achieved by clustering, that is, classifying a large amount of data into relatively few classes of similar objects.

The primary objective of clustering is to partition the given data into so-called homogenous clusters [2]. In fact, homogenous indicates that all points in the same group are close to each other and are not close to points in other groups. The clustering algorithm is then used to build pattern classes or to reduce the size of a set of data while retaining relevant information. The separation of a cluster is a fuzzy notion, and the representation of clusters by fuzzy sets may seem more appropriate in certain situations. The captured data exhibits some imprecision owing to subjective and systematic occurrences during the measurements.

When determining the presence or absence of association, interaction, or interconnection between the elements of the kinetic, kinematics, and electromyographic data and the gait phases, the mapping is usually non-fuzzy [2]. But when the severity level of an element of the gait functions must be determined, then fuzzy sets are often involved in the mapping.

Granulation of information is an inherent and omnipresent activity of human beings carried out with the intent of better understanding of the problem and coming up with an efficient problem-solving strategy. In particular, granulation of information is aimed at splitting the problem into several manageable chunks.

In this way, we partition the problem into a series of well-defined sub-problems (modules, mappings) of a far lower computational complexity than the original one we have started with.

We granulate information over time by forming information granules over predefined time intervals. This gives rise to the notion of temporal granulation [2, 65, 68]. Information granules need to be "stable" meaning that they have to retain their identity in spite of some small fluctuations occurring within the experimental data.

The application of cluster analysis is centered on the formulation of relationships for the variables under consideration. Fuzzy clustering provides a richer description of the geometric structure of the data set in many cases, and has a lesser tendency to get stuck in local minima. Memberships can be interpreted as degrees of typicality of degrees of sharing [2].

Information granules are defined in a finite universe of discourse being multidimensional experimental data sets $X = \{Fx\ Fy\ Fz\}$, etc.

1. A fuzzy mapping is established between the 3-dimensional ground reaction forces (GRFs) depicted as set X, and the seven gait phases depicted as set Y.

   - The 3-dimensional ground reaction forces as elements of x out of a set of X:

   $$X = \{FxFyFz\};\text{and}$$

   - The functional phases in a gait cycle whose attributes include:

   $$Y = \{p1p2p3p4p5p6p7\}$$

   This relationship may be viewed as a rule or rule-base that may be used to provide a model. This crisp relationship, $R\ (x,\ y)$, depicts a feature space that describes the association, interaction, or interconnection between the elements of the force plate data and the seven gait phases. This table is then expressed in the matrix form *R (x, y)* as:

$$R_{ref}(x,y) = \begin{matrix} & P1 & \cdots & P7 \\ \begin{matrix} Fx \\ Fy \\ Fz \end{matrix} & \multicolumn{3}{c}{\begin{pmatrix} x1y1 & \cdots & x1y7 \\ \vdots & \vdots & \vdots \\ x3y1 & \cdots & x3y7 \end{pmatrix}} \end{matrix} \tag{8.1}$$

2. In a similar manner, the establishment of the fuzzy mapping, *S (y.z)* describes the relationship between.
   - The muscle activity whose attributes include:
   - *Z* = {Sol TA LG VL RF BF Gmed ES};

   Where
   Sol indicates soleus;
   TA indicates the Tibialis Anterior;
   LG indicates Gastrocnemius Lateralis;
   VL indicates Vastus Lateralis;
   RF indicates Rectus Femoris;
   BF indicates Biceps Femoris, and.
   Gmed indicates Gluteus Medius.
   And

   - The functional phases in a gait cycle: $y \in Y$.

   Whose attributes include:

$$Y = \{p1p2p3p4p5p6p7\}.$$

This relationship may be viewed as a rule or rule-base that may provide a model. This crisp relationship, *S (y.z),* depicts a feature space that describes the association, interaction, or interconnection between the elements of the muscle activity and the gait phases. The table is then expressed in the matrix form, *S (y.z),* as:

$$S(y,z) = \begin{matrix} & \text{Sol} & TA & LG & VL & RF & BF & \text{Gmed} & ES \\ \text{phase1} & y_1z_1 & & & \cdots & & & & y_1z_8 \\ \text{phase2} & & & & & & & & \\ \text{phase3} & & & & & & & & \\ \text{phase4} & \vdots & & & \vdots & & & & \vdots \\ \text{phase5} & & & & & & & & \\ \text{phase6} & & & & & & & & \\ \text{phase7} & y_7z_1 & & & \cdots & & & & y_7z_8 \end{matrix} \tag{8.2}$$

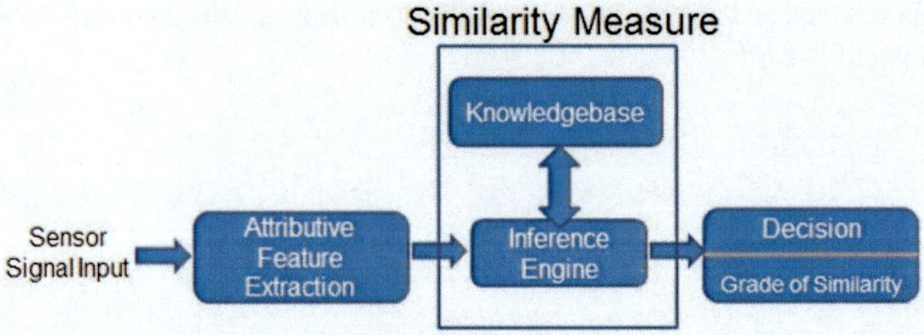

Fig. 8.14 Principle of determination of variances in the gait parameters

Equation (8.1) illustrates the strong relationship between the elements of the ground reaction forces and the seven gait phases, described in a matrix form, whereas Eq. (8.2) illustrates strong relationship between the activities of the various muscle within the seven gait phases. In fact, numerous relationships amongst various kinematic parameters as function of the seven gait phases are possible (these two cases have been explicitly described in order to explain the phenomenon involved in the analysis).

The next important stage involves the determination of any functional deficits of a subject (blinded input) and the corresponding severity of the impairment [2, 68]. In that context, the principle depicted in Fig. 8.14 is applied.

The measured reference feature pattern of the able-bodied subject (reference pattern) is compared with the measured pattern of the impaired subject (blinded input pattern) within the fuzzy rule base. This procedure offers an aggregation between the input and reference feature patterns, and this is expressed through the fuzzy similarity algorithm [2]. This procedure is illustrated in expanded form in Fig. 8.15.

By using systems analysis approach, mathematical methods, biomechanics principles, and computational techniques, it is possible to understand the neuro- and pathophysiology of the human locomotor system, the knowledge, and insight that will improve the quality of diagnosis, prognosis, and follow-up for individual patients suffering from neurological and degenerative diseases, and even assist in countermeasures associated with microgravity during and after space flights [102, 118].

Notion of Neuromorphic Engineering

Neuromorphic computing constitutes a form of process that mirrors the structure and functionality of the human brain. It is a method of computer engineering in which elements of a computer are modeled after systems in the human brain and the nervous system [49].

Furthermore, neuromorphic engineering refers to the design of both hardware and software computing elements. It draws from several disciplines including

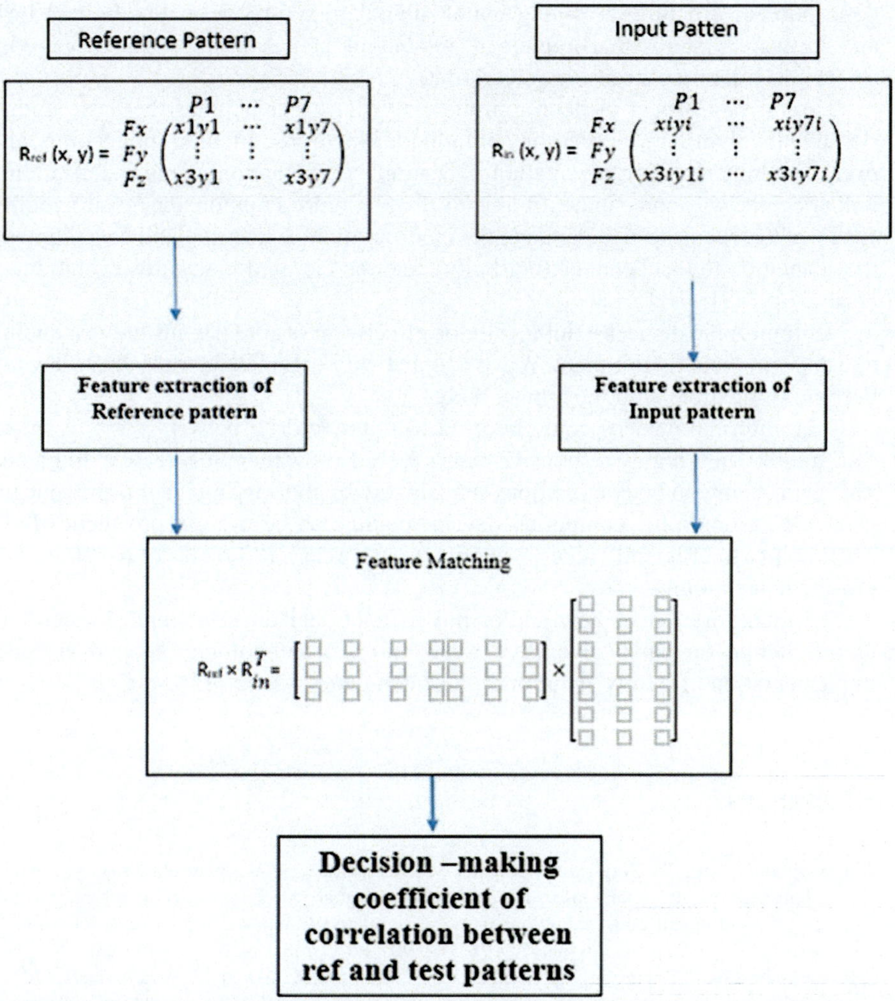

Fig. 8.15 Concept of the fuzzy similarity measurement

computer science, biology, mathematics, electronic engineering and physics, etc. towards the creation of bio-inspired computer systems and hardware. Neuromorphic architectures are most often modeled after neurons and synapses since neurons are often considered as the fundamental units of the brain [47–50].

Neurons use chemical and electronic impulses to transmit information between different regions of the brain and the rest of the nervous system. They use synapses to connect to one another [48, 50]. Neurons and synapses are far more versatile, adaptable, and energy-efficient information processors than traditional computer systems.

A neuromorphic computer/chip is any device that uses physical artificial neurons to perform computations. The expression of "neuromorphic" has emerged as a description of structures involving analog, digital, mixed-mode analog/digital VLSI, and software systems that implement models of neural systems (for perception, motor control, or multisensory integration).

A significant aspect of neuromorphic engineering constitutes the basic understanding involving the morphology of individual neurons, circuits, applications, and overall architectures for the creation of desirable computations. This understanding will affect the way and manner of the information representation, the resulting influences on the robustness to damage, the incorporation of learning and development, the adaptation to local change (plasticity), and the facilitation towards evolutionary change [48, 49].

Neuromorphic devices exhibit extreme effectiveness at receiving and responding to data from their environment. When coupled with organic materials, these devices become compatible with the human body.

Neuromorphic devices could be used to improve drug delivery systems in the near future. Their highly responsive nature would allow them to release a drug upon sensing a change in body conditions (e.g., involving the varying insulin and glucose levels). Neuromorphic computing devices could also be used in prosthetics [48]. Their ability to efficiently receive and process an external signal is of benefit to this prosthetic technology.

The incorporation of new and/or novel functional materials and devices into unique architectures may give rise to a revolutionary technological leap towards the implementation of a fully "neuromorphic" computer system [42].

References

1. Sarkodie-Gyan, T., Huiying, Y., Bogale, M., Hernandez, N. V., & Pirela-Cruz, M. (2017). Application of multiple sensor data fusion for the analysis of human dynamic behavior in space: Assessment and evaluation of mobility-related functional impairments. *Journal of Biomedical Science and Engineering, 10*, 182–203. https://doi.org/10.4236/jbise.2017.104015
2. Sarkodie-Gyan, T., Huiying, Y., Alaqtash, M., Bogale, M. A., Moody, J., & Brower, R. (2013). Application of fuzzy sets for assisting the physician's model of functional impairments in human locomotion. *Journal of Intelligent Fuzzy Systems, 25*, 1001.
3. Alaqtash, M., Huiying, Y., Brower, R., Abdelgawad, A., & Sarkodie-Gyan, T. (2011). Application of wearable sensors for human gait analysis using fuzzy computational algorithm. *Elsevier Journal Engineering Applications of Artificial Intelligence, 24*, 1018.
4. Mergner, T. (2002). The Matryoshka Dolls principle in human dynamic behavior in space: A theory of linked references for multisensory perception and control of action. *Cahiers de Psychologie Cognitive/Current Psychology of Cognition, 21*, 129.
5. Nishikawa, K., Biewener, A. A., Aerts, P., Ahn, A. N., Chiel, H. J., Daley, M. A., Daniel, T. L., Full, R. J., Hale, M. E., Hedricks, T. L., Kristopher Lappin, A., Richard Nichols, T., Quinn, R. D., Satterlie, R. A., & Azymik, B. (2006). *Neuromechanics: An integrative approach for understanding motor control*. From the symposium 'Biomechanics and Neuromuscular Control' presented at the annual meeting of the Society for Integrative and Comparative Biology; January 4–8, 2006.
6. Nuffield Council on Bioethics. (2013). *Novel neurotechnologies: Intervening in the brain*. ISBN: 978-1-904384-29-8.

7. Taylor, B., & Marsiske, M. (2019). Central nervous system. https://doi.org/10.1007/978-3-319-69892-2-668-1.
8. Brain facts: A primer on the brain and nervous system. In *The society for neuroscience* (4th ed), pp 53. ISBN: 10-916110-00-1.
9. Martin, T. P., Edgerton, V. R., & Grindeland, R. E. (1988). Influence of spaceflight on rat skeletal muscle. *Journal of Applied Physiology, 65*, 2318–2325.
10. Philippe, A. S., Landrock, C. K., & Blaber, A. P. Reducing incapacitating symptoms during space flight: Is the postural deficiency syndrome an applicable model? *Hippokratia, 12*, 41.
11. Riley, D. A., Ellis, S., Slocum, G. R., Sedlack, F. R., Bain, J. L. W., Krippendorf, B. B., Lehman, C. T., Macias, M. Y., Thompson, J. L., Vijayan, K., & De Bruin, J. A. (1996). In-flight and post-flight changes in skeletal muscles of SLS-1 and SLS-2 spaceflown rats. *Journal of Applied Physiology, 81*, 133–144.
12. Cavanagh, P. & Bloomfield, S. (2015). *Dcrements in bone, connective tissue and muscle function with prolonged spaceflight: mechanisms and countermeasures*. Bone/Muscle Division of Space Life Sciences, USRA.
13. Oman, C. M., & Bloomberg, J. J. *Sensorimotor adaptation*. Research summary of the Sensorimotor Adaptation Team.
14. Wood, S. J. (2009). *Sensorimotor adaptation following exposure to ambiguous inertial motion cues*. Project description of the Sensorimotor Adaptation Team.
15. Young, L. R. (2012). *Sensorimotor displays and controls to enhance the safety of human/machine cooperation during lunar landing*. Project Description of the Sensorimotor Adaptation Team.
16. Bloomberg, J. J. (2011). *Development of countermeasures to enhance sensorimotor adaptation*. Project description of the Sensorimotor Adaptation Team.
17. Oman, C. M. (2008). *Advanced displays for efficient training and operation of robotic systems*. Project description of the Sensorimotor Adaptation Team.
18. Harris, B. A., Jr., Bilica, R. D., Bishop, S. L., Blackwell, T., Layne, C. S., Harm, D. L., Sandoz, G. R., & Rosenow, E. C., III. (1997). Physical exam during space flight. *Mayo Clinic Proceedings, 72*, 301–308.
19. MacNeilage, P., & Angelaki, D. E. *Psychophysics and modeling of spatial orientation perception*. In Project description of the Sensorimotor Adaptation Team.
20. Small, R. L. (2011). *Modeling and mitigating spatial disorientation in low-gravity environments*. Project description of the Sensorimotor Team.
21. Songer, J. E., Merfeld, D. M., & Eaton, R. A. (2009). *Investigating high-frequency function: a potential flight diagnostic*. Project description of the Sensormotor Adaptation Team.
22. Moore, S.T. (2008). *Galvanic vestibular stimulation as an analog of post-flight sensorimotor dysfunction*. Project description of the Sensorimotor Adaptation Team.
23. Chi, W., Chen, S., Ishioka, A., Long, R., Chun, Y., Yan, S., Zong-Chun, Y., & Feng-Yuan, Z. (2008). Changes of muscle-related genes and proteins after spaceflight in Caenorhabditis elegans. *Progress in Biochemistry and Biophysics, 35*(10), 1195–1201.
24. Gopalakrishnam, R., Genc, K. O., Rice, A. J., Maender, C., Evans, H., & Cavanagh, P. R. (2005). Effects of long-term space flight on muscle volume. In *ISB XXth congress-ASB 29th annual meeting,* July 31–August 5, 2005, p. 922.
25. Macho, L., Kvetnansky, R., Ficova, M., Popova, I. A., & Grigoriev, A. (2001). Effects of exposure to space flight on endocrine regulations in experimental animals. *Endocrine Regulations, 35*, 101–114.
26. Levine, B. D., Pawelczyk, J. A., Ertl, A. C., Cox, J. F., Zuckerman, J. H., et al. (2002). Human muscle sympathetic neural and haemodynamic responses to tilt following spaceflight. *Journal of Physiology, 5389*(1), 331–340.
27. Professor Gilles Clement. (2005). The maintenance of physiological function in humans during spaceflight. *International SportMed Journal, 6*(4), 185–198.
28. Iwasaki, K.-i., Levine, B. D., Rong, Z., Zuckerman, J. H., & Pawelczyk, J. A. (2007). Human cerebral autoregulation before, during and after spaceflight. *The Journal of Physiology, 579*(3), 799–810.

29. The World Health Organization (WHO). *Neurological disorders: Public health challenges.*
30. Hausdorff, J. M. (2007). Gait dynamics, fractals and falls: Finding meaning in the stride-to-stride fluctuations of human walking. *Human Movement Science, 26*, 555–589.
31. Dingwell, J. B., Cusumano, J. P., Cavanagh, P. R., & Sternad, D. (2001). Local dynamic stability versus kinematic variability of continuous overground and treadmill walking. *Journal of Biomechanical Engineering, 123*, 27.
32. Steimer, T. (2002). The biology of fear- and anxiety-related behaviors. *Dialogues in Clinical Neuroscience, 4*, 231.
33. Matsumoto, D., Hwang, H. S., & Frank, M. G. The role of emotion in predicting violence. *FBI Law Enforcement Bulletin*, 1–11.
34. Ko, T. (2008). *A survey on behaviour analysis in video surveillance applications*. Raytheon Company. www.intechopen.com
35. American Public Transportation Association. (2009). *Identifying suspicious behavior in mass transit*. APTA SS-SRM-RP-009-09 Approved October 15, 2009.
36. Yang, W., Yang Wang, A. V., & Mori, G. (2012). Kernel latent svm for visual recognition. In NIPS, 2012, p. 6.
37. National Defense Research Institute (RAND). (2013). *Using behavioral indicators to help detect potential violent acts*. RAND Corporation Copyright 2013.
38. Steinwart, I., Hush, D., & Scovel, C. (2005). A classification framework for anomaly detection. *Journal of Machine Learning Research, 6*, 211–232.
39. Newell, M. E. (2005). *The connection between emotion, brain lateralization, and heart rate variability*. MS thesis, Uniformed Services University of the Health Sciences, F. Edward Hebert School of Medicine.
40. Johnson, R. R., & Aaron, J. L. (2013). Adults' beliefs regarding nonverbal cues predictive of violence. *Criminal Justice and Behavior, 40*, 881–894.
41. Anderson, D., Campos, J. J., Witherington, D. C., Dahl, A., Rivera, M., He, M., Uchiyama, I., & Barbu-Roth, M. (2013). The role of locomotion in psychological development. *Frontiers in Psychology, 4*, 440.
42. Pahapill, P. A., & Lozano, A. M. (2000). The pedunculopontine nucleus and Parkinson's disease. *Brain, 123*(Pt 9), 1767–1783.
43. One Team, One Mission, Securing Our Homeland U.S. Department of Homeland Security Strategic Plan Fiscal Years 2008–2013.
44. Department of Homeland Security Strategic Plan; Fiscal Years 2012–2016; February 2012.
45. Chang, H.-C., Grossberg, S., & Cao, Y. (2014). Where's Waldo? How perceptual, cognitive, and emotional brain processes cooperate during learning to categorize and find desired objects in a cluttered scene. *Frontiers in Integrative Neuroscience*, 1–46.
46. Wolpert, D. M., & Ghahramani, Z. (2000). *Computational principles of movement neuroscience*. Nature America Inc. http://neurosci.nature.com
47. Greengard, S. (2020). Neuromorphic chips take shape. *Communications of the ACM, 63*(8), 9–11.
48. Tuchman, Y., Mangoma, T. N., Gkoupidenis, P., van de Burgt, Y., John, R. A., Mathews, N., Daly, S. R., Malliaras, G. G., & Salloe, A. (2020). Organic neuromorphic devices: Past, present, and future challenges. *MRS Bullletin, 45*(8), 619–630. https://doi.org/10.1557/mrs.2020.196
49. Lutkevich, B. (2020) What is neuromorphic computing? *TechTarget*. https://searchenterpriseai.techtarget.com/definition/neuromorphic-computing
50. Pelé, A. (2020, April 21). Neuromorphic vision sensors eye the future of autonomy. *EETimes* (April 21). https://www.eetimes.com/neuromorphic-vision-sensors-eye-the-future-of-autonomy/
51. Lotte, N. S., Struijk, A., Bentsen, B., Lontis, E. R., Caltenco, H. A. (2009). Fully integrated wireless inductive tongue computer interface for disabled. In *Proceedings; annual intern conference of the IEEE engineering in medicine and biology society*.
52. Boettiger, J. P. (2020). *A comparative evaluation of the detection and tracking capability between novel event-based and conventional frame-based sensors*. Theses and dissertations.

53. Hausdorff, J. M. (2005). Gait variability: Methods, modeling and meaning. *Journal of Neuroengineering and Rehabilitation, 2*(19), 1–9.
54. Grossberg, S. (1987). *The adaptive brain I: Cognition, learning, reinforcing, and rhythm*. Elsevier Science Publishers.
55. Grossberg, S. (1988). *The adaptive brain II: Vision, speech, language, and motor control*. Elsevier Science Publishers.
56. Grossberg, S., & Kuperstein, M. (1989). *Neural dynamics of adaptive sensory-motor control* (Expanded Edition). McGraw-Hill.
57. Keller, A., & Vosshall, L. B. (2004). Human olfactory psychophysics. *Current Biology, 14*(20).
58. Narens, L., & Mausfeld, R. (1992). On the relationship of the psychological and the physical in psychophysics. *Psychological Review, 99*(3), 467–479.
59. Preuveneers, D., & Ilie-Zudor, E. (2017). The intelligent industry of the future: A survey on emerging trends, research challenges and opportunities in industry. *Journal of Ambient Intelligence and Smart Environments*, 1–12.
60. Augusto, J. C., Callaghan, V., Cook, D., Kameas, A., & Satoh, I. (2013). Intelligent environments: A manifesto. *Human-centric Computing and Information Sciences, 3*(1), 1–18. https://doi.org/10.1186/2192-1962-3-12
61. LeChun, Y., Bengio, Y., & Hinton, G. (2015). Deep learning. *Nature, 521*, 436–444.
62. Zadeh, L. A. (1994). Fuzzy logic, neural networks and soft computing. *Communications of the ACM, 37*, 77–84.
63. Zadeh, L. A. (1998). Roles of soft computing and fuzzy logic in the conception, design and deployment of information/intelligent systems. In O. Kaynak, L. A. Zadeh, B. Tiirksen, & I. J. Rudas (Eds.), *Computational intelligence: Soft computing and fuzzy-neuro integration with applications* (pp. 10–37). Springer.
64. Sarkodie-Gyan, T., Yu, H., Alaqtash, M., Spier, E., & Brower, R. (2009). Recognition and decision-making in human locomotion based on the principles of fuzzy reasoning. In *Proceedings of the IEEE international conference on robotics and biomimetics*, pp. 529–534.
65. Yu, H., Alaqtash, M., Spier, E., & Sarkodie-Gyan, T. (2010). Analysis of muscle activity during gait cycle using fuzzy rule-based reasoning. *Elsevier Journal Measurement, 43*(2010), 1106–1114.
66. Brown, R. J. C. (2021). Measuring measurement – What is metrology and why does it matter? *Elsevier Journal Measurement, 168*, 108408.
67. Yu, H. (2010). *Categorization of functional impairments in human locomotion using the methods of the fusion of multiple sensors and computational intelligence*. PhD dissertation at the Department of Electrical and Computer Engineering, University of Texas at El Paso.
68. Alaqtash, M. (2012). *The application of fuzzy granular computing for the analysis of human dynamic behavior in 3D space*. PhD dissertation at the Department of Electrical and Computer Engineering, University of Texas at El Paso.
69. Wellings, D. J., & Unsworth, J. (1997). Fortnightly review: Environmental control systems for people with a disability: An update. *British Medical Journal, 315*, 409–412.
70. Platts, R., & Fraser, M. (1993). Assistive technology in the rehabilitation of patients with high spinal cord lesions. *Paraplegia, 313*, 280–287.
71. Göpel, W., Hesse, J., & Zemel, J. N. (1989). Magnetic sensors. In *Sensors, a comprehensive survey* (Vol. 5). VCH Verlagsgesellschaft.
72. Hayashi, R., Tsuga, K., Hosokawa, R., et al. (2002). A novel handy probe for tongue pressure measurement. *The International Journal of Prosthodontics, 15*(4), 385–388.
73. Andreasen, L. N. S. (2006). An inductive tongue computer interface for control of computers and assistive devices. *IEEE Transactions on Biomedical Engineering, 53*(12), 2594.
74. Huo, X. (2008). A magneto-inductive sensor based wireless tongue-computer interface. *IEEE Transactions on Neural Systems and Rehabilitation Engineering, 16*(5), 497–504.
75. Struijk, L. N. S. A. (2006). An inductive tongue computer interface for control of computers and assistive devices. *IEEE Transactions on Biomedical Engineering, 53*(12), 2594–2597. PubMed: 17152438.

76. Lu, W., Yan, G., Li, J., Gong, S., Han, Y., & Li, X. (2017). FlexFlow: A flexible dataflow accelerator architecture for convolutional neural networks. In *2017 IEEE international symposium on high performance computer architecture*.
77. Desmedt, J. E., & Godaux, E. (1977). Fast motor units are not preferentially activated in rapid voluntary contractions in man. *Nature, 267*(5613), 717–719.
78. Colledge, N. R., Cantley, P., Peaston, I., Brash, H., Lewis, S., & Wilson, J. A. (1994). Ageing and balance: The measurement of spontaneous sway by posturography. *Erontology, 40*, 273–278.
79. Straube, A., Boetzel, K., Hawken, M., Paulus, W., & Brandt, T. (1988). Postural control in the early: differential effects of visual, vestibular and somatosensory input. In *Posture and gait: Development, adaptation and modulation*. Proceedings of the 9th International symposium on postural and gait research.
80. Diener, H. C., & Dichgans, J. (1986). Long loop reflexes and posture. In W. Bles & T. Brandt (Eds.), *Disorders of posture and gait* (pp. 41–76). Elsevier.
81. Allum, J. H. J., Honegger, F., & Pfaltz, C. R. (1989). The role of stretch and vestibule-spinal reflexes in the generation of human equilibrating reactions. *Progress in Brain Research, 80*, 399–409.
82. Runge, C. F., Schupert, C. L., Horak, F. B., & Zajac, F. E. (1988). Role of vestibular information in initiation of rapid postural responses. *Experimental Brain Research, 122*, 403–412.
83. Shumway-Cook, A., & Woollacott, M. H. (1995). *Motor control: Theory and practical applications* (1st ed., pp. 119–142). Williams & Wilkens.
84. Nashner, L. M. (1997). Physiology of balance, with special reference to the healthy elderly. In J. C. Masdeu, L. Sudarsky, & L. Wolson (Eds.), *Gait disorders of aging: Falls and therapeutic strategies* (pp. 37–53). Lippincott-Raven.
85. Massion, J. (1994). Postural control systems. *Current Opinion in Neurobiology, 4*, 877–887.
86. Horak, F. B. (1987). Clinical measurement of postural control in adults. *Physical Therapy, 67*, 1881–1885.
87. Fukuoka, Y., Nagata, T., Ishida, A., & Minamitani, H. (2001). Characteristics of somatosensory feedback in postural control during standing. *IEEE Transactions on Neural Systems and Rehabilitation Engineering, 9*(2), 145–153.
88. Nashner, L. M. (1972). Vestibular postural control model. *Kybernetic, 10*, 106–110.
89. Ishida, A., & Imai, S. (1980). Responses of the posture control system to pseudorandom acceleration disturbances. *Medical & Biological Engineering & Computing, 18*, 433–438.
90. Johansson, R., Magnusson, M., & Akesson, M. (1988). Identification of human postural dynamics. *IEEE Transactions on Biomedical Engineering, 35*, 858–869.
91. Ishida, A., & Miyazaki, S. (1987). Maximum likelihood identification of a posture control system. *IEEE Transactions on Biomedical Engineering, BME-34*, 1–5.
92. Ishida, A., Imai, S., & Fukuoka, Y. (1997). Analysis of the posture control system under fixed and sway-referenced support conditions. *IEEE Transactions on Biomedical Engineering, 44*, 331–336.
93. Fukuoka, Y., Tanaka, K., Ishida, A., & Minamitani, H. (1980). Characteristics visual feedback in postural control during standing. *IEEE Transactions on Rehabilitation Engineering, 7*, 427–434.
94. Johansson, R. S., & Vallbo, A. B. (1980). Spatial properties of the population of mechanoreceptive units in the glabrous skin of the human hand. *Brain Research, 184*, 353–366.
95. Magnusson, M., Enbom, H., Johansson, R., & Pyykko, I. (1990). Significance of pressor input from the human feet in anterior-posterior postural control: The effect of hypothermia on vibration-induced body-sway. *Acta Oto-Laryngologica, 110*, 182–188.
96. Toppila, E., & Pyykko, I. (2000). Chaotic model of postural stability: A position and velocity dependent system? *Automedica, 19*, 115–134.
97. Brandt, T., Paulus, W., & Straube, A. (1986). Vision and posture. In W. Bless & T. Brandt (Eds.), *Disorders of posture and gait* (pp. 157–175). Elsevier Science Publishers B.V.

98. Baloh, R. W., Jacobson, K. M., & Socotch, T. M. (1993). The effect of aging on visual-vestibulaoocular responses. *Experimental Brain Research, 95*, 509–516.
99. Naboack, C. R., & Demarest, R. J. (1981). *The human nervous system, basic principles of neurobiology*. McGraw-Hill.
100. Sarkodie-Gyan, T. (2005). *Neuro-rehabilitation devices: Engineering design, measurement, and control*. McGraw-Hill Publishing. ISBN: 0-07-144830-6.
101. Bamman, M. M., Clarke, M. S. F., Feeback, D. L., Talmadge, R. J., Stevens, B. R., Lieberman, S. A., & Greenisen, M. C. (1998). Impact of resistance exercise during bed rest on skeletal muscle sarcopenia and myosin isoform distribution. *Journal of Applied Physiology, 84*, 157–163.
102. Bonin, G. (2005). *Physiological issues in human spaceflight: Review and proposed countermeasures* (MAAE 4906/MECH 5801). Biomedical Engineering and Biomechanics.
103. Peterka, R. J. (2002). Sensorimotor integration in human control. *Journal of Neurophysiology, 88*, 1097–1118.
104. Berthoz, A., Lacour, M., Soechting, J. F., & Vidal, P. P. (1979). The role of vision in the control of posture during linear motion. *Progress in Brain Research, 50*, 197–210.
105. Bronstein, A. M. (1986). Suppression of visually evoked postural responses. *Experimental Brain Research, 63*, 665–658.
106. Dijkstra, T. M. H., Schoner, G., & Gielen, C. C. A. M. (1994). Temporal stability of the action-perception cycle for postural control in a moving visual environment. *Experimental Brain Research, 97*, 477–486.
107. Lee, D. N., & Lishman, J. R. (1975). Visual proprioceptive control of stance. *Journal of Human Movement Studies, 1*, 87–95.
108. Lestienne, F., Soechting, J., & Berthoz, A. (1977). Postural readjustments induced by linear motion of visual scenes. *Experimental Brain Research, 28*, 363–384.
109. van Asten, W. N. J. C., Gielen, C. C. A. M., & van der Gon Denier, J. J. (1988). Postural adjustments induced by simulated motion of differently structured environments. *Experimental Brain Research, 73*, 371–383.
110. Allum, J. H. J. (1983). Organization of stabilizing reflex responses in tibias anterior muscles following ankle flexion perturbations of standing man. *Brain Research, 264*, 297–301.
111. Jeka, J. J., Schoner, G., Dijkstra, T., Ribeiro, P., & Lackner, J. R. (1977). Coupling of fingertip somatosensory information to head and body sway. *Experimental Brain Research, 113*, 475–483.
112. Kavounoudias, A., Gilhodes, J. C., Roll, R., & Roll, J. P. (1999). From balance regulation to body orientation: Two goals for muscle proprioceptive information processing? *Experimental Brain Research, 124*, 80–88.
113. Day, B. L., Severac Cauquil, A., Bartolomei, L., Pastor, M. A., & Lyon, I. N. (1977). Human body-segment tilts induced by galvanic stimulation: A vestibularly driven balance protection mechanism. *Journal of Physiology (London), 500*, 661–672.
114. Hlavacka, F., & Nijiokiktjien, C. (1985). Postural responses evoked by sinusoidal galvanic stimulation of the labyrinth. *Acta Otolaryngology (Stockholm), 99*, 107–112.
115. Johansson, R., Magnusson, M., & Fransson, P. A. (1995). Galvanic vestibular stimulation for analysis of postural adaptation and stability. *IEEE Transactions on Biomedical Engineering, 42*, 282–292.
116. Nashner, L. M., & Wolfson, P. (1974). Influence of head position and proprioceptive cues on short latency postural reflexes evoked by galvanic stimulation of the human labyrinth. *Brain Research, 67*, 255–268.
117. Mergner, T., & Rosemeier, T. (1998). Interaction of vestibular, somatosensory and visual signals for postural control and motion perception under terrestrial and microgravity conditions – A conceptual model. *Brain Research Reviews, 28*(1–2), 118–135.
118. Hausdorff, J., Forman, D., Pilgrim, D., Rigney, D., & Wei, J. (1992). A new technique for simultaneous monitoring of electrocardio and walking cadence. *The American Journal of Cardiology, 70*, 1064–1071.

119. Hausdorff, J. M., Forman, D. E., Ladin, Z., Goldberger, A. L., Rigney, D. R., & Wei, J. Y. (1994). Increased walking variability in elderly persons with congestive heart failure. *Journal of the American Geriatrics Society, 42*, 1056–1061.
120. Yasaka, K., Akai, H., Abe, O., & Kiryu, S. (2018). Deep learning with convolutional neural network for differentiation of liver masses at dynamic contrast-enhanced CT: A preliminary study. *Radiology, 286*, 887–896.
121. Gulshan, V., Peng, L., Coram, M., et al. (2016). Development and validation of a deep learning algorithm for detection of diabetic retinopathy in retinal fundus photographs. *Journal of the American Medical Association, 316*, 2402–2410.
122. Anwar, S. M., Majid, M., Qayyum, A., Awais, M., Alnowami, M., & Khan, M. K. (2018). Medical image analysis using convolutional neural networks: A review. *Journal of Medical Systems, 42*(11), 1.

Index

© The Editor(s) (if applicable) and The Author(s), under exclusive license to Springer Nature Switzerland AG 2023

T. Sarkodie-Gyan, H. Yu, *The Human Locomotor System*,
https://doi.org/10.1007/978-3-031-32781-0